Electroporation-based therapies for cancer: From basics to clinical applications

Related titles:

An introduction to biotechnology
(ISBN 978-1-907568-28-2)

Concepts and techniques in genomics and proteomics
(ISBN 978-1-907568-10-7)

Fed-batch fermentation
(ISBN 978-1-907568-92-3)

Woodhead Publishing Series in Biomedicine:
Number 49

Electroporation-based therapies for cancer: From basics to clinical applications

Raji Sundararajan

ELSEVIER

AMSTERDAM • BOSTON • CAMBRIDGE • HEIDELBERG •
LONDON
NEW YORK • OXFORD • PARIS • SAN DIEGO
SAN FRANCISCO • SINGAPORE • SYDNEY • TOKYO
Woodhead Publishing is an imprint of Elsevier

WP
WOODHEAD
PUBLISHING

Woodhead Publishing is an imprint of Elsevier
80 High Street, Sawston, Cambridge, CB22 3HJ, UK
225 Wyman Street, Waltham, MA 02451, USA
Langford Lane, Kidlington, OX5 1GB, UK

British Library Cataloguing-in-Publication Data
A catalogue record for this book is available from the British Library

Library of Congress Control Number: 2014931623

ISBN 978-1-907568-15-2 (print)
ISBN 978-1-908818-29-4 (online)

For information on all Woodhead Publishing publications visit our website at http://store.elsevier.com/

Typeset by RefineCatch Limited, Bungay, Suffolk

Printed and bound in the United Kingdom

Dedication

To dearest God, mother Mrs. Dhanalakshmi R. Sundararajan, father, Mr. C Sundararajan, sisters, brothers, teachers and friends, who helped and supported in all the ways possible.

Epigraph

Diamond is a chunk of coal, made good under pressure and time

-Anonymous

Contents

List of figures and tables

Figures

Tables

Acknowledgments

As it takes a village to raise a child, it took several people and Institutions to teach me and advance me to this level. I am extremely grateful to

Dr. Mike Arnold, IRL, New Zealand for introducing me to this exciting field, and helped me a lot by explaining and clarifying many aspects,

Dr. Leslie Tung, Biomedical Engineering, School of Medicine, Johns Hopkins University, USA, for teaching/training me electroporation, because of which I could grow to this level,

Dr. Shesha Jayaram, EE Dept., University of Waterloo, Canada, for introducing me to the ESA conference where many papers were presented in their initial stage and many valuable comments and appreciations were obtained, which led to publications in the Journal of Electrostatics, and for all her constant and continuous help/support/encouragement,

Dr. McHenry, Dean, College of Technology & Applied Sciences, Arizona State University East for all his help, support, encouragement to work in this topic,

Dr. David Dean, Northwestern Medical school, Chicago (currently at University of Rochester Medical School) for giving me the great opportunity of spending eight great months in his lab, working with him and other surgeons on electro-gene-therapy for acute lung disease and to learn so many assays and produce several publications in the electroporation-related topic, enhancing and enriching my knowledge in this topic,

Dr. Mark Prausnitz, Georgia Tech for giving me the great opportunity to work on electroporation of prostate cancer cells,

Mr. Basudev Choudhery, Purdue University, for his help with some of the cells, and all other help,

The two former chairs of ECET Dept.-Prof. Robert Herrick, and Prof. Neal Widmer, and my current Chair, Dr. Ken Burbank and the former Dean, Dr. Dennis Depew and he current Dean Dr. Gary Bertoline of College of Technology, Purdue University, for their continuous and constant help, support, and encouragement,

Dr. Robert Nowlin, former Chair, ECET Dept, Arizoan State University East for editing almost all of my papers and book chapters,

Dr. Ken Burbank also for excellent editing,

and to all of my and my collaborators' students who pitched in conducting several experiments and building the first nanosecond pulser of my research group, which led to the 2nd one presented in Chapter 8, and all the technical and nontechnical help from so many others (staff, postdocs, and students) in the various labs, including the Bindley Bioscience Lab and the BircK Nanotechnology Center of Purdue University, and to

Dr. Glyn Jones, the Publisher of this book for his constant and continuous patience, support, and encouragement, to bring this to print and all of his team members, who are also so kind with me, including Ed Gibbons, Annette Wiseman, Jo Egre, Ian Hutchins and Harriet Clayton and the designer for the beautiful cover.

About the editor

Raji Sundararajan, PhD in Electrical Engineering is a Professor at the Electrical and Computer Engineering Technology Department, Purdue University. She has served at Arizona State University Electronics and Computer Engineering Technology Department (1996–2005) and The Department of Space, India (1982–86). She earned a BE (Hons) with gold medal and the University of Madras endowment prize in EEE from Government College of Engineering, Salem, India, an ME in High Voltage Engineering from Indian Institute of Science, Bangalore, India, and a PhD in Electrical Engineering from Arizona State University, Tempe, Arizona, USA. She obtained her initial training on electroporation at Johns Hopkins Medical School, using neo-natal cardiac primary culture cells and spent eight months at Northwestern Medical School working on electroporation-based gene delivery for acute lung diseases along with several surgeons. She spent a summer at Georgia Tech Bioengineering Department, working on electroporation of prostate cancer cells and attended several electrochemotherapy surgeries for melanoma and chest wall breast carcinoma patients at the University of Padova Medical School, Italy, and the Cork Cancer Research Centre, Ireland and has been doing research on electroporation for the past several years. In addition to the 'Teacher of the Year' award and the 'Researcher of the Year' awards at the Arizona State University East Campus, she was the recipient of the 2010 Indiana 'Women & Hi-Tech' award for 'Distinguished use of technology in healthcare Life Science' and won a prestigious, one-year (2010/11) fellowship to study regenerative medicine therapy using optoinjection and electroporation techniques at Purdue. She is a reviewer of NIH, NSF (including CAREER awards), US International Science & Technology Center, and US National Research Council proposals and various scholarly journals, including *International Journal of Cancer, Molecular Biotechnology, Journal of Biomedical Microdevices, Journal of Anticancer Drugs, Sensors, Computers in Biology*, etc. She is also a reviewer of a number of IEEE Transactions including *IEEE Trans. Plasma Science, Access, Dielectrics and Electrical Insulation, Power Delivery,*

Education, etc. and the *Journal of Electrostatics*. She is continually invited for keynote, workshop, panel moderator, invited talks, etc. for many international conferences. Dr Sundararajan, an IEEE Senior Member, is very active in various IEEE societies, including serving as Associate Editor of two IEEE Transactions. She was also the chair of the Task Force and lead author of two IEEE Standards. A fellow of the Institution of Engineers, India, Dr Sundararajan has numerous publications and is the first lady Conference Chair for IEEE International Conference on Electrical Insulation and Dielectric Phenomena (since the 1920s) and the first lady President of Electrostatics Society of America (since the 1970s), both with sessions on biomedical applications of electrical phenomena.

About the contributors

Poornima Agoramurthy obtained her BE (2008) in Electrical and Electronics Engineering from Visvesvaraya Technological University, Bangalore, India. She completed her MSc in the Department of Electrical and Computer Engineering Technology at Purdue University, USA. Her research focuses on electric field studies in human tissues and electroporation as a method for cancer treatment.

Michela Basso, MD is a radiotherapist at the Veneto Institute of Oncology of Padova, Italy. She has a special interest in treating thyroid and breast cancer.

Ignacio G. Camarillo is an Associate Professor in the Department of Biological Sciences at Purdue University. He received his BSc from St Mary's College in Orchard Lake, MI and his PhD in the Department of Physiology from Wayne State University. As a PhD student he studied 'Prolactin signal transduction in the mammary gland' and was supported by the Minority Biomedical Research Support (MBRS) Program. He continued his training as a National Cancer Institute Postdoctoral Fellow in the Department of Molecular Cellular and Developmental Biology at the University of California Santa Cruz, where he studied 'The protective effect of pregnancy against breast cancer'. Dr Camarillo is a member of the Purdue Center for Cancer Research and serves on the Executive Committee for the Obesity and Cancer Discovery group. His current research focuses on the mechanisms involved in the relationship between diet, obesity and breast cancer. He promotes diversity in STEM careers through leadership roles with Purdue SACNAS (Society for the Advancement of Chicanos and Native Americans in Science), SROP (Summer Research Opportunity Program) and LSAMP (Louis Stokes Alliance for Minority Participation) programs.

Luca G. Campana, MD, PhD is a surgical oncologist at the Sarcoma and Melanoma Unit of the Veneto Institute of Oncology of Padova, Italy. He received his doctorate in Oncology and Surgical Oncology from the

University of Padova, where he completed his general surgical training at the Department of Oncological and Surgical Sciences. His clinical interests lie in the prevention and therapy of malignant melanoma, soft tissue sarcoma and breast cancer. His research interest is focused on locoregional therapies in surgical oncology and, in particular, on the clinical application of electrochemotherapy in cancer patients.

Vanna Chiarion-Sileni, MD is a staff specialist in medical oncology at the Veneto Institute of Oncology of Padova, Italy. She has a special interest in the treatment of patients with malignant melanoma and esophageal cancer. She is a leading expert in the systemic treatment of malignant melanoma and participates in several ongoing clinical trials.

Christy L. Cooper, MS is a Bioanalytical Chemist in the laboratories of James F. Leary, located in the Birck Nanotechnology Center at Purdue University. With intensive cell culture and molecular biology experience, she offers a unique perspective on the interface of chemistry, molecular biology and nanomedicine. Ms Cooper is currently working in collaboration with Dow AgroSciences to evaluate the efficacy of transient fluorescent reporter gene expression in maize and soybean crops. She is also pursuing a PhD in Basic Medical Sciences in rare leukemia stem cell detection via flow cytometry and subsequent treatment using a targeted nanomedical approach. She completed an MSc in Chemistry at Indiana University-Purdue University Indianapolis (IUPUI) in 2005.

Fabrizio Dughiero, PhD is Professor of Electrical Engineering at University of Padova, Italy. His activities deal with electroheat and the application of induction and RF heating to hyperthermia and industrial processes. His research interests are focused on numerical methods for electromagnetic and thermal problems and optimization techniques.

Cristina Falci, MD is a staff specialist in medical oncology at the Veneto Institute of Oncology of Padova, Italy. She has a special interest in treating breast cancer.

Steve Haltiwanger, MD, CCN is the Health and Science Director of LifeWave, Inc., where he is involved in setting up research studies on LifeWave technology. Present responsibilities include coordinating research projects at universities and colleges, and writing up research protocols, research studies and background papers on LifeWave technology. He also formulates nutritional compounds to support cellular energy production and cell repair. For 29 years he has presented at many national conferences and international conferences in 12 countries. He earned a BSc in Professional Chemistry in 1975 from Valdosta State College in Valdosta, Georgia. He received his medical degree from the Medical College of Georgia in 1980. While there, he completed medical school and a one-year fellowship in Pathology. He completed a Psychiatry and Neurology residency at the William S. Hall Psychiatric Institute/ University of South Carolina School of Medicine in 1984. In 1985, he was board certified in Psychiatry and Neurology. In 2003, he became a CCN (Certified Clinical Nutritionist). He is currently a member of the scientific counsel of the International and American Association of Clinical Nutritionists. Dr Haltiwanger's past research of combining electromagnetic and phototherapy treatments with nutritional supplementation led to his current position as Health and Science Director of LifeWave. He has done research on the effects of LED phototherapy on health and the effects of magnetic resonance with Jacobson Resonators.

Jason D. Harper is currently an Electrical Engineer in the Electric Vehicle Smart Grid Interoperability Center at Argonne National Laboratory. Jason received his B.S. (ECET) and M.S. (ECET) from Purdue University in 2006 and 2009 respectively. Jason's work experience includes the avionics, medical device, and chemical instrumentation R&D industries. Jason holds two US patents on novel chemical instrumentation, has published numerous peer-reviewed technical papers and is the author of a technical book involving chemical instrumentation.

Priyadarshan Kathirvelu is a graduate student at the AU-KBC Research Centre of Anna University, Chennai. She works with her thesis supervisor Dr Kavitha Sankaranarayanan on pacemaker function. She has a BSc, MSc and MPhil in Microbiology and has been awarded various research scholarships, including the Junior Scholarship by Lady Tata Memorial Trust, Bombay and the Anna Centenary Research Fellowship by Anna University, Chennai. Her research interests include understanding mechanisms in stem cells, cancer, etc.

James F. Leary, PhD is a named Professor of Nanomedicine, and Professor of Basic Medical Sciences and Biomedical Engineering at Purdue University. He is the original inventor of high-speed flow cytometry/cell sorting, a technology important in the evolution of rare stem/progenitor cell isolation and circulating tumor cell enumeration for cancer diagnostics and minimal residual disease monitoring. He is also an early researcher in the field of nanomedicine and now devises new nanomedical devices for advanced drug/gene delivery and for multimodal non-invasive molecular imaging.

S Madhivanan earned his MBBS (equivalent to MD), MCH (Master of Child Health), MS (General Surgery) and MCh (Neurosurgery) degrees in India. He is serving as a Professor of Surgery at the KAPV Government Medical College, Trichy, India and, as a consultant, performs surgery at several other hospitals. He is the winner of the Dr D. K. Varadhachariar Memorial Prize for Surgery, and the recipient of the Best Outgoing Student in Medicine. He also is the winner of TB Association prize and earned second rank in Pharmocology. He is very interested in electrochemotherapy and its applications.

Arutselvan Natarajan, with a PhD in Industrial Chemistry, is a senior research scientist at the Stanford University, USA. His current research interests include developing novel cancer diagnostics, imaging probes and therapeutics; and the development of innovative therapeutic and diagnostic tools based on nanotechnology. He also has rich experience in synthetic peptides, PEG technology, antibodies, protein scaffolds, small molecules and natural compounds, including anti-microbial, anti-oxidants and anti-cancer. He has extensive pharmaceutical R&D experience.

Maxine Nichols, PhD was the Manager/Research Scientist at the Biological Sciences Department, Purdue University. She has over 25 years experience in study design strategies and development of animal models in both rats and athymic mice, including feeding behavior, cancer cachexia, drug therapy and drug targeted delivery systems. She also has conducted extensive cell culture work. She has experience in performing necropsies, histology and immuno-cytochemistry. Dr Nichols has also extensive technical and managerial laboratory experience and her efforts are critical for the proposed work.

Gerald O'Sullivan, MD, FRCS qualified in medicine from UCC in 1969. Following surgical residency in Ireland and the USA, he was appointed consultant surgeon in 1985 and to a chair in surgery at University College Cork in 1999. He was a Fellow of the Royal College of Surgeons in Ireland (RCSI), and an Honorary Fellow of the Royal College of Physicians and Surgeons in Glasgow, of the Royal College of Surgeons Edinburgh, of the American College of Surgeons and of the College of Surgeons of East Central and Southern Africa. He had received numerous international awards and has published over 200 papers in peer reviewed publications including *Nat Medicine, J. Ex. Med, Gastroenterology, Annals of Surgery, Molecular Therapy*, etc. He has presented numerous video recordings of surgical operations to international societies, including the American College of Surgeons, The Association of Surgeons of Great Britain and Ireland, and the Endoscopic Surgery Associations. He has pioneered three new treatments from laboratory to full clinical usage. Professor O'Sullivan is the past President of the Irish Society of Gastroenterology, past President of the RCSI and the immediate past President of the European Surgical Association. He has been involved in electrochemotherapy of cancer since 2000 and his group contributed to a multicentre/multinational study in the development of the European Standard Operating Procedure of Electrochemotherapy, which is now the standard therapy. Currently research by his group involves the development of devices for electrochemotherapy to deep seated tumors and internal organs and the conduct of a prospective randomised trial of electrochemotherapy versus surgery of basal cell carcinomas of the facial region.

Kevin J. Otto PhD is the Director of the NeuroProstheses Research Laboratory at Purdue University. He received his BSc in Chemical Engineering from Colorado State University, and MSc and PhD in Bioengineering from Arizona State University. He then completed a Postdoctoral Fellowship at the University of Michigan. For over 16 years his work has focused on neural engineering, sensory neuroprostheses, brain-machine interface systems, implantable devices and cochlear implants. He is currently an Associate Professor in the Department of Biological Sciences and Biomedical Engineering. His current research interests include neuroprostheses, systems neuroscience, neurotechnologies and diagnostic systems.

Raja Prabu Ramachandran PhD received his PhD in High Voltage Engineering from College of Engineering, Guindy, Anna University, Chennai, India. Currently, he is working as Dean (Academic Research) and Professor & Head, Department of EEE, BS Abdur Rahman University (formerly Crescent Engineering College), Chennai. He is a member of IEEE, CIGRE, I.E (I) and ISTE. His research interests include pulsed electric fields, outdoor polymeric insulators and micro-grid protection. He is a Chartered Engineer. He has been awarded various fellowships by AICTE, DST, CSIR and TNSCST. He has generated more than 50 research publications.

Anita Ramesh, DCH(UK), MD, DipNB, DM (Medical Oncology), MSc Oncology (UK), MBA (Health Service Management) is a Professor of

Medical Oncology at the Department of General Medicine, Sri Ramachandra Medical College and Research Institute, Sri Ramachandra University, Porur, Chennai, India. She is a visiting consultant for Kanchi Kamakoti Child Trust Hospital, Sankara Nethralaya Medical Research Institute and Madras Mission Hospital in Chennai. Her education qualifications are long and varied. Her education in Oncology started at Dr. MGR Medical University, Chennai, where she passed D.M. (Medical Oncology) exam. She completed MSc in Oncology from the University of Nottingham, UK. She also possesses a Diploma in Child Health from Loyal College of Paediatrics & Child Health, London and a European certification in Medical Oncology from European Society of Medical Oncology. In addition, she acquired a MBA degree in Health Management from Anna University, Chennai. She passed with distinction and has received recognition through innumerable awards, including the Yong Scientist award in 2001, Best paper award at ASCO PANASIA Conference in 2005. She was also awarded international scholarships to attend and present papers in various conferences of world-renowned Institutes since 2001. She is a Member of SIOP, ESMO, UKCCSG, ASCO, IAP, Stem cell research Committee and Pharmacy, etc. Of Sri Ramachandra Medical College. She has about 24 articles in publications at International level, 10 publications in Indexed Journals and 13 in National level. Her ongoing research include Late effects clinic, Epidemology Childhood Cancer, Retinoblastoma project Cancer Registry and Tissue Banking, Clinical Trials, etc.

Lisa Reece is the Biosafety Research Engineer at the Birck Nanotechnology Center at Purdue University, West Lafayette, IN, USA, where she provides supervision and training of the BNC bionano lab users. She is also the Lab Director for the Indiana CTSI (Clinical Translational Sciences Institute) BioNanotechnology Core Facility, which includes all nine bionano labs within Birck. She is available for scientific consultation on bionanotechnology, BioMEMS and BioNEMS projects. Her goal is to help make the bionanotechnology sections of Birck an integrated set of labs having the best technology and expertise for generating, characterizing and testing nanomaterials for biomedical research. Furthermore, she was the Technical Editor for *Nanotechnology for Dummies*, 2nd Edition (July 2011) and is currently a PhD Candidate in the Department of Basic Medical Sciences, College of Veterinary Medicine at Purdue. She holds a BSc in Microbiology from the University of Texas at El Paso, USA. She has four years of industrial microbiology experience with divisions of Baxter Scientific, Inc., Baltimore Spice Co. and Promega, Inc. She also has over 20 years of academic experience in immunology, oncology and infectious disease research with UT El Paso, William Beaumont Army Medical Center, Ft Bliss, TX, UT MD Anderson Cancer Center, Houston, TX and UT Medical Branch, Galveston, TX, which includes both basic and clinical research.

Carlo R. Rossi, MD is Professor of General Surgery at the University of Padova, Italy. He is the Chief of the Sarcoma and Melanoma Unit at the Veneto Institute of Oncology of Padova, with main research interests in the treatment of patients with malignant melanoma, soft tissue sarcomas and peritoneal carcinomatosis.

Therese S. Salameh earned her PhD in the Department of Biological Sciences at Purdue University. She received her BA from Clark University in Worcester, MA and her MA from Purdue University, West Lafayette, IN. Her current research focuses on the underlying biological mechanisms involved in the relationship between obesity and breast cancer. She examines changes in tissue fatty acid composition in relation to various stages of tumor growth and development, to identify links between diet and breast cancer.

Kavitha Sankaranarayanan PhD obtained her PhD from the Tata Institute of Fundamental Research (TIFR) – National Centre for Biological Sciences (NCBS) (Advisor: Prof M. K. Mathew). She has since undergone a Department of Biotechnology, Government of India sponsored postdoctoral training at NCBS-IISc (Advisors: Prof K. S. Krishnan/Prof P. Balram). After working with eminent cardiothoracic surgeon Padmashree Dr K. M. Cherian for about two years, she is currently heading the Ion Channel Biology Laboratory at the AU-KBC research centre. She has more than a decade of experience in the areas of electrophysiology, molecular biology and cell biology from leading labs across the globe. She has been trained in advanced techniques in electrophysiology, including the Cut-open Oocyte Vaseline Gap technique, different types of Patch recordings at UCLA, California, USA, Hong Kong University of Science and Technology, The Max-Planck Guest Laboratory, Institute of Biochemistry and Cell Biology, Shanghai and other international labs. Her research interests revolve around the role of ion channels in health and disease, in addition to its role in development. She employs a biophysical approach in understanding the basics of cellular function and the role of channels in the same. Her lab aims to dissect out the functions of these channels in various physiological and pathophysiological states, including cardiac pacemaker function, stem cell proliferation and differentiation, cancer cell proliferation, autoimmune diseases. The lab also works on channels and their role in abiotic stress and development in plants and also on insect ion channels.

Elisabetta Sieni, PhD received her Degree in Electronic Engineering in 2002 and her PhD in Bioelectromagnetics and Electromagnetic Compatibility and in Information Engineering, Bioengineering, from Padova University in 2006 and 2011, respectively. She is currently working at the Department of Industrial Engineering of Padova University as a post-Doc. Her research interest is focused on biomedical application of electromagnetic fields.

Raji Sundararajan, PhD is Professor at the Electrical and Computer Engineering Technology Department, Purdue University. She obtained her initial training on electroporation at Johns Hopkins Medical School and has been doing research on electroporation for the past several years. She was the recipient of the 2010 Indiana 'Women & Hi-Tech' award for 'Distinguished use of technology in healthcare Life Science' and won a prestigious, one-year (2010/11) fellowship to study regenerative medicine therapy using optoinjection and electroporation techniques at Purdue. She is a reviewer of NIH, NSF, US International Science & Technology Center, and US National Research Council proposals, and various scholarly journals including *International Journal of Cancer, Molecular Biotechnology, Journal of Biomedical Microdevices, Journal of Anticancer drugs*, and several IEEE Transactions.

Gowri Sree Varadarajan, PhD received her doctorate in High Voltage Engineering in 2008 from the College of Engineering, Guindy, Anna University, India. She is presently an Associate Professor in the Department of Electrical and Electronics Engineering, College of Engineering, Guindy, Anna University, India, with over 19 years of teaching experience. Her current research topics are application of high voltage engineering for cancer treatment and assisted germination of paddy seeds. She is also interested in water processing and preservation of liquid foods by pulsed electric fields. She also has extensive experience in testing of high voltage power apparatus.

Ganesh Venkatraman PhD is an Associate Professor in the Human Genetics Department of the Sri Ramachandra University in Chennai, India. He obtained his PhD in Biochemistry from the Cancer Institute, University of Madras, Chennai in 2005. His PhD work involved development of monoclonal antibodies to cancer cell surface antigens for targeted diagnosis and therapy using radiolabeling techniques. After completing his PhD, he joined the University of Nebraska Medical Center, USA, for his postdoctoral training. He worked on radiolabeled single chain antibody fragments for radio-immunotherapy and radio-immunodiagnosis of solid tumors. He also worked on molecular studies on the role of mucins in gastric cancers and on combination chemotherapy for prostate cancer. Subsequently, he joined Fels Institute of Cancer Research, Temple University in Philadelphia, where he worked on signaling pathways implicated in metastasis of ovarian cancer cells. On completing his postdoctoral training in the USA, he joined Sri Ramachandra University as Assistant Professor in 2008. He has several ongoing Government of India funded projects on targeted drug delivery, new drug discovery and molecular oncology.

Brent Vernon, PhD is an Associate Professor of Biomedical Engineering at Arizona State University (ASU) School of Biological and Health Systems. He is the director of the Center for Interventional Biomaterials in the Fulton School of Engineering at ASU. He has more than 50 publications in the area of injectable, *in situ* gelling biomaterials from research supported by National Institutes of Health, American Heart Association, Arizona Biomedical Research Commission and Arizona State University.

Funian Xiao obtained his BS from Purdue University with a major in Physics and a minor in EE. He completed his MS in the Department of Electrical and Computer Engineering Technology at Purdue University, USA. His thesis was on enhanced anticancer drug delivery using electrical pulses, which resulted in several publications, including journal and conference papers and book chapters, and second prize from the 2009 Electrostatics Society of America (ESA) Annual Meeting Student Paper Competition. His research interests include applications of science and engineering for various medical practices.

Introduction

Motivation:

While the existing cancer treatments, such as surgery, radiation, and chemotherapy, save millions of lives, they are not serving all patients. For example, cancer kills one American every minute (about 1500 deaths per day and 550,000 per year). Worldwide, millions die each year due to cancer. The pitiful survival rate of 5–10% for pancreatic cancer and the mere one year median survival of glioblastoma cancer tell us that current treatments are inadequate. Every three minutes, a woman in the US is diagnosed with breast cancer and every 13 minutes one dies of it. Worldwide, every minute, one woman dies due to breast cancer. Only 21 to 41% respond to the premiere hormone drug for breast cancer, Tamoxifen, among the Estrogen Receptor positive women. There are inoperable, recurrent, and radio- and chemo-resistant cancers that do not respond well to the current standard of cure. Thus, there is a critical need for safe, effective, yet affordable, alternate physical treatment techniques for the various cancers, and the electrical-pulse-mediated, enhanced molecular (drug/gene/DNA/vaccine/siRNA) delivery technique, known as electroporation, is one of the ways to effectively treat these patients.

Electroporation or Electropermeabilization (EP) is the application of high intensity, short duration electrical pulses for enhanced drug uptake. When an electrical pulse is applied across cell membranes, charges accumulate on the leaky dielectric plasma membranes of the cells; when the voltage reaches about 0.5V to 1V, the electric field across the membrane enhances to a very high value (such as 0.5V/5nm = 1x109V/m) causing dielectric breakdown, opening up pores, enhancing the permeability, allowing generally impermeable macro drug/gene/DNA molecules to enter into the cells and do their therapeutic action as evidenced by various clinical, in vivo and in vitro studies with a very small dose of the drug.

Unlike other cancer therapies, electroporation has a very short treatment period (just a few minutes) and a low rate of injury to cells and tissues. It can be out-patient-based. It is an efficient technique to enhance drug delivery efficacy of cancer treatment, gene transfer and other similar applications in medicine, biology, and biotechnology. It involves minimal side effects to patients with local, site-specific influence to healthy cells due to the small dose of drug needed. In addition, some unstable medical drugs cannot be delivered through in vivo by other methods, while electroporation has the potential to deliver these in cancer therapy. Therefore, electroporation is a promising technology to directly deliver drugs into living cancer cells for better treatment outcomes.

Electroporation-mediated chemotherapy, known as Electrochemotherapy (ECT) is a viable alternative to conventional cancer treatments as evidenced by successful clinical trials for various cancers, including chest wall breast carcinomas, lymphomas, squamous cell carcinomas, testicular carcinomas, and malignant pleural effusions. When surgery, radiotherapy, and/or chemotherapy did not work, ECT was found to be effective for the treatment of chemo-resistant, recurrent melanoma and chest wall breast carcinoma. This method allows enhanced entry of poorly permeant chemotherapeutic drugs into cells at specific tissue locations without affecting neighboring regions to a great extent. It is applicable to all histologies of cancers and has been shown to offer up to 1000 fold improvement in therapeutic benefits compared to using a drug alone.

In this book, the various authors report their positive experiences with electroporation-based clinical and in vitro studies.

Organization of the chapters

Chapter 1 presents the need for new/alternate cancer treatment techniques and elucidates the benefits, advantages and the potential of electrochemotherapy.

Chest wall recurrence is a type of breast carcinoma that returns after a mastectomy. A chest wall recurrence may involve skin, muscle and fascia beneath the site of the original breast tumor, as well as lymph nodes. A chest wall recurrence may be classified as a localized recurrence, or it may be linked to distant metastasis. There are several inoperable cases and those that do not respond to systemic chemotherapy. For these cases, ECT is an attractive alternative which has worked well for several patients. Chapter 2 presents the positive experiences of Dr. Luca G Campana and his group at the Sarcoma and Melanoma Unit of the Veneto Institute of Oncology of Padova, Italy.

Chapter 3 presents the successful clinical trial results of the treatment of various melanoma cases, including aggressive, superficial extreme melanomas that are inoperable, recurrent and/or chemo-refractive. In these treatments, eight 1200V/cm, 100μs pulses at 5000Hz frequency (0.2ms interval between pulses) were applied after intra-tumoral Bleomycin (a FDA approved chemo drug in the USA and approved by respective medical agencies in other countries) administration.

In the clinical trials, typically, 8 pulses of intensity 1200V/cm, and duration 100μs each, are applied. However, it is always safe to use lower voltages, but longer duration pulses, that have enough energy to cause transient dielectric breakdown of the plasma membrane that enables the permeance of the impermeable or poorly permeable drug molecules. Thus, the efficacy of low voltage long duration pulses, along with the above high voltage, short duration pulses and intermediate values were studied in Chapter 4. It presents the results of a study using an in vitro model of human breast carcinoma MCF-7 cells which showed that low, medium and high voltage pulses were effective in uploading several anticancer drugs into these cells. Usually Bleomycin is the go to drug for ECT in clinical practice; occasionally Cisplatin is used. This group studied the effect of bleomycin and other FDA approved breast cancer drugs, including Paclitaxel (Taxol) and Doxirubicin (Dox) using both low and high voltage pulses. In addition, they studied the effect of Tamoxifen, the endocrine (hormone) drug and Curcumin, the yellow pigment of the 6000 years old, famous Indian spice, turmeric. This research is very relevant as these are the most commonly used breast cancer chemo and hormone drugs and have severe side effects and a life time dosage limit, thus ideal candidates for electrochemotherapy, as only reduced dosage is used in this therapeutic technique.

Chapter 5 discusses the electrical properties of cancer cells and how various functions of the cells are electrical and electronic, and why/how electroporation is a viable technique for cancer treatment.

Chapter 6 presents the results of the study on adult human mesenchymal stem cells using various pulse parameters and Bleomycin. These primary culture stem cells were isolated from the sternal bone marrow of cardiac patients undergoing surgery and their response to electroporation in proliferation control was studied. Another objective of this study was the use of these stem cells as vehicles to carry other therapeutic molecules to the tumor, as they have the propensity to move to the tumor tissue. Colony forming units and viability assays were used to characterize these cells.

The feasibility of the use of electroporation for various cancer cell lines, such as human leukemia cell lines and cervical cancer cell line along with Bleomycin is discussed in Chapter 7 and it again shows the potential of using electroporation for uploading chemo drugs into various cancer cells.

In all the above studies, only micro and millisecond pulses were utilized. It is also possible to use nanosecond pulses with very low energy to do the same. Chapter 8 furnishes the design and building of a compact nanosecond pulse generator using state-of-the-art, radio frequency, fast, MOSFETs. The effect of nanosecond pulses at low intensities using Tamoxifen has been studied.

Chapter 9 discusses the study of electroporation using ovarian cancer cells with no drug. This indicates the proliferation control using electrical pulses only. This study was extended to other cell lines, such as lung cancer, and breast cancer cells and investigated the effective proliferation control using pulses only and no drug. This technique, known as Irrversible Electroporation is discussed in Chapter 10.

Chapter 11 reviews the uploading of siRNA into cells using electroporation. These are difficult to transfect molecules and their uploading using electroporation is appropriate.

The efficacy of electroporation is based on the electric field of the applied pulses. The intensity and distribution of the electric field plays a critical role in its use as a therapeutic tool. This electric field distribution and intensity was studied using industry-standard, simulation software. The electric field distributions of various configurations of breast tumors were studied in Chapter 12. It presents electric field distribution of malignant and normal tissues and the influence of electrode geometry and placement on the electric field profile.

Audience

Many potential drugs that have been developed to treat cancer have found limited success due to lack of safe and efficient delivery systems that allow molecules to cross the hydrophilic and hydrophobic lipid bilayers of cell plasma membranes which are normally impermeable. The use of electrical pulses for active delivery of drug/gene molecules to the interior of cells offers exciting prospects. It is safe, effective, economical, and repetitive. The electroporation technique relies on the cell's natural ability to heal itself after electroporation. This is because membrane disruption and resealing is common in vivo as cells normally heal smaller wounds naturally in day to day life. Using this inherent nature of cells has the advantage of uploading molecules of our choice as desired. And since one size does not fit all, we can use a combination of electro-chemo-therapy and other conventional modalities to offer better treatment with increased efficacy and low cytotoxicity.

We hope that these scholarly chapters of our book will be very useful to basic and clinical researchers, medical professionals, biomedical and biotechnology professionals in academia and industry, and other researchers, post-docs and students.

1

Electrochemotherapy – A novel cancer treatment

Gerald C. O'Sullivan

DOI: 10.1533/9781908818294.1

Abstract: There are so many cancer patients for whom the current standard of cure doesn't work. Electrochemotherapy could serve as an attractive alternative to many of these patients who have inoperable, chemo-and radio-resistant tumors. This technique is safe, simple, effective and budget-friendly.

Key words: electrochemotherapy, bleomycin, cisplatin, palliative, cancer.

1.1 Why electrochemotherapy?

There is a pressing need for new treatments for cancer – currently, so many afflicted patients are incurable at presentation and the required palliative treatments are often invasive, ineffectual and unbeneficial. Also many patients have neither the resources nor the support systems for them to be able to afford expensive hospital-based therapies. While new treatments may fall below expectations, their use should still inform and be a force for ongoing research and developments in cancer control.

Electrochemotherapy is the local application of pulses of electric voltage to tumour tissue, to render the cell membranes permeable to otherwise impermeant or poorly permeant (hydrophilic) anticancer drugs (i.e. bleomycin or cisplatin), thereby facilitating a potent localized cytotoxic effect [1–5]. Intracellular concentrations of bleomycin are increased by almost 1000-fold, resulting in cell death by an apoptosis-like phenomenon. Consequently there is little in the way of inflammatory reactions and because the treatment is confined to the electric field, there is little collateral injury of tissues.

To date, experience with electrochemotherapy has mainly been confined to cutaneous or sub-cutaneous cancers, but the results have been encouraging and show that it is a safe treatment – 85 major cancer centres across Europe now make it available for treatment of skin based cancers. Given the positive benefit, particularly as a palliative treatment, there is the promise of new and exciting developments. Any new developments in generator technologies will permit low voltage systems and ECG triggered treatments, thus allowing safe application close to the heart.

Current developments in electrode technologies and devices by our group will allow application to deep seated and internal cancers. Electroporation techniques will also find application in the delivery of non-viral vectors into tumours and tissues for gene therapy, a promising modality for neoadjuvant and adjuvant treatment.

(Excerpts from the preface to the International workshop on Electro-Chemo-Therapy for Effective cancer treatment, July 27–28, 2011, Chennai, India.)

1.2 References

1. Larkin J O, Collins C G, Aarons S, Tangney, M, Whelan M, et al. (2007), 'Electrochemotherapy: Aspects of preclinical development and early clinical experience', *Annals of Surgery*, **245**(**3**): 469–79.
2. Sadadcharam M, Soden D M and O'Sullivan G C (2008), 'Electrochemotherapy: An emerging cancer treatment', *International Journal of Hyperthermia*, **24**(**3**): 263–73.
3. Soden D M, Larkin J O, Collin C G, Tangney M, Aarons S, et al. (2006), 'Successful application of targeted electrochemotherapy using novel flexible electrodes and low dose bleomycin to solid tumors', *Cancer Letters*, **232**(**2**): 300–10.
4. Whelan M C, Larkin J O, Collins C G, Cashman J, Breathnach O, et al. (2006), 'Effective treatment of an extensive recurrent breast cancer which was refractory to multimodal therapy by multiple applications of Electrochemotherapy', *European Journal of Cancer Supplements*, **4**(**11**): 32–4.
5. Marty M, Sarsa G, Garbay J R, Gehl J, Collins C G, et al. (2006), 'Electrochemotherapy – an easy, highly effective and safe treatment of cutaneous and subcutaneous metastases: Results of ESOPE (European Standard Operating Procedures of Electrochemotherapy) study,' *EJC Supplements*, **4**(**11**): 3–13.

2

Clinical electrochemotherapy for chest wall recurrence from breast cancer

Luca G. Campana, Cristina Falci, Michela Basso, Elisabetta Sieni and Fabrizio Dughiero

DOI: 10.1533/9781908818294.3

Abstract: Chest wall recurrence (CWR) after mastectomy is a disabling condition affecting 2–15% of breast cancer (BC) patients. When tumor skin infiltration occurs, the main disease-related complaints often require additional medication, advanced wound dressings and repeated hospitalization. Therefore, even in the presence of visceral disease, an optimal locoregional treatment can significantly reduce the morbidity of uncontrolled local disease spread. A number of BC patients with CWR are not suitable for standard locoregional treatments (mainly radiotherapy) or are refractory to systemic chemotherapy. The benefits of electrochemotherapy (ECT) are the shortness of treatment, ease of application, brief hospitalization, high activity rate, repeatability, possibility of safe combination with other treatments and negligible systemic toxicity. In the few early clinical experiences of CWR, ECT proved an active antitumor therapy and ensured sustained superficial tumor control. Larger clinical trials are awaited to corroborate these promising results and improve clinical outcome of BC patients receiving ECT.

Key words: breast cancer, chest wall recurrence, electroporation, electrochemotherapy, bleomycin.

2.1 Introduction

Worldwide, breast cancer (BC) is the most common cancer diagnosed in women, with approximately 1.3 million new cases each year. In American women, BC is the most common cancer and the second leading cause of cancer death. BC is also the most common cancer in females in Europe, with approximately 140,000 deaths in 2008. Chest wall recurrence (CWR) after mastectomy is an uncommon but serious and disabling condition affecting a subgroup of BC patients, with tumor invasion ranging from skin or subcutaneous tissue to the underlying muscles or bones (Figure 2.1). After the primary surgical treatment, BC can recur locally and/or regionally, sometimes also in association with distant metastases. A *local recurrence* is defined as the reappearance

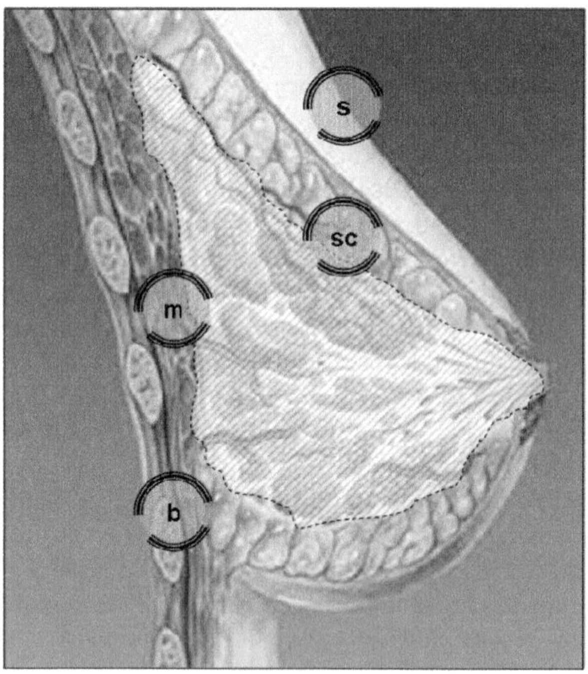

Figure 2.1 Degrees of chest wall infiltration from breast cancer. Tumor spread can range from invasion of superficial soft tissues, e.g. skin (s) or subcutaneous layer (sc) to deep deposits in the muscles (m) or bones (b). The hatched area indicates the mammary parenchyma that is removed during mastectomy

of cancer on the ipsilateral breast (after conservative surgery) or the chest wall (CW) (after mastectomy). A *regional recurrence* is defined as the involvement of the regional lymph nodes, usually ipsilateral axillary or supraclavicular, and less commonly infraclavicular or internal mammary nodes.

The optimal treatment approach for BC patients who develop a CWR in the field of previous mastectomy is widely debated. Because the optimal management of these patients typically requires a combined modality therapy, both multidisciplinary assessment and planning are essential. With the exception of some patients with favorable prognostic factors, CWR from BC has generally a dismal prognosis. However, even in the presence of visceral disease, an optimal locoregional treatment can significantly reduce the morbidity of uncontrolled local disease spread. In recent years, there has been considerable interest in introducing electrochemotherapy (ECT) as a new treatment opportunity for patients with superficial CWR from BC.

In this chapter, the incidence, clinical presentation and disease-related complaints of patients with CWR from BC are presented, along with the results of the conventional treatments. The characteristics of patients suitable for ECT application, at least with the available technology, are illustrated. The preliminary clinical results of ECT in the management of CWR from BC are reported, focusing on the pros and cons of this treatment approach. In the preliminary clinical experiences, the electroporation-driven administration of two anticancer agents (bleomycin and cisplatin) has proved active in the treatment of refractory skin metastases from BC and ensured satisfactory superficial tumor control. Finally, some technical considerations from the engineer's point of view will be treated, in order to delineate further possible advancements in the clinical application of ECT in patients with recurrent BC.

2.2 Scope of the problem

2.2.1 Incidence and clinical presentation of CWR from BC

Local recurrence rates of BC after mastectomy alone have been reported to be as high as 45% for patients with an advanced primary tumor [1]. This high rate of relapse is reduced to 2–15% by the addition of postmastectomy radiotherapy (RT) [2]. Between 50 and 70% of postmastectomy locoregional recurrences are isolated to the CW, the mastectomy scar and the surrounding skin flaps being the most often involved sites [3]. A *local* plus *regional* recurrence in the supraclavicular, axillary or internal mammary nodes occurs in 30–40% of these patients. Approximately one-third of patients who present with a postmastectomy *locoregional* recurrence also have synchronous distant metastases [3, 4]. The median interval to a locoregional recurrence after mastectomy is 2 to 3 years, and 90% arise within 5 years [5, 6].

The two major risk factors for *local recurrence* are the number of positive axillary lymph nodes and the size of the primary tumor [7]. Other factors that may help to predict locoregional recurrence include age at diagnosis, hormone receptor status, the presence of lympho-vascular invasion or extracapsular nodal extension, and positive deep surgical margins [8, 9]. Postmastectomy CW irradiation reduces the rate of recurrence by 65–75%. In modern randomized trials, the rate of locoregional recurrence in women receiving postmastectomy RT ranges between 8 and 13%, compared to 30 and 35% in the non-irradiated control groups [9, 10]. Available clinical practice guidelines recommend postmastectomy RT for patients with tumor size greater than 5 cm, tumor invasion of the skin, the pectoral muscle or the CW, 4 or more positive lymph nodes, and for those with positive surgical margins [11].

The degree of CW infiltration can range from the involvement of superficial soft tissues (skin and subcutaneous layer) to the invasion of underlying muscles and bone (ribs, sternum) (Figure 2.1). Later in this chapter, the term CWR will refer exclusively to the disease spread on the superficial soft tissues of the CW. In this setting, CWR after mastectomy most often presents as one or more palpable nodules near the mastectomy scar (Figure 2.2(a)), but the skin may also be diffusely involved with inflammatory changes without a discernible mass (Figure 2.2(b)). In more advanced patients, widespread ulceration, sometimes secondary to the treatments for recurrence, (Figure 2.2(c)) or diffuse soft tissues infiltration with woody induration, are often present (Figure 2.2(d)). The main disease-related complaints reported by the patients with CWR are not only pain, burning sensations, skin ulceration, bleeding and soft tissue infections, which often require analgesics and repetitive wound dressings, but also a constant level of anxiety and an impairment of their social relations and quality of life.

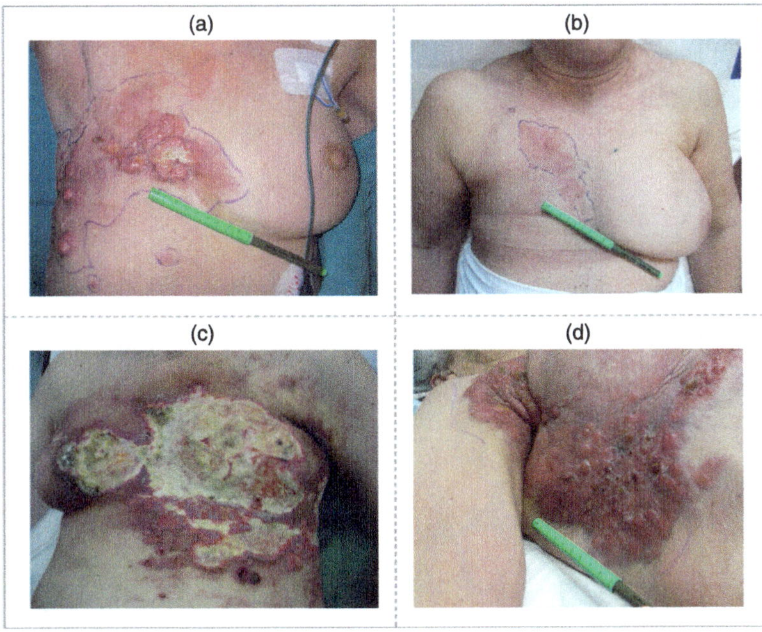

Figure 2.2 Different clinical presentations of superficial CWR from BC. Multiple palpable nodules disseminated from the mastectomy scar to the lateral portion of the CW, the skin of the ipsilateral axilla and the abdominal wall (a). In this patient, the skin of the CW presents inflammatory changes without a discernible mass (b). In more advanced patients, the CW may be diffusely involved with widespread ulceration (c). Alternatively, diffuse soft tissues infiltration with woody induration beyond the limits of standard surgical or radiation boundaries may be present (d)

2.2.2 Prognostic factors

After the appearance of CWR, the most documented prognostic factors are lymph node status at the time of primary tumor, time interval between BC diagnosis and recurrence, site and volume of disease recurrence [5, 6, 12–25] (Table 2.1).

Table 2.1 Significant prognostic factors in BC patients after locoregional recurrence

Reference	Prognostic factors									
	Size of primary tumor	Node status of primary tumor	Histological grade	Vascular invasion	Disease extension/ size	Disease-free interval	ER/PR status	HER-2 status[a]	Adjuvant RT	Adjv hormonal therapy
[22]	–	–	–	–	⊗[b]	⊗	–	–	–	–
[18]	–	–	–	–	⊗	⊗[g]	–	–	–	–
[20]	–	–	–	–	⊗	⊗	–	–	–	–
[25][c]	⊗	⊗	–	–	⊗[d]	⊗[e]	–	–	ns	ns
[13]	⊗	⊗	–	⊗	⊗[f]	⊗[g]	–	–	–	–
[21]	ns	⊗	⊗	–	ns[h]	⊗	⊗	–	–	–
[16]	–	–	–	–	–	⊗[g]	⊗	⊗	–	–
[23]	–	⊗	–	–	–	⊗[g]	–	–	⊗	–
[24]	–	–	–	–	⊗[d]	–	–	–	–	–
[17]	⊗		⊗	–	–	⊗[e]	⊗[i]		⊗	⊗
[19]	ns	ns	ns	⊗	⊗[l]	⊗[m]	ns	–	ns	ns

ER: oestrogen receptor; PR: progesterone ≥ receptor; HER-2: Human Epidermal Growth Factor Receptor 2: Adjv, adjuvant; ns: non significant.

[a] Positive HER2 status was associated with poorer locoregional tumor control. [b] In this study also the number of recurrences had prognostic significance. [c] In this study, also patients age (>50 years), local control in recurrent site and presence of necrosis in primary tumor showed additional independent influence on survival. [d] Isolated CWR. [e] Cut-off for disease-free interval was 12 months. [f] Limited recurrence: single metastasis and ≤3 cm; extensive recurrence: >3 cm, multiple/diffuse metastases. [g] Cut-off, ≤2 years versus >2 years. [h] Local vs. regional recurrence. [i] In this study, also p53 and Ki67 status, in primary as well as in recurrent tumors, were associated with a poorer survival. [l] The number of metastases was categorized as follows: 1 vs. ≥2. [m] Cut-off for disease-free interval, 6 years.

2.2.3 Patient management

The patients who experience postmastectomy CWR should undergo complete clinical and instrumental evaluation in order to define the real extent of the disease, according to the present guidelines by the international societies of medical oncology [26]. A radionuclide bone scan is generally recommended, even in asymptomatic patients, as bone is the most common site of metastatic disease. Computed tomography (CT) scan of chest is useful for differentiating sites of recurrence in the skin, subcutaneous fat, pectoralis muscles and brachial plexus from postoperative and postradiation changes. Since the lymphatic drainage patterns may be substantially altered following primary treatment, magnetic resonance imaging (MRI) may also delineate treatment-related fibrosis from recurrent BC.

2.3 Treatment options for chest wall recurrence

The available treatment options for the patients with CWR include surgery, re-irradiation (when feasible, according to the previous radiation exposure) and/or systemic treatments (endocrine-, chemo- or targeted-therapy). A CWR after mastectomy represents a difficult clinical problem for a number of reasons. First is the need to assess the status of the regional lymph nodes. While in the past nearly all patients with recurrent disease had already undergone axillary dissection as a component of upfront surgical treatment, at present many women have undergone sentinel lymph node biopsy which, if negative, is not followed by the axillary clearance. This has the implication to re-stage the axilla at the time of disease recurrence. Second, because of the increasing use of postmastectomy RT, many CWR are located in previously irradiated fields. Finally, decisions on local treatment must be balanced with the need of systemic therapies. For instance, for the patients with a CWR that would require large, demolitive surgical resection, it is preferable to start with a systemic therapy, not only to facilitate possible locoregional treatment, but also to treat possible distant undetected metastases.

2.3.1 Surgery

The utility of salvage surgery in achieving local control has been reported in several small retrospective series, but unfortunately the majority of patients remain unsuitable for resection. Whenever possible, a wide local excision of all gross disease is recommended. The likelihood of local control after surgery alone is highest when:

a) the recurrence pattern is a single and small (generally <3 cm) recurrence as compared to multiple nodules; and
b) when the relapse-free interval is long (2–10 years, according to different authors) [25, 27].

In a very few selected patients, a full-thickness CW resection and prosthetic reconstruction may be necessary in order to obtain negative margins, particularly when treating extensive locoregional recurrences or when the recurrent tumor involves the ribs and/or the sternum [27–32]. Although, in recent experience, the morbidity of such an aggressive approach is contained, extensive surgery is generally avoided and should only be performed if other options are not available. Moreover, despite an aggressive surgical treatment, also in selected patients resected with a curative intent [28], the prognosis remains severe, 5-year overall survival ranges from 18–46% (27–32) and the incidence of a second local recurrence is around 20–30% [29, 33].

2.3.2 Radiotherapy management

General considerations

Radiotherapy management has the potential to provide long-term disease control for patients who develop CWR after mastectomy. However, the recurrences in previously irradiated areas present a challenge, as the re-irradiation dose that can be given without a high risk of unacceptable toxicity is lower than that considered adequate. Furthermore, the poorly functioning microvasculature, caused by previous radiation, may render the tumor less sensitive to the effects of radiation and chemotherapy. The radiotherapeutic management of locoregional recurrence from BC generally involves treatment of the entire CW and draining lymph nodes, even if the recurrence is confined to a small area; in fact, limited RT fields can increase the risk of additional failures during follow-up [34, 35].

Table 2.2 Results of radiotherapy in the treatment of isolated locoregional recurrence from BC after mastectomy

Reference	No. of patients	Median follow-up (mo)	Local control	5-year OS (%)
[13]	67	74	62% (5 yrs)	58
[41]	120	24	66% (10 yrs)	74
[43]	130	68.5	78% (crude)	48
[16]	113	122	79% (10 yrs)	46
[9]	535	216	47% (5 yrs)	36

In general, long-term local control is achieved in up to 61–86% of patients, particularly in those with favorable selection criteria [5, 6, 17, 34–44]. Unfortunately, only about one-third of the treated patients remain free from distant metastases and the 5-year overall survival rate ranges from 35–82%. Local outcomes are most favorable in patients who undergo RT for isolated recurrence or after complete surgical excision (Table 2.2). For patients with an unresectable CWR, re-irradiation alone is sometimes attempted to provide local palliation, but these patients are more often initially treated with systemic therapies, followed by locoregional treatment.

Radiation techniques

External beam RT is the standard irradiation technique for CWR, if the patient did not receive postmastectomy RT as part of the initial BC treatment. In most patients, the CW is irradiated with tangential photon beams. An alternative technique is to use one or more anterior electric fields of appropriate energy, according to the CW thickness. The treatment of the CW along with supraclavicular nodes, when indicated, is traditionally accomplished by matching a supraclavicular field to the tangents treating the CW. In conformal radiation therapy, CT-based treatment planning is adopted. The patient is scanned in a standard supine position and dedicated software generates a 3-D treatment planning, where the target volume and the critical anatomical structures that should be spared from irradiation (organs at risk, OAR, e.g. lungs and heart) are contoured on CT slices (Figure 2.3). Conventional RT is typically administered once daily, at 1.8–2.0 Gy per fraction, 5 times weekly. If the CW was not already irradiated, total radiation dose ranges

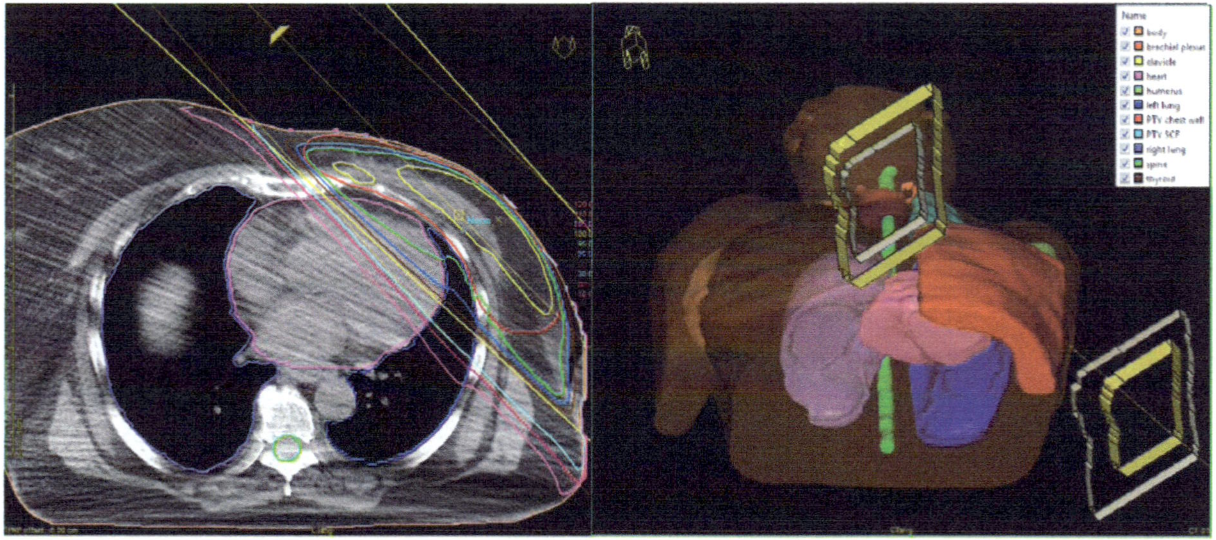

Figure 2.3 Radiotherapy pre-treatment planning. An example of tangential fields technique used to treat the CW in patients with locoregional recurrence from BC. Axial slice demonstrating contours of the CW target (red line), contours of normal organs and dose distribution. The two fields consist of a medial and lateral opposite photon fields that include the chest wall while sparing the radiation dose to the lung and the heart. CT-based treatment planning is applied to determine the appropriate dose distribution (such that the CW is covered by the 90% isodose curve) and to respect the dose constraints to OAR of radiation-induced toxicity (left). Three-dimensional reconstruction with the CW target (red volume) (right). Normal organ and beam shapes are projected in virtual planes. PTV: Planned Target Volume

from 45–50 Gy, with a boost of 10–20 Gy to the areas of gross residual disease and scar sites. By means of modern treatment techniques, treatment-related toxicity is contained, skin desquamation and erythema being the most frequent but self-limited complaints.

2.3.3 Non-conventional fractionation regimens

Limited data exist on CW re-irradiation [45–49] (Table 2.3). For previously irradiated patients, full dose re-irradiation is used with caution, due to concerns for long-term toxicity. In general, limited field re-irradiation, by means of tailored conformal therapy techniques and/or non-conventional fractionation regimens, may be safely employed. 'Hyperfractionated' RT consists of multiple daily treatments with smaller than conventional fraction sizes. By decreasing the size of each fraction, it is possible to apply a high total dose without increasing late morbidity. 'Accelerated fractionation' RT involves the same or similar total dose as in conventional treatment, but it is given in a reduced overall time by giving conventional sized fractions more than once daily. The aim of accelerated fractionation RT is to prevent tumor repopulation during a course of RT.

Despite the use of different radiation techniques, fractionation schemes and irradiation fields, a lower locoregional control after recurrence was observed in re-irradiated patients compared to patients without prior radiation [45–48]. In general, if re-irradiation is undertaken, radiotherapists try to limit the dose to critical structures, including the ribs, sometimes using a more superficial type of beam (e.g. electrons) or

Table 2.3 **Clinical results of CW re-irradiation for BC recurrence**

Reference	No. of patients	Local DFS
[46]	13	62% at 20 months
[47]	44	67% at 1 year
[48]	13	85% at 2 years
[45]	81	66% at 1 year[a]

Abbreviations: DFS: disease-free survival. [a]The 1-year DFS rate for patients with gross disease was 53% compared with 100% for those without gross disease.

hyperfractionated schemes (i.e. twice a day irradiation) to reduce the risk of late side effects. 'Hyperfractionated accelerated' RT was explored in a small series of 80 patients with locoregionally recurrent BC after mastectomy [49]. Five- and 10-year local control was achieved in 68 and 55% of patients, respectively. Although this local control was no better than that achievable with conventional fractionation RT, this approach is sometimes used because of the potential for a more favorable long-term toxicity profile.

2.3.4 Hyperthermia and photodynamic therapy

Hyperthermia

Hyperthermia is an anticancer treatment in which the exposure of body tissues to a supra-physiological level of temperature (in the range of 40–45 °C) by means of different techniques, can kill or selectively sensitize cancer cells to radiation or some anticancer agents. Based on the results obtained in recurrent BC with re-irradiation alone, some investigators advocate combining radiation and hyperthermia. Due to its radiosensitizing properties, hyperthermia lowers the radiation dose needed to achieve durable tumor control and therefore long-term toxicity. A meta-analysis of 5 clinical trials, comparing RT with and without hyperthermia in the setting of locoregionally recurrent BC, showed that the addition of hyperthermia improved the complete response rate from 41–59%, but had no impact on survival [50]. The National Comprehensive Cancer Network (NCCN) has included in its 2007 Breast Cancer Guidelines, hyperthermia therapy and radiation as combination treatment for CWR [26]. The addition of concurrent chemotherapy to hyperthermia and RT, constituting thermo-chemo-RT, has been evaluated only in few early clinical trials, with moderate success [51].

Photodynamic therapy

Photodynamic therapy (PDT) uses a photosensitizing agent that accumulates in cancer tissues. Tumor exposure to a laser light induces a phototoxic reaction and selective tumor destruction [52–57]. Although promising results have been achieved in a variety of skin cancers, the most recent clinical experiences with PDT for CWR rely on very limited monocentric series [56–57].

2.3.5 Systemic treatments

Systemic treatment of BC includes cytotoxic chemotherapy, endocrine manipulations, targeted agents and supportive therapies (Table 2.4). These drugs are used in the neoadjuvant (e.g. before surgical excision), adjuvant (e.g. after surgical excision) and metastatic setting. In general, systemic treatments are active at the beginning of therapy; however, after a variable period of time, progression occurs and resistance to therapy is a common finding. The therapeutic approach to the patients with CWR relies on patients' characteristics and disease extension.

Table 2.4 **Different systemic therapies used in current clinical practice of recurrent or metastatic BC**

Cytotoxic chemotherapy	Endocrine therapy	Targeted therapy	Supportive therapy
Anthracyclines Doxorubicine Epirubicine	**Estrogen receptor modulators** Tamoxifen	**Anti HER2** Trastuzumab Laptinib	**Bisphosphonates** Pamidronate Zoledronate Clodronate
Taxanes Paclitaxel Docetaxel	**Estrogen receptor down regulator** Fulvestrant	**Angiogenesis inhibitors** Bevacizumab	
Other microtubule inhibitors Vinorelbine Eribulin	**Nonsteroidal aromatase inhibitors** Anastrozole Letrozole	**PARP inhibitors** Olaparib Iniparib	
Other agents Capecitabine Gemcitabine Cyclophosphamide Fluorouracil Methotrexate	**Steroidal aromatase inhibitors** Exemestane		
Platinum agents Cisplatin Carboplatin			

Abbreviations: HER2L: human epidermal growth factor receptor 2; PARP: Poly ADP-ribose polymerase: (a family of proteins involved in the process of DNA repair and programmed cell death).

Patients with local or locoregional disease

After radical local treatment, patients should be proposed for systemic therapies aimed to reduce the risk of further local or distant relapse. Available strategies include several antiblastics, multiple hormonal treatments and some targeted agents (Table 2.4). The choice between chemotherapy and endocrine therapy is driven by clinical and immunohistochemical tumor characteristics [58]. Patients with unfavorable predicting factors such as extensive or multisite disease, node involvement, short disease-free interval after diagnosis of BC, high tumor grade, endocrine-negative or hormone-refractory disease (e.g. relapse during endocrine treatment) and young age, generally deserve chemotherapy. If patients have been previously exposed to antiblastics, different drugs should be chosen, unless an interval of 24 months or more has gone after previous chemotherapy; in this case, a rechallange with the same drugs is considered.

However, in the presence of limited disease, hormone-responsiveness, long disease-free interval and advanced age, the endocrine manipulation represents the best treatment option. Endocrine therapies in post-menopausal women include different agents (Table 2.4) and the choice is driven by previous treatments and patients' characteristics [58]. In the presence of a Human Epidermal Growth Factor Receptor-2 (HER-2) positive tumor, the humanized monoclonal antibody trastuzumab should be introduced, even if already administered in the adjuvant setting, in addition to chemotherapy. Some evidence also supports the use of trastuzumab associated to endocrine-therapy in older patients. However, one of the significant side effects of trastuzumab is on the heart. A cardiac dysfunction has been reported in 2–7% of treated patients and approximately 10% are unable to tolerate the treatment because of pre-existing heart problems.

Patients with skin metastases associated to distant disease

Initial palliation with endocrine therapy should be the primary consideration for patients with metastatic hormone receptor-positive tumors. Cytotoxic chemotherapy is appropriate for those with hormone-refractory disease, rapidly progressive visceral disease, or early relapse after adjuvant therapy. If a tumor over-expresses HER2, targeted treatment with trastuzumab or lapatinib is possible. Consequently, accurate determination of the status of these predictive markers in tumor tissue, possibly from a recurrence site, is recommended.

Notably, each patient with advanced BC absolutely deserves a specifically-tailored and multidisciplinary treatment plan. Simultaneously to systemic anticancer therapies, the best supportive care, including for example bisphosphonates and palliative RT for bone metastases or ECT for symptomatic CW skin infiltration, should be prescribed in order to preserve patients' quality of life [59].

2.4 Clinical experience with electrochemotherapy

2.4.1 Patients' selection and preparation

Patients with CWR may be considered for ECT independently of the presence of visceral metastases, provided they have an acceptable performance status to undergo general sedation. To be deemed appropriate candidates for ECT treatment, they generally have to meet both general and procedure-specific inclusion criteria (Table 2.5).

2.4.2 Nodules registration

In the great majority of BC patients, there is no need for additional radiological imaging. Superficial metastases can be simply counted and measured by means of a millimetre scale ruler.

2.4.3 ECT administration

Drug injection

According to the current standard operating procedures, ECT may be performed either with bleomycin (BLM) (both using the intratumoral and/or the intravenous route) or cisplatin (CDDP) (administered exclusively intratumorally, according to the guidelines) [60]. In clinical practice, CDDP can be considered an alternative agent for selected patients who develop drug resistance or cannot receive BLM. However, according to the clinical experience that led to the European Standard Operating Procedures on Electrochemotherapy (ESOPE), CDDP administration has been restricted exclusively to the intralesional route, thus limiting its application to the patients with a limited number of

Table 2.5 Patients selection criteria for ECT of CWR from BC

Selection/exclusion criteria	
General	**Procedure-specific**
Previous mastectomy[a]	Superficial (skin, dermis, hypodermis) metastases suitable for electrode application[b]
Exclusion from radical resection	Exclusion of allergy to bleomycin or cisplatin
Contraindication to CW irradiation	No history of epilepsy
Resistance or intolerance to systemic treatments	No presence of: – cardiac pace-maker – mammary expander or prosthesis – subcutaneous port-a-cath for chemotherapy
Absence of systemic and soft tissue infections	
Adequate Performance Status[c]	
Absence of severe organ impairment[c]	
Adequate life expectancy (>3 months)	
Multidisciplinary discussion	

[a] ECT is also technically applicable in BC patients with skin metastases, who did not undergo previous mastectomy, but at present there is not enough clinical evidence to support such an approach out of a clinical trial.
[b] With the available pulse generators for clinical ECT, deep soft tissue or bone metastases cannot be treated; muscle and bone metastases are the specific target of ECT application in the frame of dedicated, ongoing clinical trials in Italy (Rizzoli Orthopaedics Institute, Bologna) and Slovenia (Ljubljana Institute of Oncology).
[c] ECT candidates are often frail and heavily-treated patients. Generally, to be considered for ECT they have to be suitable for a mild general sedation. Moreover, an accurate and realistic balance of the possible pros and cons of treating superficial metastases is mandatory for the treating oncologist.

metastases. The selection of the anticancer agent, BLM or CDDP, is at the discretion of the physician who takes into account the ESOPE, but also the patient's characteristics and previous treatments.

Since an extensive tumor skin infiltration is often present, most of the patients undergoing ECT for CWR from BC are preferentially treated with BLM, administered by means of a peripheral endo-venous catheter. In fact, the systemic route of administration allows a more homogeneous drug distribution and thus an adequate tumor treatment. The bolus of BLM (at a dosage of 15 IU/m^2 of body surface area) is not administered before verifying that the pulse generator is regularly charged and ready to deliver the electric pulses. BLM is given in 50–60' to the patient through the peripheral venous access. Over the following 8 minutes the physician prepares the patients for the application of electric pulses. The anatomical region to be electroporated is thoroughly disinfected and edged with sterile drapes. If necessary, the operator also proceeds with tissue infiltration with a mixture of short- and long-acting local anesthetics.

Electroporation

In general, the choice of the electrode type for ECT administration depends on the tumor location, size and depth, as well as on the patient's anatomy. Generally, CWR are best electroporated by applying a needle electrode, available in different sizes and geometries. The maximum length of the available needle electrodes for the routine clinical practice is 30 mm. However, in most BC patients who have previously received mastectomy, the thickness of the tumor-involved CW soft tissues is by far less than 30 mm, therefore ECT can also be performed with shorter electrodes.

However, the application of a longer needle electrode could be indicated in selected cases, when treating large, fungating tumors. In most BC patients, ECT can be safely and effectively administered by means of a 7-needle, hexagonal array, needle electrode (Figure 2.4) that constitutes an acceptable compromise between invasiveness and possibility to treat a large skin surface during a single treatment session. Considering that the hexagonal-array needle electrode covers a surface of at least 3 cm^2 and that the maximum number of electrode applications, in the 20-minute interval after the administration of BLM bolus, is around 100–120, therefore in theory, the amount of body surface area that can be electroporated during a single course of ECT ranges from 300–360 cm^2. The possibility to additionally administer some BLM (or CDDP) by the intratumoral route during the same procedure, allows effective treatment of additional, albeit small, tumor nodules [61]. If a patient presents with extensive CW disease that cannot be encompassed in a single treatment, the involved area can be divided into adjacent fields, and the treatment planned in multiple ECT sessions with a minimum 1–2 week interval.

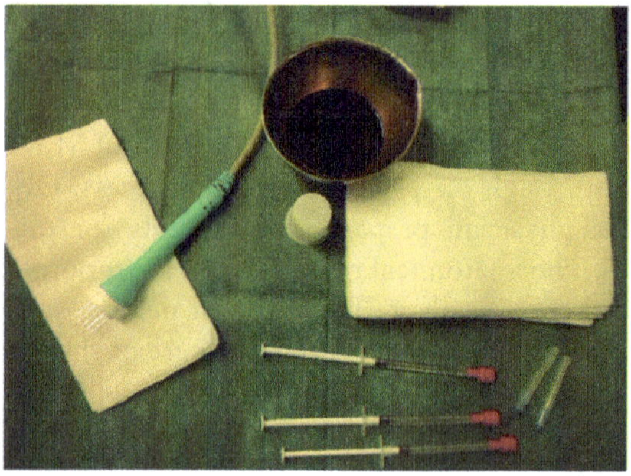

(a)

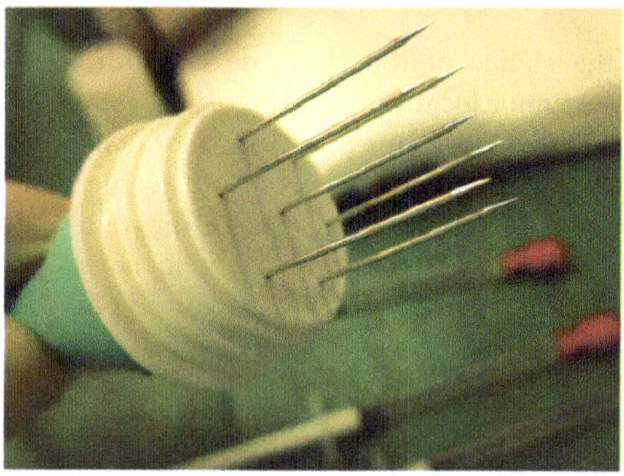

(b)

Figure 2.4 The instruments for an ECT treatment prepared on the table of the operating room. Gauze, skin disinfectant and the electrode connected to the appropriate handle (a). A magnification of the electrodes of the hexagonal-array needle (b)

2.4.4 Follow-up (response assessment and quality of life)

After discharge from hospital, a number of patients do not require specific skin care, but simply to take some analgesics and to clean and dress the electroporated area. Conversely, after the treatment of an extensive CW surface, the patients are seen at least weekly, to assess possible dermatological toxicity, verify the adequacy of the prescribed analgesics and manage the treated area with proper dressings. Local tumor response is assessed at the 1-month visit and confirmed at the 2-month control, according to the Response Evaluation Criteria In Solid Tumors (RECIST) [62]. Alternatively, the World Health Organization (WHO) measurement criteria can be used, but they require a bi-dimensional tumor assessment of each target lesion. Since they proved to be no more accurate than RECIST, they are also more impractical than the uni-dimensional assessment required by RECIST criteria [63].

At the end of ECT, patients are referred to the treating medical oncologist for follow-up and further systemic treatments, as clinically indicated. Generally, when the oncologists deal with advanced BC patients who underwent ECT for the occurrence of CWR, the local tumor response is still an important endpoint, but the duration of such a response (i.e. local tumor control, expressed alternatively by the endpoints of 'local progression-free survival' or 'local-disease-free survival') becomes the primary objective. Finally, because ECT is often applied in a palliative setting, the clinical benefit of this treatment could not be claimed without a rigorous assessment of patients' quality of life outcomes by means of dedicated instruments (validated questionnaires on quality of life and patients' reported outcomes).

2.4.5 Review of clinical trials

Recently, the comprehensive ECT clinical experience on CWR from BC has been reviewed by Sersa et al. [64]. The authors analyzed eight clinical trials from 1996 to 2009, none of which was specifically designed for BC. Comparisons of retrospective series from different institutions are always problematic. From the collected data, the overall response rate was 89%, complete 59%. Although encouraging, these results were obtained by means of different ECT protocols, either for the chemotherapeutic agents (BLM or CDDP) or their delivery route (intravenous or intratumoral). Finally, only 37 out of 49 patients considered in this review were treated according to the ESOPE [60].

2.4.6 Personal experience

The experience on ECT treatment of BC patients with CWR relies on a cohort study of a mixed series of different cancer patients with skin metastases, and a BC-specific phase-II study on BC patients with CWR after mastectomy [65,66]. From 2006, a total of 51 BC patients (median age 70 years, range 38–88) underwent ECT for CWR at the Veneto Region Oncology Research Institute of Padova. They received a median of 2 ECT courses (range 1–5). Local tumor response was assessed, according to RECIST, on 212 target lesions (median size 20 mm, range 10–320). Two-month objective response was as follows: 43.2% complete (22/51 patients), 47.5% partial (24/51), 9.8% no change (5/51). Treatment-related morbidity was mild, worsened with the increasing number of ECT applications and consisted primarily of pain ('moderate'/'severe' in 6, 13 and 17% of patients 1 month after the first, second and third ECT, respectively) and dermatological toxicity (acute clinically relevant skin ulceration in 14, 20 and 33% of patients, respectively). Median follow-up was 32 months (range, 6–53) and the 3-year local control rate of 71%. However, 33 of 51 patients developed new lesions (NL) in non-electroporated areas after a median time of 6.6 months (range 2.3–29.5), therefore 28 patients received repetitive treatments (up to 5 ECT cycles).

2.5 Electrochemotherapy: the engineer's point of view

2.5.1 Description of electric signal in ECT treatment

The most widely adopted pulse generator for clinical ECT in Europe (the Cliniporator™ device, Igea, Modena, Italy) records the voltages and electric currents in tissues when a rectangular voltage pulse is applied between two electrodes. An example of voltage and current data as a function of time is shown in Figure 2.5, where 10 electrode applications of an ECT treatment using an hexagonal electrode are reported. The upper graph indicates the applied voltage in which the maximum amplitude is imposed by the pulse generator. The bottom graph is the corresponding current through the electrodes. Vertical lines divide the pulses in groups of 8, and each 8-pulse group represents a single electrode application. It is noteworthy that, during ECT treatment, given the same applied voltage by the pulse generator, the values of electric current change in the single applications, probably because during the procedure, the electrodes are progressively moved in order to treat all the target area.

Considering the hexagonal-array electrode (Figure 2.4), that is composed of seven needles, the voltage is progressively applied to

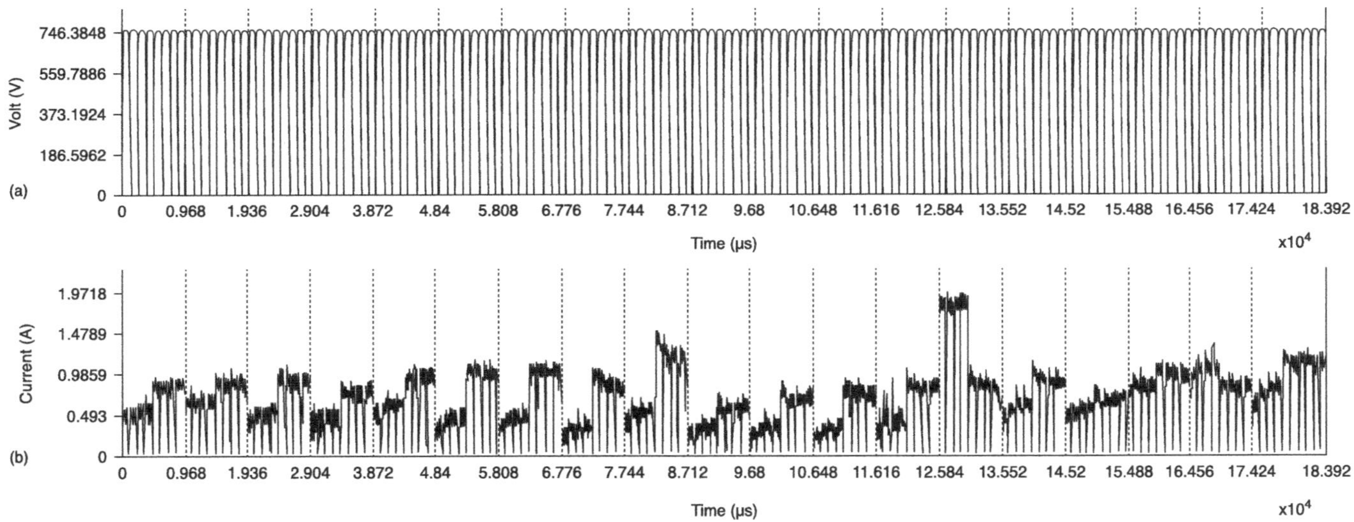

Figure 2.5 Voltage (a) and current (b) recorded by the software of the pulse generator during an ECT treatment of a BC patient with CWR. Ten needle electrode applications by means of an hexagonal-array needle electrode. Each 8-pulse group represents a single electrode placement

different electrode pairs. The differences in the current amplitude that can be observed in each 8-pulse group are because each group can be subdivided into two subsets of four pulses that correspond to the voltage application to different tissue area. The possible variations of the current amplitude in a subset of four pulses are due to the change in tissue electric conductivity derived by the cell membrane electro-permeabilization [65, 66].

In general, since the electro-permeabilization process is improved by applying more consecutive pulses [66], then the fourth pulse of each group can be considered the one that corresponds to the best cell membrane permeabilization and should be the one considered in the treatment analysis. The variability of the current values can be related to the difference of tissue electrical characteristics [67–69], but also to the depth at which the needle electrodes are inserted in the tissues. In general, the deeper the electrodes are inserted, the more the current flow.

In clinical practice, the variability in the depth of electrode insertion depends on the treated anatomical area and tissue stiffness (e.g. it is harder to insert a needle into fibrous tissues than into intact tissues). For instance, in the tissue of limbs (e.g. a forearm or thigh), the needles are more easily placed than in the CW, especially after mastectomy. The system of two electrodes implanted in biological tissues and supplied by a pulse generator can be represented as the parallel combination of a resistance and capacitance (Figure 2.6). This simple circuit describes the behavior of tissues when a voltage pulse is applied [70–74]. To compute the circuit resistance, the constant part of the ratio of the pulse voltage and current (after that the current transient due to capacity effect has

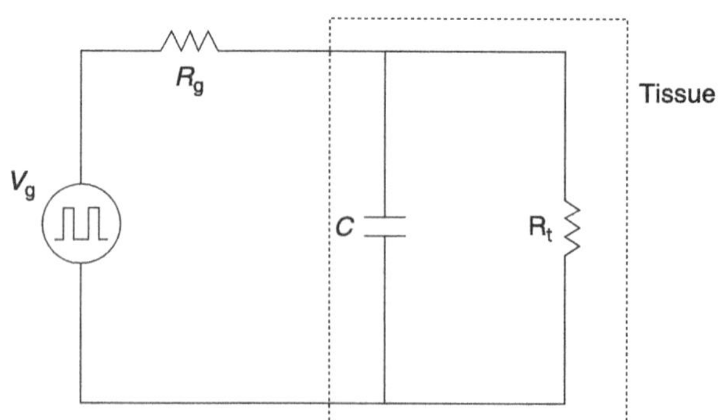

Figure 2.6 Equivalent electric circuit describing the electrical behavior of tissues during an ECT treatment

vanished) is used. The circuit in Figure 2.6 shows a steady state network and the resistance value R_t is the ratio between voltage and current.

2.5.2 Analysis of ECT data

The treatment of tumors with ECT can be considered in terms of voltage and current time sequences. The corresponding resistance values for each pulse can be computed. For simplicity, only one pulse for each ECT application (e.g. the last pulse of the second subset of four pulses corresponding to each ECT application) was considered. The resistance values of different electrode applications in the same as well as different treatments have been compared according to tumors' histology and their anatomical location. In the clinic, the effectiveness of pulse delivery and therefore of the process of electro-permeabilization is evaluated in real time by the clinician in terms of the current between the electrodes, since the pulse generator device visualizes on a screen the current value of each pulse. Generally, for an hexagonal-array needle electrode, an electric current of 1.5 A identifies a successful pulse delivery that corresponds to cell electroporation [61].

It is a common finding in clinical ECT that the recorded current widely changes during the same and different treatments. For instance, in Figure 2.7(a) the current values reported during the ECT treatment of patients with melanoma metastases on a lower limb are compared with those of a BC patient with CWR. Although in the two patients the applied voltage was the same, different levels of electric current have been recorded. In this example, given a voltage pulse of about 700 V (Figure 2.5) of several tens of μs at a few kHz, the current registered during the ECT of the melanoma patient is in the order of some Ampère (A) and can be close to 6 A, whereas in the ECT data of the BC patient, the current is by far lower, close to 1 A.

Moreover, it is noteworthy that each 8-pulse group corresponds to a different treated area, since the tumor can have an extension of several cm^2 and the 7-needle electrode has a treatment area of approximately $1.3\,cm^2$. Therefore, several electrode placements are needed to treat tumors infiltrating large skin surfaces. In fact, also analyzing the ECT treatment record of a single patient, some differences in the values of electric current for each electrode application can be noted (Figure 2.7(a)). For each 8-pulse series, the electrical resistance corresponding to the 8th pulse has been computed. Then, for each patient that has been treated with N electrode placements to cover the whole tumor area, N resistance

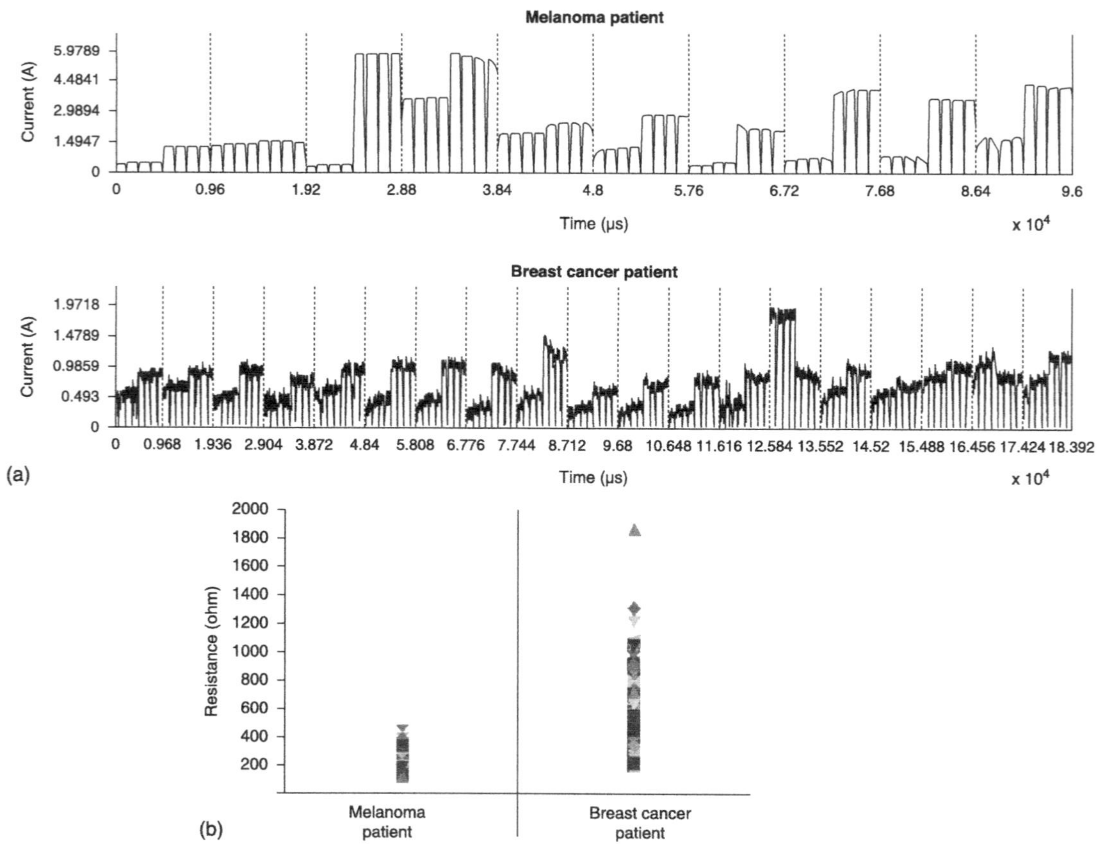

Figure 2.7 Electric current and resistance during clinical ECT. Values of the electric current (in Ampere, A) recorded during an ECT treatment of in-transit metastases from melanoma located in a lower limb and of CWE from BC after mastectomy (a). Electrical resistance values computed from voltage and current value of the two patients (b)

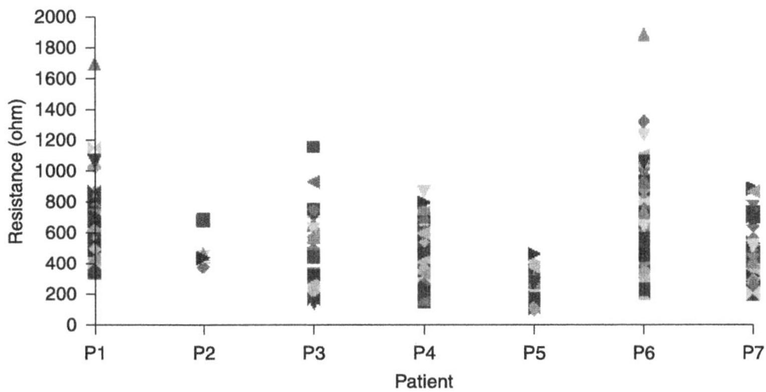

Figure 2.8 Resistance values (in Ohms, Ω) computed for 7 BC patients (P1–P7) treated with ECT for CWR

values have been computed. It is evident that the resistance evaluated in the ECT treatment of the BC patient is higher than the resistance computed for the melanoma patient (Figure 2.7(b)). Moreover, the resistance values during the ECT for BC have a larger interval of variability (200–1900 Ω in this example) with respect to the ones of the lower limb melanoma (150–400 Ω in this example). Since higher values of resistance correspond to lower currents flowing between the electrodes, this observation could be relevant to improve ECT activity in different cancer types or in the same tumor type, but localized in different anatomical regions.

Comparing the resistance data from different BC patients, their values and range were similar (Figure 2.8). In these patients, the values of the electric resistance displayed a similar pattern, even in the presence of a high intra-patient variability. This finding could be related to the inhomogeneity of the treated tissues (presence of fibrous tissues and surgical scars on the CW) and consequently to the difficulty to properly place the electrodes in the tissues.

2.6 Conclusions and perspectives

ECT represents a new promising treatment option to be considered by oncologists when dealing with patients with skin involvement from BC, particularly in the setting of CWR after mastectomy. The use of locally-delivered electrical pulses is well tolerated by the patients and magnifies the anticancer effect of BLM, administered at standard dosages, thanks to the mechanism of reversible cell permeabilization and drug entrapment. The mechanism of electroporation also allows the enhancement of CDDP activity, but its clinical application is limited by the intratumoral administration, according to the available guidelines [60]. The preliminary clinical results of the various combinations (i.v. BLM plus electroporation, i.t. BLM plus electroporation, i.t. CDDP plus electroporation) of ECT treatment in metastatic BC are encouraging and highlight the possibility of ECT application in different clinical scenarios of recurrent BC. Although promising, the exact role of this therapeutic approach in the multimodality management of BC patients is still under investigation and further clinical studies are warranted to confirm the preliminary results of treatment activity, explore its toxicity (especially in case of multiple applications), confirm its palliative value and find possible shared indications in BC patients.

Nevertheless, it is a common finding in clinical practice that a number of patients with CWR are not suitable for standard locoregional treatments, mainly RT, due to previous radiation exposure or comorbidities. In this setting, the benefits of ECT are the shortness of treatment duration (the procedure lasts a maximum of 20–30 minutes), easiness of application, brief hospitalization (generally 1–2 days), high activity rates, repeatability, possibility of safe combination with other treatments and limited systemic toxicity. In light of the favorable preliminary results from published clinical experiences and the technical possibility of an improved delivery of the electric voltages at the tumor site by means of new technological solutions (e.g. tailored electric voltages), it is conceivable that ECT might gain a role in the multidisciplinary management of recurrent BC and enlarge the therapeutic armamentarium for metastatic BC.

2.7 Acknowledgments

The authors wish to acknowledge the valuable input from Carlo Riccardo Rossi, Sarcoma and Melanoma Unit of the Veneto Region Oncology Research Institute, Padova, Italy and the insightful discussions with Luis Mir, Institute Gustave-Roussy, Villejuif Cédex, France. They also wish to thank Guido Sotti and Luigi Corti, radiotherapists, and Davide Canonico, health physicist, from the Radiotherapy Section of the Veneto Region Oncology Research Institute, Padova, Italy, for logistic support, graphic processing and radiotherapy pretreatment planning.

2.8 References

1. Buchanan C L, Dorn P L, Fey J, et al. (2006), 'Locoregional recurrence after mastectomy: incidence and outcomes', *J Am Coll Surg*, **203**: 469.
2. Schmoor C, Sauerbrei W, Bastert G and Schumacher M (2000), 'Role of isolated locoregional recurrence of breast cancer: results of four prospective studies', *J Clin Oncol*, **18**: 1696.
3. Andry G, Suciu S, Vico P, et al. (1989), 'Locoregional recurrences after 649 modified radical mastectomies: incidence and significance', *Eur J Surg Oncol*, **15**: 476.
4. Pisansky T M, Ingle J N, Schaid D J, et al. (1993), 'Patterns of tumor relapse following mastectomy and adjuvant systemic therapy in patients with axillary lymph node-positive breast cancer. Impact of clinical, histopathologic, and flow cytometric factors', *Cancer*, **72**: 1247.
5. Recht A, Gray R, Davidson N E, et al. (1999), 'Locoregional failure 10 years after mastectomy and adjuvant chemotherapy with or without tamoxifen without irradiation: Experience of the Eastern Cooperative Oncology Group', *J Clin Oncol*, **17**: 1689.
6. Hsi R A, Antell A, Schultz D J and Solin L J (1998), 'Radiation therapy for chest wall recurrence of breast cancer after mastectomy in a favorable subgroup of patients', *Int J Radiat Oncol Biol Phys*, **42**: 495.
7. Willner J, Kiricuta I C and Kölbl O (1997), 'Locoregional recurrence of breast cancer following mastectomy: always a fatal event? Results of univariate and multivariate analysis', *Int J Radiat Oncol Biol Phys*, **37**: 853.
8. Katz A, Strom E A, Buchholz T A, et al. (2000), 'Locoregional recurrence patterns after mastectomy and doxorubicin-based chemotherapy: implications for postoperative irradiation', *J Clin Oncol*, **18**: 2817.
9. Nielsen H M, Overgaard M, Grau C, et al. (2006), 'Loco-regional recurrence after mastectomy in high-risk breast cancer-risk and prognosis. An analysis of patients from the DBCG 82 b&c randomization trials', *Radiother Oncol*, **79**: 147.
10. Cheng S H, Horng C F, Clarke J L, et al. (2006), 'Prognostic index score and clinical prediction model of local regional recurrence after mastectomy in breast cancer patients', *Int J Radiat Oncol Biol Phys*, **64**: 1401.
11. Overgaard M, Hansen P S, Overgaard J, et al. (1997), 'Postoperative radiotherapy in high-risk premenopausal women with breast cancer who receive adjuvant chemotherapy. Danish Breast Cancer Cooperative Group 82b Trial', *N Engl J Med*, **337**: 949.
12. Taylor M E, Haffty B G, Rabinovitch R, et al. (2009), 'CR appropriateness criteria on postmastectomy radiotherapy', *Int J Radiation Oncology Biol Phys*, **73**: 997–1002.
13. van Tienhoven G, Voogd A C, Peterse J L, et al. (1999), 'Prognosis after treatment for loco-regional recurrence after mastectomy or breast conserving therapy in two randomised trials (EORTC 10801 and DBCG-82TM). EORTC Breast Cancer Cooperative Group and the Danish Breast Cancer Cooperative Group', *Eur J Cancer*, **35**: 32.

14. Kamby C and Sengelov L (1997), 'Pattern of dissemination and survival following isolated locoregional recurrence of breast cancer', *Breast Canc Res Treat*, **45**: 181–92.
15. Juan O, Lluch A, de Paz L, et al. (1999), 'Prognostic factors in patients with isolated recurrences of breast cancer (stage IV-NED)', *Breast Canc Res Treat*, **53**: 105–112.
16. Haffty B, Hauser A, Ho Choi D, et al. (2004), 'Molecular markers for prognosis after isolated postmastectomy chest wall recurrence', *Cancer*, **100**: 252–63.
17. Carreno G, del Casar J M, Corte M D, et al. (2007), 'Local recurrence after mastectomy for breast cancer: analysis of clinicopathological, biological and prognostic characteristics', *Breast Canc Res Treat* **102**: 61–73.
18. Halverson K J, Perez C A, Kuske R R, et al. (1992), 'Survival following locoregional recurrence of breast cancer: univariate and multivariate analysis', *Int J Radiat Oncol Biol Phys*, **23**: 285.
19. Raimbault M, Lavoué V, Morcel K, et al. (2011), 'Isolated skin recurrence following salvage mastectomy for intramammary recurrence (after initial breast conservation therapy): is it a fatal event?', *The Breast*, **20**: 380–4.
20. Haffty B, Fischer D, Beinfield M, et al. (1991), 'Prognosis following local recurrence in the conservatively treated breast cancer patient', *Int J Radiation Oncol Biol Phys*, **21**: 293–98.
21. Schmoor C, Sauerbrei W, Bastert G, et al. (2000), 'Role of isolated locoregional recurrence of breast cancer: Results of four prospective studies', *J Clin Oncol*, **18**: 1696–708.
22. Bedwinek J, Lee J, Fineberg B, et al. (1981), 'Prognostic indicators in patients with isolated local-regional recurrence of breast cancer', *Cancer*, **47**: 2232–5.
23. Chapgar A, Meric-Bernstam F, Hunt K K, et al. (2003), 'Chest wall recurrence after mastectomy does not always portend a dismal outcome', *Ann Surg Oncol*, **10**: 628–34.
24. Bhandari T, Dizon DS, Taneja C, et al. (2007), 'Clinical characteristics of women presenting with skin-only recurrence of breast cancer', *Am J Surg*, **194**: 494–6.
25. Willner J, Kiricute I C, Kolbi O (1997), 'Locoregional recurrence of breast cancer following mastectomy: always a fatal event? Results of univariate and multivariate analysis', *Int J Rad Oncoly Biol Phys*, **37**: 853–63.
26. National Comprehensive Cancer Network (NCCN) guidelines. Available at: *www.nccn.org* (accessed April 20, 2012).
27. Santillan A A, Kiluk J V, Cox J M, et al. (2008), 'Outcomes of locoregional recurrence after surgical chest wall resection and reconstruction for breast cancer', *Ann Surg Oncol*, **15**: 1322–29.
28. van der Pol C C, van Geel A N, Menke-Pluymers M B E, et al. (2009), 'Prognostic factors in 77 curative chest wall resections for isolated breast cancer recurrence', *Ann Surg Oncol*, **16**: 3414–21.
29. Faneyte I F, Rutgers E J, Zoetmulder F A (1997), 'Chest wall resection in the treatment of locally recurrent breast carcinoma: indications and outcome for 44 patients', *Cancer*, **80**: 886–91.

30. Pfannschmidt J, Geisbusch P, Muley T, et al. (2005), 'Surgical resection of secondary chest wall tumors', *Thorac Cardiovasc Surg*, **53**: 234–9.
31. McCormack P M, Bains M S, Burt M E (1989), 'Local recurrent mammary carcinoma failing multimodality therapy. A solution', *Arch Surg*, **124**: 158–61.
32. Friedel G, Kuisper T, Dippon J, et al. (2008), 'Full-thickness resection with myocutaneous flap reconstruction for locally recurrent breast cancer', *Ann Thorac Surg*, 85: 1894–900.
33. Flook D, Webster D J, Hughes L E and Mansel R W (1989), 'Salvage surgery for advanced local recurrence of breast cancer', *Br J Surg*, **76**: 512–4.
34. Halverson K J, Perez C A, Kuske R R, et al. (1990), 'Isolated local-regional recurrence of breast cancer following mastectomy: Radiotherapeutic management', *Int J Radiat Oncol Biol Phys*, **19**: 851.
35. Bedwinek J M, Fineberg B, Lee J and Ocwieza M (1981), 'Analysis of failures following local treatment of isolated local regional recurrence of breast cancer', *Int J Radiat Oncol Biol Phys*, **7**: 581–5.
36. Aberizk W J, Silver B, Henderson I C, et al. (1986), 'The use of radiotherapy for treatment of isolated locoregional recurrence of breast carcinoma after mastectomy', *Cancer*, **58**: 1214.
37. Schwaibold F, Fowble B L, Solin L J, et al. (1991), 'The results of radiation therapy for isolated local regional recurrence after mastectomy', *Int J Radiat Oncol Biol Phys*, **21**: 299.
38. Mendenhall N P, Devine J W, Mendenhall W M, et al. (1988), 'Isolated local-regional recurrence following mastectomy for adenocarcinoma of the breast treated with radiation therapy alone or combined with surgery and/ or chemotherapy', *Radiother Oncol,* **12**: 177–85.
39. Chu F C H, Lin F J, Kim J H, et al. (1976), 'Locally recurrent carcinoma of the breast. Results of radiation therapy', *Cancer*, **37**: 2677–81.
40. Janjan N A, McNeese M D, Buzdar A U, et al. (1986), 'Management of locoregional recurrent breast cancer', *Cancer*, **58**: 1552–6.
41. Haylock B J, Coppin C M, Jackson J, et al. (2000), 'Locoregional first recurrence after mastectomy: prospective cohort studies with and without immediate chemotherapy', *Int J Radiat Oncol Biol Phys*, **46**: 355.
42. Kuo S H, Huang C S, Kuo W H, et al. (2008), 'Comprehensive locoregional treatment and systemic therapy for postmastectomy isolated locoregional recurrence', *Int J Radiat Oncol Biol Phys*, **72**: 1456.
43. Chagpar A, Kuerer H M, Hunt K K, et al. (2003), 'Outcome of treatment for breast cancer patients with chest wall recurrence according to initial stage: implications for post-mastectomy radiation therapy', *Int J Radiat Oncol Biol Phys*, **57**: 128.
44. Chen K K, Montague E D and Oswald M J (1985), 'Results of irradiation in the treatment of locoregional breast cancer recurrence', *Cancer*, **56**: 1269.
45. Wahl A O, Rademaker A, Kiel K D, et al. (2088), 'Multi-institutional review of repeat irradiation of the chest wall and breast for recurrent breast cancer', *Int J Radiat Oncol Biol* Phys, **70**: 477–84.
46. Laramore G E, Griffin T W, Parker R G, et al. (1978), 'The use of electron beams in treating local recurrence of breast cancer in previously irradiated

fields', *Cancer*, **41**: 991–5.
47. Phromratanapongse P, Steeves R A, Severson S B, et al. (1991), 'Hyperthermia and irradiation for locally recurrent previously irradiated breast cancer', *Strahlenter Onkol*, **167**: 93–7.
48. Wagman R, Katz M, Hong L, et al. (2002), 'Re-irradiation of the chest wall for recurrent breast cancer', *Int J Radiat Oncol Biol Phys*, **54**: S237–8.
49. Ballo M T, Strom E A, Prost H, et al. (1999), 'Local-regional control of recurrent breast carcinoma after mastectomy: Does hyperfractionated accelerated radiotherapy improve local control?', *Int J Radiat Oncol Biol Phys*, **44**: 105.
50. Vernon C C, Hand J W, Field S B, et al. (1996), 'Radiotherapy with or without hyperthermia in the treatment of superficial localized breast cancer: Results from five randomized controlled trials. International Collaborative Hyperthermia Group', *Int J Radiat Oncol Biol Phys*, **35**:7 31.
51. Feyerabend T, Wiedemann G J, Jäger B, et al. (2001), 'Local hyperthermia, radiation, and chemotherapy in recurrent breast cancer is feasible and effective except for inflammatory disease', *Int J Radiat Oncol Biol Phys*, **49**: 1317.
52. Schuh M, Nseyo U O, Potter W R, et al. (1987), 'Photodynamic therapy for palliation of locally recurrent breast carcinoma', *J Clin Oncol*, **5**: 1766.
53. Sperduto P W, DeLaney T F, Thomas G, et al. (1991), 'Photodynamic therapy for chest wall recurrence in breast cancer', *Int J Radiat Oncol Biol Phys*, **21**: 441.
54. Koren H, Alth G, Schenk G M and Jindra R H (1994), 'Photodynamic therapy – an alternative pathway in the treatment of recurrent breast cancer', *Int J Radiat Oncol Biol Phys*, **28**: 463.
55. Khan S A, Dougherty T J, Mang T S (1993), 'An evaluation of photodynamic therapy in the management of cutaneous metastases of breast cancer', *Eur J Cancer*, **29A**: 1686.
56. Wyss P, Schwarz V, Dobler-Girdziunaite D, et al. (2001), 'Photodynamic therapy of locoregional breast cancer recurrences using a chlorin-type photosensitizer', *Int J Cancer*, **93**: 720.
57. Cuenca R E, Allison R R, Sibata C, et al. (2004), 'Breast cancer with chest wall progression: treatment with photodynamic therapy', *Ann Surg Oncol*, **11**: 322–27.
58. Guarneri V and Conte P (2009), 'Metastatic breast cancer: Therapeutic options according to molecular subtypes and prior adjuvant therapy', *The Oncologist*, **14**: 645–56.
59. Schmid-Büchi S, Halfens R J, Müller M, Dassen T and van den Borne B (2013), 'Factors associated with supportive care needs of patients under treatment for breast cancer', *Eur J Oncol Nurs*, **17(1)**: 22–9.
60. Mir L M, Gehl J, Sersa G et al. (2006), 'Standard operating procedures of the electrochemotherapy: Instructions for the use of bleomycin or cisplatin administered either systemically or locally and electric pulses delivered by the cliniporator by means of invasive or non-invasive electrodes', *EJC Suppl*, **4**: 14–25.
61. Marty M, Sersa G, Garbay J, Gehl J, Collins C, Snoj M, et al. (2006), 'Electrochemotherapy – an easy, highly effective and safe treatment of

cutaneous and subcutaneous metastases: results of ESOPE (European Standard Operating Procedures of Electrochemotherapy) study', *EJC Suppl*, **4**: 3–13.
62. Therasse P, Arbuck S G, Eisenhauer E A (2000), 'New guidelines to evaluate the response to treatment in solid tumors. European Organization for Research and Treatment of Cancer, National Cancer Institute of the United States, National Cancer Institute of Canada', *J Natl Cancer Inst*, **92**: 205–16.
63. Miller A B, Hoogstraten B, Staquet M and Winkler A (1981), 'Reporting results of cancer treatment', *Cancer*, **47**: 207–14.
64. Sersa G, Cufer T, Paulin S M, Cemazar M and Snoj M (2012), 'Electrochemotherapy of chest wall breast cancer recurrence', *Cancer Treat Rev*, **38(5)**: 379–86.
65. Campana L G, Mocellin S, Basso M, Puccetti O, De Salvo G L, et al. (2009), 'Bleomycin-based electrochemotherapy: clinical outcome from a single institution's experience with 52 patients', *Ann Surg Oncol*, **6**: 191–9.
66. Campana L G, Valpione S, Falci C, et al. (2012), 'The activity and safety of electrochemotherapy in persistent chest wall recurrence after mastectomy: a phase II study', *Breast Cancer Res Treat*, **134(3)**: 1169–78.
67. Pavlin M, Kandušer M, Reberšek M, Pucihar, Hart F X, et al. (2005), 'Effect of cell electroporation on the conductivity of a cell suspension', *Biophys J*, **88**: 4378–90.
68. Pliquett U, Langer R, and Weaver J C (1995), 'Changes in the passive electrical properties of human stratum corneum due to electroporation', *Biochimica et BiophysicaActa (BBA) – Biomembranes*, **1239**: 111–21.
69. Miklavčič D, Pavšelj N, and Hart F X (2006), 'Electric properties of tissues', in *Wiley Encyclopedia of Biomedical Engineering*, New York: John Wiley & Sons, Inc.
70. Pavšelj N and Miklavčič D (2008), 'Numerical models of skin electropermeabilization taking into account conductivity changes and the presence of local transport regions', *Plasma Science, IEEE Trans*, **36**: 1650–8.
71. Pavšelj N, Préat V and Miklavčič D (2007), 'A numerical model of skin electropermeabilization based on *in vivo* experiments', *Annals Biomed Eng*, **35**: 2138–44.
72. Rubinsky B (2010), *Irreversible Electroporation*. Berlin: Heidelberg: Springer.

3

Clinical electrochemotherapy for advanced superficial melanoma

Luca G. Campana, Raji Sundararajan, Vanna Chiarion-Sileni and Carlo R. Rossi

DOI: 10.1533/9781908818294.35

Abstract: Cutaneous and subcutaneous metastases represent the target of electrochemotherapy (ECT) and occur in 2–20% of melanoma patients, depending on primary tumor histopathological features. ECT has been successfully tested in melanoma patients in the frame of several early clinical trials that confirmed its high activity on advanced, refractory disease. In 2006, the ECT technique was standardized thanks to the European Standard Operating Procedure on Electrochemotherapy (ESOPE) multicenter trial and entered into routine oncological practice, mainly in a palliative setting. Recently, some clinical experiences with ECT in the treatment of melanoma patients have shown a promising tumor control rate on electroporated tumors, although the majority of patients continue to undergo repetitive treatment applications due to disease progression. Researchers are actively investigating new ways to expand and improve ECT technology, or the combination of ECT itself with immunomodulatory agents on electroporated cancers and possibly elicit a systemic and effective antitumor response.

Key words: melanoma, in-transit metastases, electroporation, electrochemotherapy, bleomycin.

3.1 Introduction

Since the 1970s, the incidence of cutaneous melanoma has risen steadily in Western populations [1]. In 2007, the number of death from malignant melanoma exceeded 8000 in the US, with an estimated 8790 deaths in 2011 [2]. Taking into account the patients who died for melanoma between 2005 and 2009 in the US, the age-adjusted death rate was 2.7 per 100 000 per year, with a median age of 68 years [3]. Although the 10-year survival rate for patient with an early stage melanoma without unfavorable histo-prognostic factors is greater than 95% [4], the prognosis is much more severe for patients diagnosed with advanced disease [5]. When melanoma is detected in late stages (Stage III or IV, i.e. locoregional or metastatic melanoma, respectively), it carries a dismal prognosis, with a mean survival of about 6–8 months in Stage IV disease.

Malignant melanoma usually spreads both by the lymphogenous and the haematogenous route and can virtually metastasize to any organ in the body. When disease recurrence occurs, this can follow a sequential pattern (local recurrence first, then locoregional recurrence) or, alternatively, present directly as distant metastases. Local recurrence comprises cutaneous tumor nodules at the site of primary tumor, within 2 cm from the surgical scar (i.e. *satellitosis*), while locoregional recurrence includes both tumor deposits (skin and/or subcutaneous tissue) beyond 2 cm from the primary cancer (i.e. in-transit metastases) and the regional lymph node metastases. Satellitosis, in-transit metastases and lymph node metastases represent different stages of the melanoma dissemination through the dermal lymphatics. Finally, distant metastases include soft tissues (skin, subcutaneous tissue, muscles), visceral localizations (mainly lung, liver, brain and bowel) and bone.

At present, cutaneous and subcutaneous metastases represent the tumor localizations that can be targeted using electrochemotherapy (ECT) and occur in 2–20% of melanoma patients, depending on primary tumor histopathological features. They can occur either during the early or late phase of disease.

The treatment options for patients with metastatic melanoma are few and generally based on low-level evidence. Although most patients with metastatic disease receive systemic treatments, until recently no therapeutic regimen has been shown to prolong survival in large, randomized, phase III trials. The combination of different therapeutic approaches is based on the concept that agents with distinct mechanisms of action may have a synergistic effect, but none of these combinations have been shown to improve survival over the single-agent chemotherapy with dacarbazine [6–7]. More recently, both the anti-CTLA-4 monoclonal antibody Ipilimumab and the selective BRAF inhibitor Vemurafenib have shown a survival benefit in metastatic melanoma patients in the frame of phase III trials [8–9]. However, some economic concerns (the cost of Ipilimumab is over $120.000), the peculiar toxicity profile of the new agents and the restriction of targeted therapy only for suitable candidates (only ~50% of melanoma patients have the required BRAF mutation) make clear that new approaches for the treatment of metastatic melanoma are still required.

The electroporation-mediated delivery of anticancer drugs represents an innovative and promising approach for advanced melanoma patients who develop superficial metastases. ECT has been successfully tested since the 1990s in melanoma patients with skin metastases in the frame of several early clinical trials [10–14]. These studies, although heterogeneous, were mainly focused on improving the treatment parameters and formed the basis for the standardization of the operating procedure. The multicenter European Standard Operating Procedure on Electrochemotherapy (ESOPE) trial confirmed the activity and safety of ECT and established the operating procedures to apply this treatment in routine clinical practice [15]. Since 2006, when the operating procedures where published, several clinical experiments with ECT in melanoma patients has been conducted in different European cancer centers [16–20]. Since its early clinical applications, the clinical experience of ECT has rapidly grown and many melanoma patients have been successfully treated.

The process of electroporation/electropermeabilization also represents the physical basis for the electro-gene delivery, an alternative therapeutic strategy that has been tested in metastatic melanoma in the context of a phase I study at the H Lee Moffitt Cancer Center (Tampa, FL) [21]. This approach, although more complex than the combination of electric pulses and antiblastics, proved to be an effective method of plasmid DNA (i.e. DNA encoding for IL-12) delivery and therefore a potential therapy for melanoma). In fact, the evidence of tumor response in 'untreated' (i.e. non-electroporated) metastases in the patients of this trial support the induction of a systemic immune response after electroporation-driven DNA administration.

3.2 Therapeutic options in advanced melanoma

3.2.1 Locoregional recurrent melanoma

A substantial proportion of patients with thick melanomas on the limbs will develop local recurrence (i.e. near the scar of primary melanoma, Figure 3.1 – top left image) or in-transit metastases (localized between the site of primary melanoma and the regional lymph node basin) (Figure 3.1 – top right image). In some patients, tumor metastases in the regional lymph node basin may also occur.

Treating extensive cutaneous/subcutaneous nodules or in-transit disease is a clinical challenge, because of unresectability and relative insensitivity to conventional systemic therapies. Superficial tumor recurrence causes significant psychological distress for patients, whose everyday life is negatively affected by the disease-related symptoms (i.e. anxiety, pain, bleeding, ulceration, infections). Moreover, many patients presenting with in-transit tumor nodules also harbor clinically- or radiologically-undetectable visceral metastases and therefore have a limited life expectancy.

In patients with locoregional recurrent melanoma, the standard therapeutic approach is that the first local recurrence or in-transit metastases should be widely excised, without a skin graft, if possible. In the melanoma patients with cutaneous metastases, it is generally advisable not to resect subsequent cutaneous metastases, because they may be indicators of tumor response to systemic therapies, or useful for evaluation of new local therapies. However, the occurrence of any cutaneous metastases, either in the limbs or elsewhere, is an indication that the patient is likely to develop multiple locoregional and/or systemic metastases, therefore the patients with any further recurrence should be considered for locoregional chemotherapy and/or systemic approaches.

Isolated limb perfusion (ILP) [22] or isolated limb infusion (ILI) [23] with high doses of cytotoxic agents are currently the most effective method for the treatment of multiple local or in-transit recurrences. The advantage of these therapeutic approaches is that high-dose regional chemotherapy can be administered without causing serious systemic side effects. By means of ILP, which uses melphalan (L-PAM) in association with tumor necrosis factor alpha (TNF-α) and varying degrees of hyperthermia, a complete response can be obtained in about 50% of patients, with a median disease-free interval of 14 months [22]. Since the

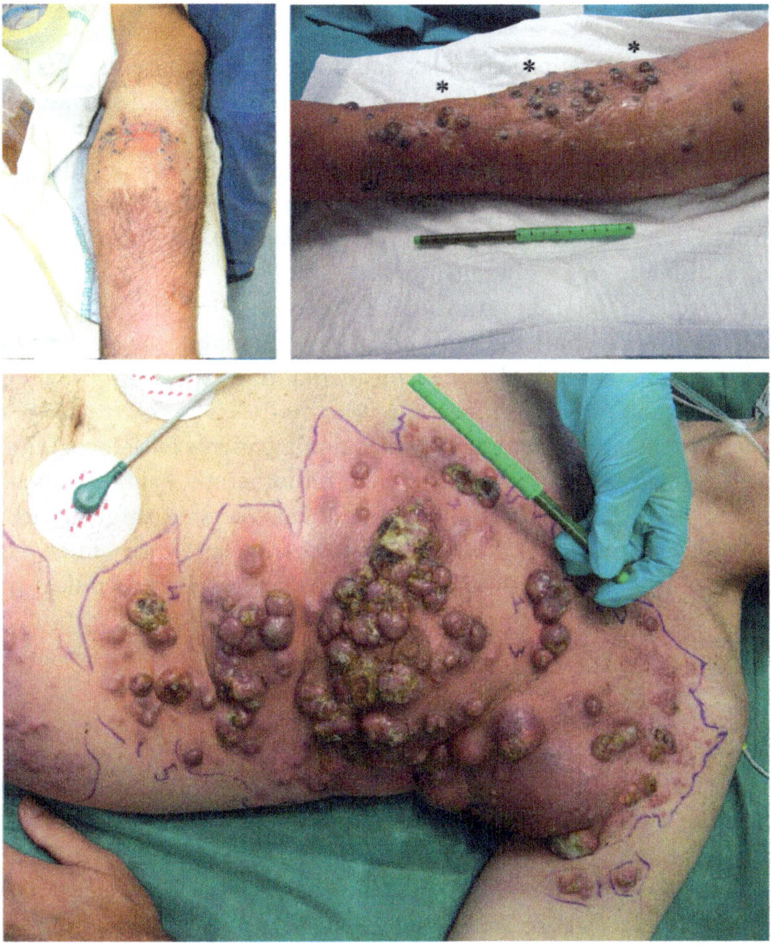

Figure 3.1 Clinical presentations of superficial recurrence from malignant melanoma. *Local recurrence* or *satellite metastases* (top left image): tumor metastases occur near the scar (arrow) of primary melanoma, located in a lower limb. *In-transit metastases* (top right image): multiple, cutaneous and subcutaneous, tumor nodules are distributed along the limb; some of them present ulceration and bleeding, but they do not diffuse outside the lower limb, beyond the inguinal lymph node basin. *Disseminated superficial melanoma* (bottom image): multiple tumor nodules have extensively infiltrated the soft tissues of the thorax, abdominal wall and reach the proximal portion of the upper limb. Massive axillary lymphadenopathy is also present

ILP technique represents a complex and invasive procedure, ILI has been developed to obtain the benefits of ILP without its major disadvantages. In fact, in the ILI technique, the catheters for drug delivery are placed percutaneously, instead of by open surgical exposure as in ILP. ILI is essentially a low-flow ILP performed under hypoxic conditions and the results obtained infusing L-PAM and actinomycin-D have been shown to be comparable to those obtained by ILP [22].

Finally, since local and in-transit recurrences are often due to lymphatic permeation, in the patients with such locoregional recurrence, lymph node dissection should also be considered if it was not previously performed at the time of primary surgery.

Other treatment options exist for melanoma patients with locoregional superficial metastases. Examples include radiotherapy (suitable in cases of limited fields of irradiation), local treatments (i.e. cryosurgery, carbon dioxide laser ablation), intralesional therapies (with interferon, Interleukin-2, bleomycin or vaccines) or the accrual into a clinical trial with novel immunotherapy or targeted agents (see following section).

3.2.2 Systemic disease

When faced with patients with superficially disseminated melanoma (Figure 3.1 – bottom image), the initiation of a systemic treatment is mandatory, regardless of the presence of visceral metastases. Until 2010, only a few chemo- and immuno-therapeutic agents were available for the treatment of metastatic melanoma: Dacarbazine, Fotemustine and interleukin-2 (IL-2) [24–30]. Recently, an immunomodulatory agent (the anti-CTLA-4 monoclonal antibody, Ipilimumab) and the BRAF inhibitor Vemurafenib have been approved for both the first- and second-line therapy of metastatic melanoma, due to the striking results in overall survival reported in randomized trials [31]. A general overview of the main anticancer drugs for the treatment of metastatic melanoma is presented in Table 3.1.

3.3 Clinical experience with electrochemotherapy

In recent years, after technique standardization [15], the effectiveness of ECT treatment has been confirmed in several small series of melanoma

Table 3.1 Current systemic treatments for metastatic melanoma

Drug	Overall Response	Response duration (months)	median survival (months)	Advantages	Side effects
Chemotherapy DTIC [24] TMZ [24]	5–12% 13.5%	5–6 months 1.9 months	6.4 7.7	Tolerability CNS[a], tolerability[b]	CNS[a] myelosuppression
Polychemotherapy DTIC + CDDP + BCNU + TMX [25]	18.5%	ns	7.7	–	Myelosuppression nausea/vomiting, fatigue
Biological agents IFN [26] (low/high dose[c]) IL-2 [27] (low/high dose)	20/55% 3% /16%	1 / 8–32[c] ns	4.2/22–38[c] 8.2/10	Tolerability Response in pts with large tumors	Fatigue, anorexia Fever, dyspnea, nausea
Biochemotherapy Chemotherapy + IFN and/or IL-2 [30]	28%	ns	No benefit (OR 0.99)	Increased tumor response	Increased toxicity
Novel immunotherapy Ipilimumab [31]	11	4	10.1	Improvement of survival in randomized studies	Immune-related adverse reactions
Targeted therapy Vemurafenib [32]	53	6.7	15.9	High (and rapid) response rate; improvement of survival in randomized studies	Cutaneous carcinoma, arthralgia, rash, photosensitivity, fatigue, alopecia

Abbreviations: DTIC: dacarbazine; TMZ: temozolomide; CDDP: cisplatin; BCNU: carmustine; TMX: tamoxifen; IL-2: interleukin-2; IFN: interferon; OR: Odds Ratio; ns: not stated.

[a] Dacarbazine is ineffective against central nervous system (CNS) metastases, while temozolomide is able to penetrate brain tumors.

[b] Temozolomide is an oral congener of dacarbazine.

[c] In this trial, High Dose Interferon was administered to stage III melanoma (i.e. palpable lymph node metastases, not distant metastases) patients as a neoadjuvant treatment before lymphadenectomy

patients with superficial metastases [18]. At present, ECT is employed routinely all over the Europe in advanced patients who are refractory to the conventional treatments, with encouraging results not only on superficial tumour control but also on palliation of disease-related symptoms and preservation of patients' quality of life [17].

3.3.1 Electrochemotherapy technique

In cases of limited disease burden, the procedure can be performed on an outpatient basis, under local anesthesia, possibly in association with a conscious sedation, according to disease anatomical site and patient performance status. Alternatively, general anesthesia may be preferred for larger or multiple tumors, or in cases of some metastases located in the head and neck region, in order to allow proper airway management.

3.3.2 Drugs

After several studies on different cytotoxic drugs, only two have been identified as the best candidates for clinical ECT: bleomycin (BLM) and cisplatin (CDDP). BLM intercalates into the cellular chromatin causing single- and double-stranded breaks in DNA, resulting in a mitotic cell death by pseudoapoptosis. However, CDDP creates an apoptotic effect on the cell. This cytolytic activity is enhanced more than 1000-fold for BLM and 100-fold for CDDP by combination with electroporation.

In 2006, the prospective multi-institutional European Standard Operating Procedures of Electrochemotherapy (ESOPE) study concluded that intravenous or intratumoral BLM were comparable when given to tumors smaller than $0.5\,cm^3$ [32]. For CDDP, the studies have shown that the intratumoral route is more effective than the intravenous route, with CR rates of 82% versus 48%, respectively [33]. For BLM administration, the intravenous route is indicated in the presence of large and widespread disease or tumor nodules located in hard and fibrotic soft tissues, thus allowing a more homogeneous drug distribution. However, the intratumoral route is more feasible in cases of limited disease spread and in the presence of less perfused tumor nodules. The dose of intravenous BLM is $15{,}000\,IU/m^2$ of body surface area, in a bolus lasting 50–60 seconds. Instead, when the intratumoral route is chosen, both BLM and

CDDP drugs are administered in a dose calculation based on tumor volume [15]. The cumulative BLM dose should not exceed 400 000 IU/m^2, due to the risk of lung fibrosis.

3.3.3 Electrodes and pulse parameters

The current pulse generators in use are the Cliniporator device (IGEA, Carpi, Italy) and the Medpulser™ (Inovio Biomedical Corporation, CA). The electric pulses may be applied to the tumors either by plate electrodes on the skin surface or by needle electrodes inserted into the lesion (Figure 3.2). The plate electrodes are more suitable for superficial, exophytic tumor nodules, while needle electrodes are used for deep seated lesions (up to a maximum depth of 3 cm). The electric field distribution is determined by the geometry of the electrodes. Regardless of the electrode type, the electric field is highest around the electrodes, so that the cancer cells in strict proximity to the electrodes are directly killed through the mechanism of irreversible electroporation. This 'ablative' effect prevents the possible risk of tumor dissemination by means of multiple electrode placements in the tumor nodules and subsequently in the surrounding healthy soft tissues. Generally, the electrical parameters during clinical ECT had the following characteristics, according to the chosen electrode type: a train of 8t rectangular pulses of 400–730 V, 910–1300 V/cm, 100 μs duration and 5 kHz repetition frequency.

The time required for performing the ECT procedure is short, with a median treatment duration of 20–25 minutes [20]. Repeated sessions are usually well tolerated by the patients [16, 17, 20] and can be safely performed with a minimum 1–2 week interval, according to disease extension and patient's tolerance.

3.3.4 Results from clinical trials

The relatively recent history of ECT application in the treatment of metastatic melanoma can be divided in two phases, the cornerstone being the ESOPE study that was published in 2006 and led to ECT standardization and dissemination in the clinical practice (Figure 3.3). Although in the pre-ESOPE era the treatment protocols and the route of drug administration were heterogeneous in the different cancer centers, ECT proved to be highly effective in many clinical trials that were conducted at the Institut Gustave Roussy (France) University of South

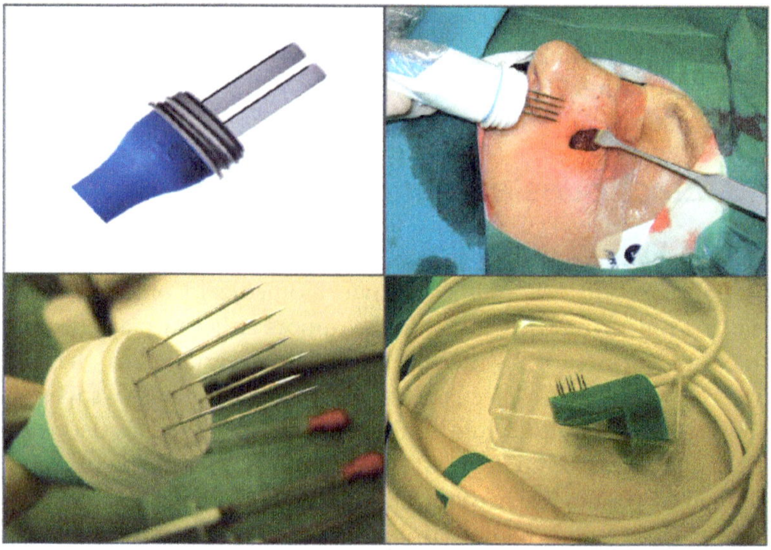

Figure 3.2 Electrode configurations for clinical ECT. The plate electrode consists of two parallel stainless-steel electrodes, used for superficial, exophytic, lesions and does not penetrate the skin (top left). Parallel-array needle electrode consists of 2 rows of eight (2 or 3 cm-long) needles with 4 mm distance between them and are used for small lesions or when treating limited disease deposits in the face, head and neck region (top right). Hexagonal-array electrode has the needles (2 or 3 cm-long) in a hexagonal configuration and is recommended for larger tumors or wider tumor spread on the skin (bottom left). The finger configuration electrode consists of 6 (0.5 or 1 cm-long) needle electrodes mounted on a small plastic support, which is easily held on the finger, in order to be easily introduced for example into the oral cavity. Both with plate or needle electrodes, during the ECT treatment the device should be placed in order to completely encompass the tumor and achieve its complete electroporation

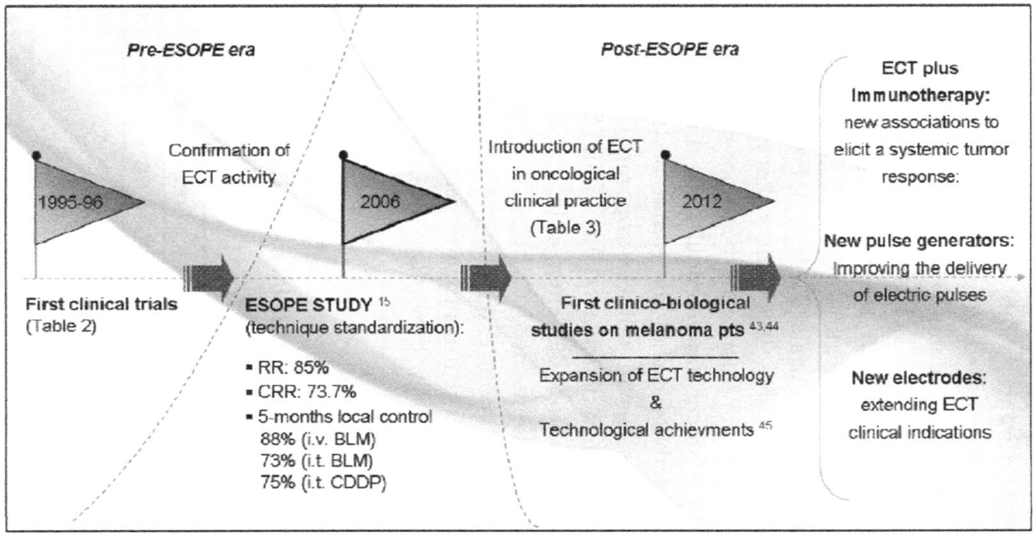

Figure 3.3 Milestones, achievements and perspectives of ECT in the treatment of melanoma. RR: response rate; CRR: complete response rate

Florida (USA), Institut de Pharmacologie et de Biologie Structurale of Toulouse (France), Institute of Oncology and Faculty of Electrical Engineering of Ljubljana University (Slovenia), Sydney Melanoma Unit (Australia) and Department of Dermatology of the Sainte Marguerite Hospital of Marseille (France) (Table 3.2). In 2006, the European multicenter study ESOPE reported an overall response rate of 80.6% on 20 melanoma patients with cutaneous metastases. Since then, the following clinical trials on ECT have been conducted in a more homogeneous fashion, according to the agreed operating procedures. In the post-ESOPE phase, the application of ECT in melanoma patients has been studied in the five clinical trials (Table 3.3). From the analysis of the results of the most recent studies, it is evident that researchers' attention has shifted from the simple evaluation of the tumor response (assessment of ECT activity on the single tumor nodules) to the patient's outcome (assessment of ECT activity on all the tumor nodules, assessment of development of new lesions in non-electroporated anatomical areas, local tumor control and also patients' quality of life) [16–20].

Another relevant observation that emerges from these recent studies, is that many melanoma patients need to be treated with multiple ECT procedures (up to 6 cycles in our experience) to maintain the superficial tumor control. Repetitive treatments are necessary because of the occurrence of new lesions in previously healthy skin, outside the

Table 3.2 Summary of the principal clinical trials on ECT for the treatment of superficial metastases from melanoma in the pre-ESOPE era

Reference	No. of pts	No. of nodules	Drug	Local tumor response (%)				
				PD	NC	PR	CR	OR
[34]	3	10	i.v. BLM	–	–	20	30	50
[35]	5	23	i.t. BLM	0	4	17	78	96
[36]	7	30	i.v. BLM	3	7	10	80	90
[37]	12	84	i.t. BLM	0	1	10	89	99
[38]	4	55	i.v. BLM	0	7	84	9	93
[39]	9	27	i.v. CDDP	–	–	37	11	48
[40]	10	82	i.t. CDDP	–	–	7	80	87
[41]	17	18	i.t. BLM	5	18	5	72	77
[42]	12	54	i.t. BLM	–	–	–	74	–

Table 3.3 **Summary of the principal clinical trials on ECT for the treatment of superficial metastases from melanoma in the post-ESOPE era**

Reference	No. of pts	No. of nodules	Drug	Local tumor response (%)					Tumor control
				PD	NC	PR	CR	OR	
[16]	14	160	i.v. BLM	0	7	43	50	93	74.5% [a] (2 years)
[17]	34	171	i.v./i.t. BLM	–	–	–	50	–	96% [b] (9 months)
[18]	9	158	i.v. BLM	8	30	39	23	62	ns
[19]	21	ns	i.v./i.t. BLM	7	11	19	60	79 [c]	ns
[20]	85	894	i.v. BLM	–	–	–	–	–	87% [d] (2 years)

a Tumor control was achieved by means of 1–3 ECT cycles
b Tumor control was achieved by means of 1–5 ECT cycles
c In 30% of the patients, tumor response was not evaluated
d Tumor control was achieved by means of a median of 3 ET cycles (range 1–6)

electroporated areas. Therefore, researchers are trying to overcome the limitations of this local treatment by means of the association of immunotherapeutic agents, hopefully capable of eliciting a systemic immune response for controlling the residual, micro-metastatic disease. Whether ECT, besides inducing local massive cell death with release of tumor antigens, can also induce systemic anti-tumor immunity is still unclear. In this search, two small but intriguing early clinical-biological studies have recently investigated the presence and prognostic role of both tumor-infiltrating lymphocytes [43] and dendritic cells [44] in melanoma metastases before and after ECT application. The early findings from these studies, together with some observations from the clinical trials [20] suggest that the combination of ECT with a biological response modifier (i.e. IL-2 or IFN) could be a suitable strategy to increase the efficacy of this local treatment approach and to prevent further tumor spreading in ECT-treated patients.

Moreover, according to the findings of the researchers from the Tuscan Tumor Institute, Florence, Italy, the high number of dendritic cells and their characterization in ECT-treated melanomas, provide the rationale for a new therapeutic approach by combining the local ECT treatment with *in situ* dendritic cells activation, for example by means of Toll-like Receptor (TLR) agonists, which are known to stimulate dendritic cell, anti-melanoma immunity and possibly induce a systemic antitumor response [44].

3.3.5 ECT in peculiar clinical conditions

Due to its intrinsic characteristics and the flexibility of the technology through which it is commonly applied, ECT also proved suitable for managing various unusual clinical presentations of patients with recurrent melanoma.

On the one hand, the development of new mini-invasive electrodes (Figure 3.2, bottom right) has made possible the treatment of tumors located in narrow body cavities (i.e. the oral mucosa or the anal canal). In our experience, the availability of such endoluminal electrodes has allowed us to treat and successfully palliate some advanced melanoma patients with symptomatic tumor metastases on the oral mucosa (Figure 3.4, middle image). Moreover, because of the availability of this

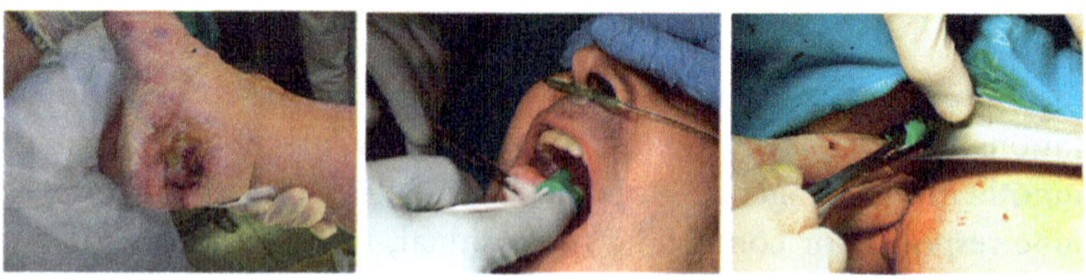

Figure 3.4 Examples of melanoma patients with cutaneous metastases who can benefit from an effective superficial tumor control obtained with ECT. An ulcerated and bleeding cutaneous metastasis, localized in the medial aspect of the heel (left). Application of the finger electrode to a patient with melanoma metastases localized in the mucosa of the oral cavity. Such a needle electrode is held on the tip of a finger and therefore can be easily maneuvered in very narrow anatomical spaces (middle). Treatment of recurrent melanoma of the anal canal by means of the electrode fingers, inserted through the retractor (right). In this patient, ECT treatment by means of a small needle electrode has enabled avoidance of a wide surgical excision that would have damaged sphincter muscles, thus causing incontinence to the patient

so-called *finger electrode*, we were able to offer a conservative (sphincter-preserving) and effective treatment strategy for some patients with recurrent melanoma of the anal canal (Figure 3.4, right image).

On the other hand, due to its various antivascular effects leading to decreased blood flow (transient vasoconstriction, endothelial damage, etc.), ECT can be applied to achieve an effective haemostasis on bleeding tumor nodules. In fact, hemorrhage from cutaneous melanoma metastases is a common clinical problem, causing considerable patient discomfort (Figure 3.4, left image).

3.4 Conclusions and perspectives

ECT, with either BLM or CDDP, is an active therapeutic option for the palliative management of unresectable, refractory superficial metastases from melanoma, either in the setting of locoregional or metastatic disease. The rate of local tumor response to ECT application ranges approximately from 62–93% in the more recent series. ECT is generally well tolerated by patients, even though many of them need to undergo multiple treatment sessions for the appearance of new lesions, as the consequence of cancer growth at the site of previously undetectable tumor deposits. Because ECT induces a local massive death on cancer cells, researchers are studying the immune system's reaction to this highly active treatment and are looking for the best combination with an immunomodulatory agent that could elicit a systemic and effective immune response toward residual melanoma deposits, hence capable of providing sustained and effective tumor control. Alternatively, other research groups are investigating the electroporation process for achieving the effective delivery of the plasmid DNA (i.e. IL-2 or IL-12 gene) as a potential systemic therapy for metastatic melanoma and are corroborated by the promising results of a phase I study [21].

At present, there are no comparative studies on ECT with other anticancer treatments. However, when taking into consideration the complexity, the costs and the learning curve of other complex treatment options (i.e. radiotherapy or locoregional chemotherapy), the ECT techniques can be considered a 'user-friendly' procedure. Since it is easy to perform, it could also be highly advantageous in countries or hospitals where other treatment modalities are not easily available or health resources are limited.

ECT technology (pulse generators and electrodes for pulse delivery) continues to evolve, thus allowing the more effective treatment of superficial tumors (i.e. by means of faster pulse generators) or the targeting of deep-seated tumors (i.e. through long needle electrodes) [45]. Due to these technological achievements, together with the continuous immunological advancements, ECT will not only consolidate its role as an excellent palliative option, but also pave the way for other effective treatment strategies for patients with recurrent melanoma [46].

3.5 References

1. Erdmann F, Lortet-Tieulent J, Schüz J, et al. (2013), 'International trends in the incidence of malignant melanoma 1953–2008 – are recent generations at higher or lower risk?', *Int J Cancer*, **132(2)**: 385–400.
2. Siegel R, Ward E, Brawley O, et al. (2011), 'Cancer Statistics, 2011. The impact of eliminating socioeconomic and racial disparities on premature cancer death', *CA Cancer J Clin*, **61**: 212–36.
3. NCI SEER Cancer Statistics Review (2012). Available at: *http://seer.cancer.gov/statfacts/html/melan.html*
4. MMMP, Molecular Map Project – An Open Access Multidatabase for Research on Melanoma Biology & Treatment (2012). Available at: *http://www.mmmp.org/MMMP/import.mmmp?page=risk_ass_progn_thin.mmmp*
5. Korn E L, Liu P Y, Lee S J, Chapman J A, Niedzwiecki D, et al. (2008), 'Metanalysis of phase II cooperative group trials in metastatic stage IV melanoma to determine progression-free and overall survival benchmarks for future phase II trials', *J Clin Oncol*, **26**: 527–34.
6. Flaherty K T (2006), 'Chemotherapy and targeted therapy combinations in advanced melanoma', *Clin Cancer Res*, **12**: 2366–70.
7. Eigentler T K, Caroli U M, Radny P, et al. (2003), 'Palliative therapy of disseminated malignant melanoma: a systematic review of 41 randomized clinical trails', *Lancet Oncol*, **4**: 748–59.
8. Hodi F S, O'Day S J, McDermott D F, Weber R W, Sosman J A, et al. (2010), 'Improved survival with ipilimumab in patients with metastatic melanoma', *N Engl J Med*, **363**: 711–23.
9. Chapman P B, Hauschild A, Robert C, Haanen J B, Ascierto P, *et al.* (2011), 'Improved survival with vemurafenib in melanoma with BRAF V600E mutation', *N Engl J Med*, **364**: 2507–16.
10. Mir L M, Glass L F, Sersa G, Teissie G, Domenge C, et al. (1998), 'Effective treatment of cutaneous and subcutaneous malignant tumours by electrochemotherapy', *Br J Canc*, **77**: 2336–42.
11. Heller R, Jaroszeski M J, Glass L F, Messina J L, Rapaport D P, et al. (1996), 'Phase I/II trial for the treatment of cutaneous and subcutaneous tumors using electrochemotherapy', *Cancer*, **1**: 964–71.
12. Glass L F, Pepine M L, Fenske N A, Jaroszeski M, Reintgen D S and Heller R (1996), 'Bleomycin-mediated electrochemotherapy of metastatic melanoma', *Arch Dermatol*, **132**: 1353–7.

13. Sersa G, Stabuc B, Cemazar M, Miklavčič D and Rudolf Z (2000), 'Electrochemotherapy with cisplatin: clinical experience in malignant melanoma patients', *Clin Cancer Res*, **6**: 863–7.
14. Byrne C M, Thompson J F, Johnston H, Hersey P, Quinn M J, et al. (2005), 'Treatment of metastatic melanoma using electroporation therapy with bleomycin (electrochemotherapy)', *Melanoma Res*, **15**: 45–51.
15. Mir L M, Gehl J, Sersa G, Collins G C, Garbay J R, et al. (2006), 'Standard operating procedures of the electrochemotherapy: instructions for the use of bleomycin or cisplatin administered either systemically or locally and electric pulses delivered by the cliniporator by means of invasive or non-invasive electrodes', *EJC Suppl*, **4**: 14–25.
16. Quaglino P, Mortera C, Osella-Abate S, Barberis M, Illengo M, et al. (2008), 'Electrochemotherapy with intravenous bleomycin in the local treatment of skin melanoma metastases', *Ann Surg Oncol*, **15**: 2215–22.
17. Campana L G, Mocellin S, Basso M, Puccetti O, De Salvo G L, et al. (2009), 'Bleomycin-based electrochemotherapy: clinical outcome from a single institution's experience with 52 patients', *Ann Surg Oncol*, **16**: 191–9.
18. Kis E, Oláh J, Ócsai H, Baltas E, Gyulai R, et al. (2011), 'Electrochemotherapy of cutaneous metastases of melanoma – a case series study and systematic review of the evidence', *Dermatol Surg*, **37**: 816–24.
19. Matthiessen L W, Chalmers R L, Sainsbury D C, Veeramani S, Kessell G, et al. (2011), 'Management of cutaneous metastases using electrochemotherapy', *Acta Oncol*, **50**: 621–9.
20. Campana L G, Valpione S, Mocellin S, Sundararajan R, Granziera E, et al. (2012), 'Electrochemotherapy for disseminated superficial metastases from malignant melanoma', *Br J Surg*, **99**: 821–30.
21. Daud A I, DeConti R C, Andrews S, Urbas P, Riker A I, et al. (2008), 'Phase I trial of interleukin-12 plasmid electroporation in patients with metastatic melanoma', *J Clin Oncol*, **20**: 5896–903.
22. Deroose J P, Grünhagen D J, van Geel A N, de Wilt J H, Eggermont A M and Verhoef C. (2011), 'Long-term outcome of isolated limb perfusion with tumour necrosis factor-α for patients with melanoma in-transit metastase', *Br J Surg*, **98**: 1573–80.
23. Kroon H M, Huismans A M, Kam P C, Thompson J F. (2014) Isolated limb infusion with melphalan and actinomycin D for melanoma: A systematic review. *J Surg Oncol*. **4**: 348–51.
24. Middleton M, Grob J, Aaronson N, et al. (2000), 'Randomized phase III study of temozolomide versus dacarbazine in the treatment of patients with advanced metastatic malignant melanoma', *J Clin Oncol*, **18**: 158–66.
25. Chapman P B, Einhorn L H, Meyers M L, et al. (1999), 'Phase III multicenter randomized trial of the Dartmouth regimen versus dacarbazine in patients with metastatic melanoma', *J Clin Oncol*, **117**: 2745–51.
26. Creagan E T, Ahmann D L, Green S J, et al. (1984), 'Phase II study of low-dose recombinant leukocyte A interferon in disseminated malignant melanoma', *J Clin Oncol*, **2**: 1002–5.
27. Moschos S J, Edington H D, Land S R, et al. (2006), 'Neoadjuvant treatment of regional stage IIIB melanoma with high-dose interferon alpha 2b induces

objective tumor regression in association with modulation of tumor infiltrating host cellular immune responses', *J Clin Oncol*, **24**: 3164–71.

28. Agarwala S S, Glaspy J, O'Day S J, et al. (2002), 'Results from a randomized phase III study comparing combined treatment with histamine dihydrochloride plus interleukin-2 versus interleukin-2 alone in patients with metastatic melanoma', *J Clin Oncol*, **20**: 125–33.
29. Atkins M B, Lotze M T, Dutcher J P, et al. (1999), 'High-dose recombinant interleukin 2 therapy for patients with metastatic melanoma: analysis of 270 patients treated between 1985 and 1993', *J Clin Oncol*, **17**: 2105–16.
30. Ives N J, Stowe R L, Lorigan P and Wheatley K. (2007), 'Chemotherapy compared with biochemotherapy for the treatment of metastatic melanoma: a meta-analysis of 18 trials involving 2621 patients', *J Clin Oncol*, **25**: 5426–34.
31. Hodi F S, O'Day S J, McDermott D F, et al. (2010), 'Improved survival with Ipilimumab in patients with metastatic melanoma', *N Engl J Med*, **363**: 711–23.
32. Marty M S G, Garbay J M, Gehl J, et al. (2006), 'Electrochemotherapy – An easy, highly effective and safe treatment of cutaneous and subcutaneous metastases: Results of ESOPE (European standard operating procedures of electrochemotherapy) study', *Eur J Canc Suppl*, **4(11)**: 3–13.
33. Sersa G. (2006), 'The state-of-the-art of electrochemotherapy before the ESOPE study: advantages and clinical uses', *Eur J Canc Suppl*, **4**: 38–44.
34. Heller R, Jaroszeski M J, Glass L F, Messina J L, Rapaport D P, et al. (1996), 'Phase I/II trial for the treatment of cutaneous and subcutaneous tumors using electrochemotherapy', *Cancer*, 77: 964–71.
35. Glass L F, Pepine M L, Fenske N A, Jaroszeski M, Reintgen D S and Heller R (1996), 'Bleomycin-mediated electrochemotherapy of metastatic melanoma', *Arch Dermatol*, **132**: 1353–7.
36. Mir L M, Glass L F, Sersa G, Teissié J, Domenge C, Miklavčič D, et al. (1998), 'Effective treatment of cutaneous and subcutaneous malignant tumours by electrochemotherapy', *Br J Cancer*, 77: 2336–42.
37. Heller R, Jaroszeski M J, Reintgen D S, Puleo C A, DeConti R C, et al. (1998), 'Treatment of cutaneous and subcutaneous tumors with electrochemotherapy using intralesional bleomycin', *Cancer*, **83**: 148–57.
38. Rols M P, Bachaud J M, Giraud P, Chevreau C, Roché H and Teissié J (2000), 'Electrochemotherapy of cutaneous metastases in malignant melanoma', *Melanoma Res*, **10**: 468–74.
39. Sersa G, Stabuc B, Cemazar M, Miklavčič D and Rudolf Z (2000), 'Electrochemotherapy with cisplatin: the systemic antitumour effectiveness of cisplatin can be potentiated locally by the application of electric pulses in the treatment of malignant melanoma skin metastases', *Melanoma Res*, **10**: 381–5.
40. Sersa G, Stabuc B, Cemazar M, Miklavčič D and Rudolf Z (2000), 'Electrochemotherapy with cisplatin: clinical experience in malignant melanoma patients', *Clin Cancer Res*, **6**: 863–7.
41. Byrne C M, Thompson J F, Johnston H, Hersey P, Quinn M J, et al. (2005), 'Treatment of metastatic melanoma using electroporation therapy with bleomycin (electrochemotherapy)', *Melanoma Res*, **15**: 45–51.

42. Gaudy C, Richard M A, Folchetti G, Bonerandi J J and Grob J J (2006), 'Randomized controlled study of electrochemotherapy in the local treatment of skin metastases of melanoma', *J Cutan Med Surg*, **10**: 115–21.
43. Quaglino P, Osella-Abate S, Marenco F, Nardò T, Gado C, et al. (2011), 'FoxP3 expression on melanoma cells is related to early visceral spreading in melanoma patients treated by electrochemotherapy', *Pigment Cell Melanoma Res*, **24**: 734–6.
44. Gerlini G, Sestini S, Di Gennaro P, Urso C, Pimpinelli N and Borgognoni L. (2013), 'Dendritic cells recruitment in melanoma metastasis treated by electrochemotherapy', *Clin Exp Metastasis*, **39(1)**: 37–45.
45. Rossi C R and Campana L G (2010), 'Clinical electrochemotherapy: the Italian Experience', in *Advanced Electroporation Techniques in Biology and Medicine*, Pakhomov A G, Miklavčič D and Markov M S (eds). Boca Raton: Taylor & Francis, 385–400.
46. Hampton T (2011), 'Electric pulses help with chemotherapy, may open new paths for other agents', *JAMA*, **305**: 549–51.

Low and high voltage electrochemotherapy for breast cancer: an *in vitro* model study

Ignacio G. Camarillo, Funian Xiao, S. Madhivanan, Therese Salameh, Maxine Nichols, Lisa M. Reece, James F. Leary, Kevin Otto, Arutselvan Natarajan, A. Ramesh, and Raji Sundararajan

DOI: 10.1533/9781908818294.55

Abstract: The death of a woman every 71 seconds in the world and every 12 minutes in the US, from breast cancer, indicate that the current standard of cure does not work. In addition, these therapies are expensive and have many side effects. This calls for safe, effective, alternate, affordable and physical therapies. Electrochemotherapy (ECT) has promising potential for treating inoperable, chemo-resistant and radio-resistant tumors that do not respond to the current standard of cure. Clinical trials show the successful applications of ECT for chest wall breast carcinomas. In this research, the efficacy of FDA approved, commonly prescribed and administered breast cancer drugs, including doxorubicin, paclitaxol, tamoxifen and bleomycin are studied using electroporation. In addition, the efficacy of curcumin is also studied using high and low voltage electroporation. MCF-7 breast cancer cells were used for this study at various pulse parameter conditions, including both low and high voltages. The positive results indicate that ECT is a good vehicle to efficiently upload these drugs intratumorally.

Key words: breast cancer, MCF-7 cells, doxorubicin, paclitaxel, tamoxifen, curcumin, turmeric, electrochemotherapy, impedance spectroscopy, flow cytometry.

4.1 Introduction

Breast cancer is the malignant tumor of the mammary gland. The mammary gland is supposed to have developed during the Jurassic Period, about 160 million years ago, when mammals were considered to evolve on our planet [1]. The breasts offered a considerable advantage, compared with their egg-laying predecessors, as these provided a superior food that gave their progeny an advantage in the struggle for survival. Breast cancer disease is frequent in female mice, dogs and humans, but is rare or unknown in other species.

In women, this is the most common cancer (Figure 4.1) [2]. For example, every 12 minutes, 6 women will develop breast cancer and one will die in

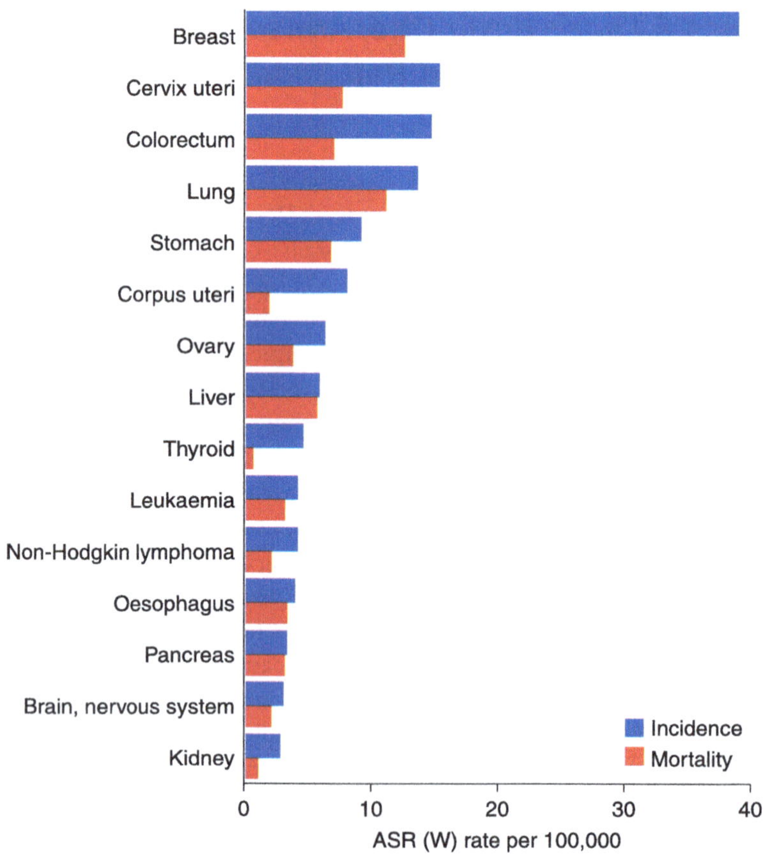

Figure 4.1 **Breast cancer is the most frequent cancer of women worldwide. With kind permission from [2]**

the US [3]. Between 1973 and 2000, the incidence rose by 60%. In Egypt, every 3 minutes a woman is diagnosed with breast cancer and is afraid that her husband will leave her [4]. Worldwide, each year, 1 million cases occur with about 440 000 deaths, about 1 every 71 seconds. Figure 4.1 illustrates the incidence and the mortality of breast cancer compared to other cancers [2].

The above deaths and the poor drug response rate (Table 4.1) [5, 6] indicate that the current chemotherapy treatments do not serve all patients. In addition, there are patients with inoperable tumors and/or chemo/radio resistant tumors. Also, conventional chemical drugs are costly and have many side effects. Sixty-two percent of bankruptcies in the US are medical-related (of which 78% had health insurance) [7]. Hence there is a critical/urgent need for alternate physical therapies.

Table 4.1 **Response rates of breast cancer drugs [5,6]**

Regimen	Response
Taxol (Paclitaxel)	Pathological: – complete 24% – partial 55% Clinical: – complete 18% – partial 82%
Adriamycin and Taxol (AT)	Pathologic complete 16% Objective clinical 89%
Flourouracil, Adriamycin, and cyclophosphamide (FAC)	Pathological: – complete 24% – partial 55% Clinical: – complete 18% – partial 82%
Tamoxifen	Objective clinical 17–36%
Aromatase inhibitors Letrozole Anastrozole Exemestane	Objective clinical: 30–55% 21–43% 41%
Herceptin (Trastuzumab-used for metastatic)	Objective: – complete 6% – partial 20%

Physical therapies are preferable, because all diseases of the body cannot be and have not been cured by treating with chemicals (drugs) only, as the biochemical processes do not explain the electrical functions [8] and the electrostatic forces and their interactions in cell regulations in deadly diseases, such as cancer. Our body possesses electrical mechanisms and uses charges and electricity to regulate and control the transduction of chemical energy and life processes. Thus, there is a critical need for alternate/additional therapies and an electrical pulse-based technique, such as electrochemotherapy (ECT) [9–13], could pave the way towards this. The combination of high intensity, short duration electrical pulses with very low dose (up to one-thousandth of the conventional dose) drugs offers other ways to treat some of the cancers that are refractory to the current standard of cure.

This is critical, as the WHO world cancer report indicates that by 2020, global cancer rates could increase by 50% to 15 million [14]. Of these, one-third are preventable (healthy life style, nutrition uptake and exercise), one-third are curable and the rest require palliative cure. ECT is a great candidate for palliative cure. In 2000, malignant tumors were responsible for 12% of about 56 million deaths worldwide from all causes. In many countries, more than 25% are attributable to cancer. Once a disease of affluent industrialized Western countries, now cancer is everywhere, affecting high, low and middle income developing countries [4,14]. The WHO report indicates that cancer is emerging as a major public health concern in developing countries, matching its effect in industrialized nations [14]. Table 4.2 shows the annual new case numbers that illustrate the seriousness of this situation [14].

The industrialized nations that have the highest overall cancer rates include the US, Germany, Italy, Netherlands, Australia, Canada and France. While it is pink ribbons and Livestrong bracelets in advanced

Table 4.2 Annual new cases of various cancers [14]

cancer	No. of new cases/year
Lung	1,200,000
Breast	1,000,000
Colorectal	940,000
Stomach	870,000
Liver	560,000

countries for breast cancer, it is not so in low and middle income developing countries. It is not even a topic of discussion, due to the culture in some of these countries. Lack of awareness, resources, socio-economic status, before the screening culture allow the cancer to grow to advanced stages [4]. Thus, there is a critical need for economical and effective alternate therapies, which are outpatient-based and use low dose drugs.

4.2 Anatomy of the breast and its cancer

The breast is a well differentiated apocrine sweat gland of the same type found in the axilla and elsewhere in the body [3]. It has three kinds of tissues: fat, glandular and normal breast tissue. It is a very dynamic organ due to the continuous change of hormones from birth to puberty to pregnancy to motherhood to menopause, explaining why breast cancer is the most common cancer among women. A constant change of state exists due to the menstrual cycle superimposed over lifecycle changes, caused by the hormonal changes. The hormones influence the normal physiological changes of the breast. The two most prominent hormones are estrogen (responsible for ductal proliferation) and progesterone (responsible for lobular proliferation and growth). The over- or under-production of the hormones is part of the reason for abnormal growth and changes in the breast. Obesity, life style and nutrition add to these changes. Figure 4.2 shows the correlation of obesity and breast cancer [4,14]. The blessing of motherhood is also the very reason for having the most frequently occurring cancer in women, and its occurrence is rising due to various environmental as well as individual lifestyle and food habit changes [14].

4.3 Drug delivery issues

Drug delivery is a challenge to both the doctor and the pharmacist, because it is possible that the drug could find other undesired and unintended paths and hence have adverse effects before reaching the desired location [15]. Orally administered drugs have several advantages, including the safest route, the most economical, non-invasive, and can be self-administered. The large surface area of the gastrointestinal tract, the mixing of its contents and the differences in pH in different parts of

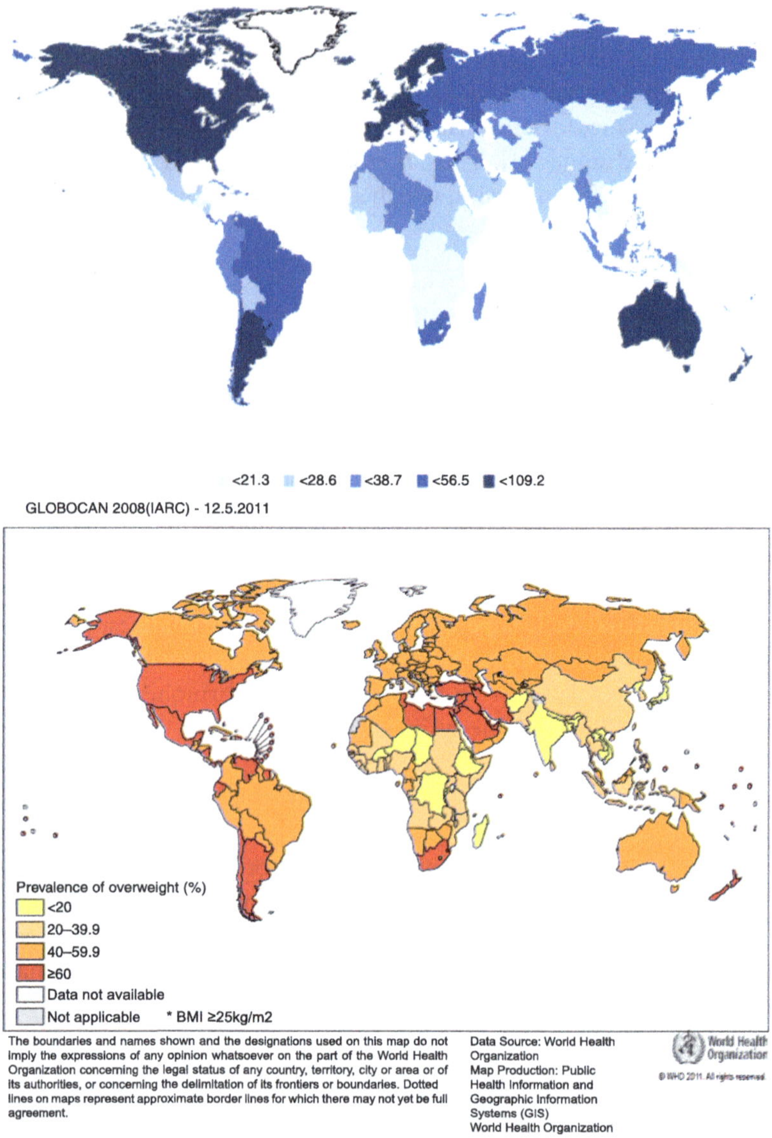

Figure 4.2 **Correlation between breast cancer (above) and obesity. With kind permission from [14,2]**

the gut facilitate effective absorption of the oral drugs. However, they also have some disadvantages. Medicines administered orally must circulate throughout the entire body before taking effect. This can pose a potential threat to the patient, as the drug can act on any part of the body causing adverse effects. It can also be broken down through enzymic degradation. Thus, the various disadvantages include:

- onset of action is slower because absorption needs time;
- irritant and unpalatable drugs cannot be administered;
- some drugs cannot be absorbed, due to certain physical and chemical characteristics;
- irritation to the gastrointestinal tract may lead to vomiting;
- there may be irregularities in absorption;
- some drugs may be destroyed by gastric juices, such as insulin.

4.4 Chemotherapy issues

Cancer is not a single disease, but a combination of a number of diseases, which makes its treatment and cure a challenge, as evidenced by the millions of deaths each year. After years of research costing billions of dollars, cancer treatment is still unsatisfactory and very costly (many bankruptcies and foreclosures in US were due to medical expenses, including cancer expenses). This is also because of cancer cells' unique nature, including capacity for uncontrolled proliferation, invasiveness, metastasis and poor differentiation [16,17]. In addition, cancer cells are like the rest of the cells in the body. Unlike microbes, the drugs that kill cancer cells also kill normal cells, due to lack of specificity. The host defense mechanisms, which help us to ward off infections are not doing so in cancers, because the cancer cells are also the host cells. Furthermore, cancer cells can remain in a resting phase during which they are not sensitive to anticancer drugs but can start multiplying later resulting in recurrence. Cancer stem cells add to the problem by being resistant to chemo drugs. These features have made treatment of cancers difficult, as evidenced by the millions of deaths each year.

In addition, most of the chemo drugs are cell cycle specific. Due to the three kinds of cells in a solid tumor, such as dividing cells that are continuously cycling, resting cells which may potentially enter cell cycle, and non-dividing cells, cell cycle phases are important for staging and scheduling chemo treatment, as some drugs work only at certain phases of the cell cycle when cells divide. Table 4.3 gives details of the cell cycle specific and nonspecific drugs [17]. Only dividing cells are susceptible to chemo drugs. The resting or cancer stem cells (a small part of tumor cells that contribute to the growth of the tumor and recur) are difficult to

Table 4.3 Cell cycle specific and nonspecific drugs [17]

Cell cycle specific	Nonspecific
S phase – Antimetabolites, Doxirubicin	Alkylating agents, anticancer antibiotics
G2 and M phases – Bleomycin	Cisplatin
M phase – Taxanes, Vinca alkaloids	Procarbazine, Camptothecins

completely eradicate in advanced tumors. Hence alternate therapies that use low dose chemotherapy, and yet are effective, are desirable.

4.5 Common adverse effects of anticancer drugs

Many anticancer drugs lack specificity; they equally are toxic to normal rapidly multiplying cells in the bone marrow, epithelial cells of skin and mucous membranes, lymphoid organs and gonads. The various common adverse effects are [5,6,17]:

- bone marrow depressing causing leucopenia, anemia, thrombocytopenia and in higher doses, aplastic anemia. Infections and bleeding are common in these patients;
- other proliferating cells:
  - GIT-stomatitis, esophagitis, glossitis and proctitis can be painful. Diarrhea and ulcers along the gut are common;
  - alopecia (loss of hair) – partial to total hair loss, but this is reversible;
  - temporary menopause for women in their reproductive period;
  - reduced spermatogenesis in men, leading to infertility in 6 months with mechlorethamine;
- immediate adverse effects – nausea and vomiting are common in about 4 to 6 hours;
- hyperuricemia – rapid tumor cell lysis can result in increased plasma uric acid levels and may lead to 'tumor lysis syndrome' and renal failure;

Table 4.4 Specific adverse effects of some anticancer drugs [17]

Drug	Specific adverse effects	Other prominent adverse effects
Doxorubicin	cardiotoxicity	bone marrow depression, loss of hair
Bleomycin	Pulmonary fibrosis, oedema of hands	stomatitis, loss of hair
Cisplatin	ototoxicity	renal dysfunction

- teratogenicity – all cytotoxic drugs are teratogenic and hence are contraindicated in pregnancy;
- carcinogenicity – cytotoxic drugs may cause secondary cancers, such as leukemias which may follow the treatment of Hodgkin's lymphoma.

In addition, long-term effects on cardiac function and cognition are also reported [18,19]. Table 4.4 gives the specific adverse effects of some anticancer drugs [17].

4.6 Anticancer drug resistance

Over time, the body becomes resistant to either single or a combination of chemo drugs [5]. This resistance may be primary or secondary [17]. When the cancer is not responsive at the first dose itself, it is known as primary resistance. Secondary resistance is when the drug response is acquired initially, and eventually not responding.

The various drug resistance mechanisms are [5,17]:

- cellular and biochemical mechanism;
- decreased drug accumulation:
  - decreased drug influx;
  - increased drug efflux;
  - altered intracellular drug trafficking;
- increased inactivation of drug or toxic intermediate;
- increased repair or tolerance to drug-increased damage to:
  - DNA;
  - protein;
  - membranes;

- decreased drug activation;
- altered drug targets;
- altered co-factor or metabolite levels;
- altered downstream effectors of cytotoxicity;
- altered signaling pathway and/or apoptotic responses to drug insult:
  - altered gene expression;
  - DNA mutation, amplification or deletion;
  - altered transcription, post-transcription processing or translation;
  - altered stability of macromolecules;
- *in vivo* mechanisms;
- pharmacological and anatomic drug barriers;
- host-drug interactions:
  - increased drug inactivation by normal tissues;
  - decreased drug activation by normal tissues;
  - relatively increased normal tissue drug sensitivity.

The various causes for multidrug resistance indicate that there are multiple ways in which our bodies could develop resistance. Table 4.5 gives the various mechanisms of drug resistance of some commonly used chemo drugs [5,17].

The mechanisms for secondary resistance could be:

- decreased amount of drug reaching the cells;
- decreased accumulation of the drug in the cells;

Table 4.5 Resistance mechanisms of some chemo drugs [17, 5]

Drug	Resistance mechanism	Pharmacological defect
Dox	Increased drug targeting Altered drug targeting Defective recognition of DNA adducts	Altered topoisomerase II Altered topoisomerase II Increased MDR expression or MDR gene amplification
Paclitaxel	Defective recognition of DNA adducts	Increased MDR expression or MDR gene amplification
Most anticancer drugs	Defective check-point function and apoptosis	P53 mutations

- transport proteins, like P-glycoprotein may expel the drug from the cell;
- conversion of the drug to active metabolites may be reduced;
- increased inactivation of the drug;
- increased production of target enzyme;
- alternate metabolic pathway used by cancer cells, as with anti-metabolites;
- altered target protein by genetic mutation.

An approach to overcoming or circumventing drug resistance is the use of novel treatment modalities and ECT fits this exactly. The goal of our research is to develop safe, efficient, novel adjuvant therapies to treat breast tumors. Towards this, to enhance drug uptake, we propose ECT using FDA approved, readily available chemo and hormone drugs, as this technique is less costly, has fewer side effects and is simple (could be performed as an outpatient treatment [10]), requiring no costly outfits such as a radiation facility.

4.7 Electroporation and electrochemotherapy

Electroporation or electropermeabilization (EP) is a viable physical technique that utilizes precisely controlled electric fields of short durations and high intensities to open up transient aqueous pathways through semi or non-permeable membranes and tissues, allowing targeted delivery of therapeutic materials, including drugs, antibodies and genes (DNA) [9–13,20]. Since the application of pulses is only for a very short duration, the cell membranes eventually reseal and force the drug molecules to act within the cell. EP offers up to 1000-fold improved therapeutic benefit compared to using the drug alone and is gaining acceptance as a viable technique to enhance the efficacy of drug delivery for cancer treatment. It is a local, site-specific, and physical technique with the least side effects, if any. This is a combination treatment that involves administering a chemotherapeutic drug followed by the delivery of electric pulses to cells or tissue.

Electroporation enhances the permeability of cell membrane. During electroporation, drug molecules can enter the cancer cells in various ways, such as diffusion, electro-osmotic and colloid-osmotic flow [21]. The magnitude of membrane potential can be defined [22] by Equation 4.1:

$$V = 1.5\ ER \qquad [4.1]$$

where V is voltage induced on the cell membrane, E is electric field intensity applied and R is the radius of the cancer cell. This equation can be modified as

$$E = 4V/(3D) \qquad [4.2]$$

where D is the diameter of cancer cell.

Equation 4.1 indicates that the E field intensity for a given potential V is inversely proportional to the diameter of the cell. It also depends on the duration and number of pulses, intervals between them, type of molecules to be uploaded and the type of tissue/cells/tumors and organ [23]. If the field is too low, there is no effect. If the field is too high, there is excessive cell death. Hence, it is necessary to optimize the electroporation pulse parameters for each cell type. It is reported that ET = k, a constant, where E is the electric field applied and T is the pulse length [24]. Figure 4.3 illustrates this concept [25]. Thus, if the electric field is high, the pulse length has to be low and if the electric field is low, the pulse length has to high. This concept is used in general for the selection of electroporation parameters.

When used to deliver chemo drugs, this technique is called Electrochemotherapy (ECT) [9–13,20]. Its success has been attested by the various clinical trials and ongoing treatments for extreme cases of melanomas, sarcomas and other skin cancers, including head and neck cancers [9–12]. When surgery, radio- and chemo-therapies did not work,

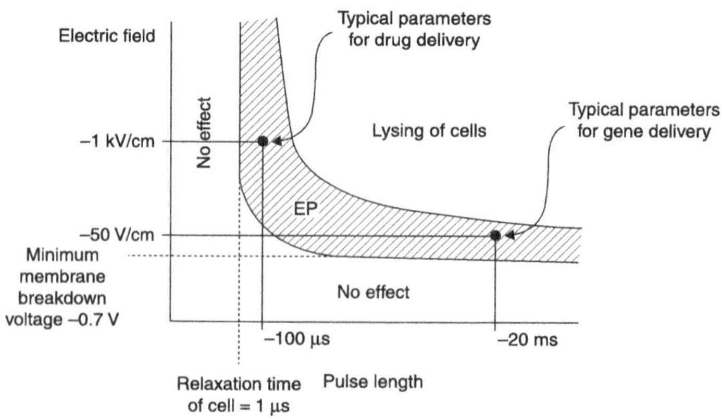

Figure 4.3 Illustration of the inverse relationship of E and T for electroporation [25]

patients suffering from chest wall breast carcinoma benefited using ECT [11,12]. Thus, electrical pulse-mediated drug delivery is an attractive option to those patients who do not respond to the current standard of cure.

Typically, 8 pulses at 1200 V/cm, with 100 μs pulses are used for ECT [10–12]. In this study, in addition, low voltage, long duration pulses, such as 200 V/cm, 10 ms were also studied to demonstrate their efficacy in reducing the proliferation of the tumor cells. Low voltage is preferred, as it is safer than high voltage for use in humans and animals [13]. It is simpler to design and build a low voltage pulse generator than a high voltage pulse generator. Towards that goal, several pulse parameters were chosen and studied.

4.8 Materials and methods

4.8.1 MCF-7 human adenocarcinoma cells

MCF-7 is a human breast cancer cell line with estrogen, progesterone and glucocorticoid receptors [26]. It is derived from the pleural effusion of a 69-year-old Caucasian metastatic breast cancer (adenocarcinoma) in 1970 by Dr Soule of the Michigan Cancer Foundation, Detroit, MI [27]. MCF-7 cells are useful for *in vitro* breast studies because they retained several ideal characteristics particular to mammary epithelium, such as the processing of estrogen, in the form of estradiol, via estrogen receptors (ER) in the cell cytoplasm [28]. It is the first hormone responding breast cancer cell line. Its unique properties are useful for experimental therapeutics. They are also sensitive to cytokeratin. When grown *in vitro*, the cells are capable of forming domes and the epithelial-like cells grow in monolayers. Figure 4.4 shows a view of the MCF-7 cells [29].

The cells were cultured in a mixture media of 90% RPMI 1640, 10% FBS serum (ATTC, Manassas, VA) and 1% Penicillin/Streptomysin (Invitrogen, Carlsbad, CA). The cells were grown in an incubator at 37 °C with 5% CO_2.

For electroporation, the cells were washed twice with 1 × PBS, whose pH was adjusted to 7.4, and left in serum-free 199 medium (Invitrogen) for 24 hours. Cells were dissociated from the incubation flask with 0.25% trypsin/EDTA (ATCC) solution. A hemocytometer was used to obtain a final concentration of 1×10^6 cells/mL. Aliquots of 750 μL in

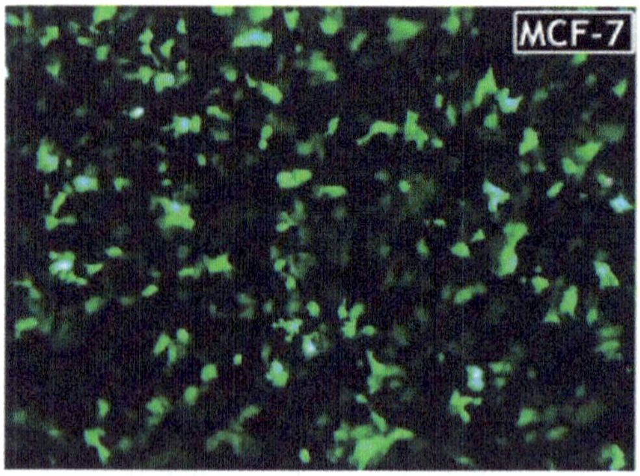

Figure 4.4 Human ER positive breast adeno-carcinoma cell line [29]

0.4 cm cuvettes were used for electroporation by adding RPMI 1640 medium with 10% charcoal stripped fetal bovine serum.

4.8.2 Chemo, hormone, and turmeric drugs

Typically bleomycin is used for ECT clinical trials. Occasionally, cisplatin is used. In our study, commonly used, commercial, FDA approved breast cancer chemo and hormone drugs, such as doxorubicin (Dox), Paclitaxel (Taxol), Bleomycin (Bleo) and Tamoxifen (Tam) were studied. They were used at very low doses compared to doses normally used in the clinics for conventional chemotherapy. All these medications have lifetime cumulative dose limits and side effects [17, 30], including fever, chills, skin reactions, hair loss, nausea, vomiting, diarrhea, muscle pain, alopecia, gastrointestinal issues, etc., thus they are ideal candidates for EP. In addition, curcumin, the yellow component of the popular Indian natural herb, turmeric, with antiseptic, anti-inflammatory, anticancerous properties is also studied.

Doxorubicin chemo drug

Doxorubicin (Dox) is a FDA approved, widely used cytotoxic drug in cancer chemotherapy. It is a natural product of a species of soil fungus, *Streptomyces*. Figure 4.5(a) shows the structure of this drug.

It is classified as an anthracycline antibiotic that works by binding to nucleic acids and intercalating DNA [31]. Anthracyclines are metabolized by reduction of a side-chain carbonyl to alcohol, resulting in some loss of

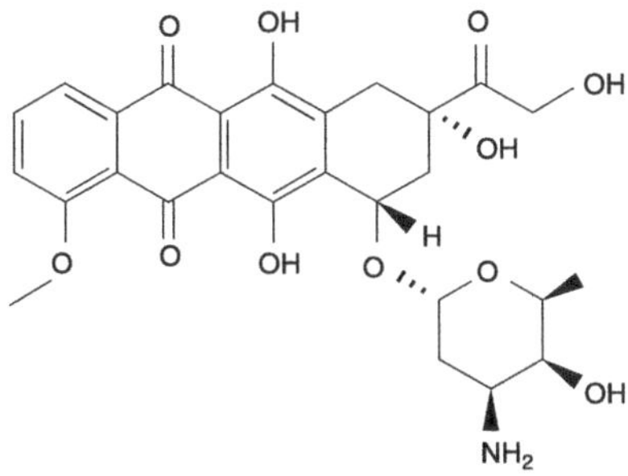

Figure 4.5 (a) Atomic structure of Doxorubicin [31]

cytotoxicity, and a one-electron reduction to a semiquinone free radical intermediate by flavoproteins, leading to aerobic production of superoxide anion, hydrogen peroxide and hydroxyl radical [5]. It kills cells (both cancer and normal cells) by working in several ways [5, 17], including activation of transduction pathways, generation of reactive oxygen intermediates, stimulation of apoptosis and inhibition of DNA topoisomerase II catalytic activity. Doxorubicin binds to DNA in cells, changing the shape of the DNA by injecting itself into some of the empty spaces in the DNA molecule. This damages the DNA, causing some unwinding of their spirals, making it impossible to form new cancer cells. Doxorubicin can damage the membranes (outer coating) of cells and may damage other parts of cells as well [17]. It is cell cycle-specific, forming a complex with DNA and topoisomerase II, inhibiting topoisomerase II and binds to DNA causing DNA breakage, ultimately leading to the generation of free radicals which oxidize DNA strands. It acts on the S phase of the cell cycle. It is given as IV infusion over 10–15 minutes.

It was approved by the FDA in 2007 for the treatment of some leukemias, lymphomas, soft tissue sarcomas, multiple myelomas, and cancers of the bladder, breast, stomach, head and neck, lung, ovaries, pancreas, prostate, uterus and thyroid. Common side effects include pain, nausea, vomiting, diarrhea, low blood counts, mouth sores and hair loss. In addition, it causes blisters on the skin (vesicants). Hence it should be carefully injected to avoid extravasation. It is metabolized by the liver, thus a dose reduction is required in the presence of hepatic dysfunction.

Anthracyclines cause cardiotoxicity with hypotension, tachycardia, arrhythmias and cardiomyopathy. Antioxidants offer significant protection against cytotoxicity by anthracyclines. They also cause vomiting, stomatitis, hair loss and bone marrow depression. They may color the urine red. An uncommon but very serious side effect is cardiotoxicity that can interfere with the pumping of the heart. For this reason there is a limit to the amount of Doxorubicin you can receive in your lifetime [32]. This makes Doxorubicin an ideal candidate for ECT.

The pharmokinetics includes the protein-binding rate that ranges from 60–70%, while its cerebro-spinal fluid (CSF)/plasma ratio is low. Dox circulates predominantly as a parent drug and 50–60% of the parent drug is eliminated by known routes. A significant portion of the compound is bound to DNA and cardiolipin in tissues [5].

Doxorubicin is a highly protein-bound drug. Its penetration in multilayered solid tumors depends up on a number of factors, including its concentration, kinetics of drug penetration, effects of tumor cell density and tissue composition. Thus drug-resistance, the inability of the drug to penetrate into the tumor, is possible. Resistance to the drug may occur to anthracyclines as a consequence of P-glycoprotein over-expression or altered topoisomerase II activities. There is increased expression of the P-170 glycoprotein related to the enhancement of drug efflux. They vary between cells and tissues. It is also reported that intermediary metabolism affects DNA cleavage and Dox resistance [5,17]. The resistance could be overcome by analogs and the time and method of delivery affects toxicity and results. An alternate physical technique, such as ECT, is one solution.

It was observed that penetration of Doxorubicin in 3-dimensional tumor cell spheroids after 1–2 hours is limited to the periphery. Similarly, a steep concentration gradient in breast tumors has been observed in patients. Hence, better methods to enhance drug delivery are needed and electroporation is one such vehicle to enhance impermeable drug uptake. It is an attractive alternative to treat patients with cutaneous or subcutaneous tumors which are recurrent, inoperable or progressive, and were refractory to systemic chemotherapy and/or radiotherapy [9–13].

Paclitaxel chemo drug

Taxol is a FDA approved taxane chemo drug for advanced breast cancer treatment. This was first discovered from the bark of the slow-growing Pacific Yew tree and has proven to be highly effective in treating women with advanced breast cancer. Figure 4.5(b) shows its chemical structure. The mechanisms of action of taxanes are high affinity binding to microtubules with enhanced microtubule formation at high drug concentrations and inhibition of mitosis. It binds to the beta-tubulin of microtubules and arrests mitosis (mitotic spindle poison).

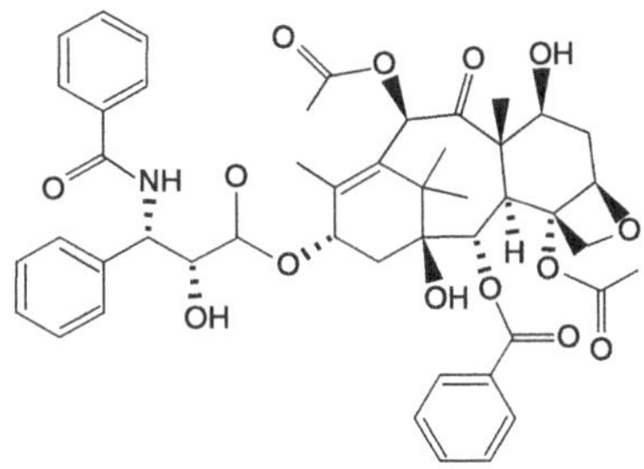

Figure 4.5 **(b) Chemical/atomic structure of Paclitaxel**

Due to metabolism, sub-micromolar concentrations of the taxanes decrease the lag time and shift the dynamic equilibrium between tubulin dimers and microtubule assembly and stabilize microtubules against depolymerization. Taxanes are eliminated predominantly by hepatic hydroxylation of cytochrome P450 enzymes and biliary excretion of metabolites. Less than 10% of each dose is eliminated via urine. About 80% of the administered dose is excreted in the feces within 7 days of treatment. It is given intravenously as an infusion over a few hours and repeated every 3 weeks. It is metabolized by the liver microsomal enzymes and excreted through the gut. Dose should be reduced in liver dysfunction.

Common side effects include low white and red blood cell counts, weakness, hair loss, fatigue, nausea, vomiting, diarrhea and muscle pain, as well as numbness, tingling and burning sensations in the arms and legs [17]. Adverse effects include myelosuppression, myalgia, allergic reactions, hypotension, arrhythmias and peripheral neuropathy. Because of this wide array of adverse effects, it is an ideal candidate for ECT, as a very low dose (9 nM) could be used (leading to reduced cost and side effects).

Bleomycin chemo drug

Bleomycin, another FDA approved anticancer chemotherapy, is a DNA targeting, cytotoxic, antitumor, antibiotic chemo drug. It is considered as a cytotoxic anticancer drug due to DNA double-stand cleavages and the loss of chromosome fragments [33]. Its structure is shown in Figure 4.5(c). It binds oxygen and redox active transition metal ions, such as Fe^{2}, Cu^{2+} and Zn^{2+}. The mechanism of action of Bleomycin is that it has accessibility into the sequence of DNA. The bithiazole and the terminal amine parts

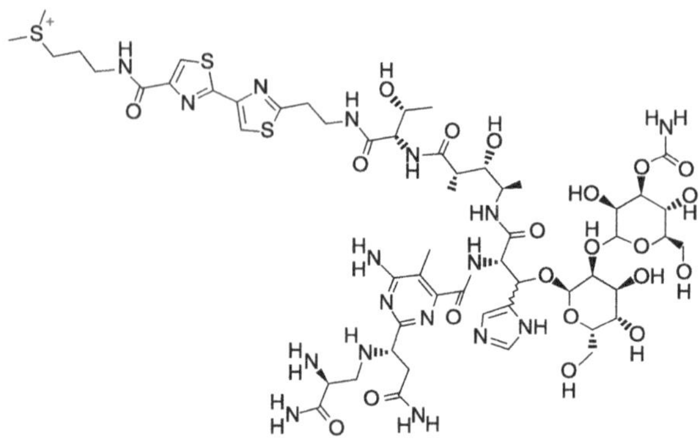

Figure 4.5 (c) Chemical/atomic structure of Bleomycin

bind stably with the DNA, leading to the formation of both double- and single-stranded breaks. It forms one double-strand for eight single-strand DNA breaks, causing more double-strand DNA breaks than would occur under normal situations. In addition, Bleomycin has specificity of cleavage on chromatin.

Bleomycin is obtained from *Streptomyces verticillus*. It binds with iron and generates free radicals and causes breakage in DNA strands. It has a unique mechanism of action and is less toxic to bone marrow. It is used in solid tumors, testicular tumors, squamous cell carcinoma of the head, neck and esophagus, melanomas, sarcomas and lymphomas [9,10]. Its most serious toxicities includes pulmonary fibrosis and cutaneous toxicity, but the trade-off is that it does not cause significant bone marrow depression like many other anticancer drugs. Its common side effects include fever, chills, skin reactions, hair loss, alopecia and gastrointestinal issues [30].

Tamoxifen

Estrogen and its receptors, α and β (ERα and ERβ), play a major role in tumor progression. Approximately two-thirds of breast cancers express these functional receptors. Because of the important role ERs play in breast cancer growth, treating breast cancer using selective estrogen receptor modulators (SERMS), such as Tamoxifen (Tam) is common, where it inhibits the ability of estrogen to bind to the ERs and induces dimerization and DNA binding to finally inactivate it [5]. This is especially true for ER^+ breast cancer patients. Tamoxifen is one of the most effective

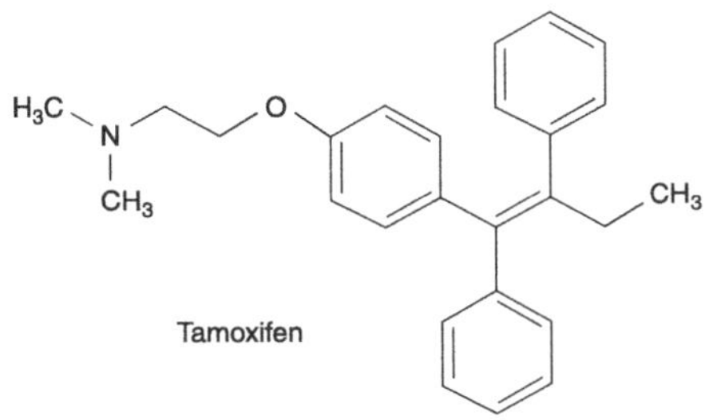

Figure 4.5 **(d) Chemical/atomic structure of Tamoxifen**

endocrine agents for treatment of breast cancer [34]. Its structure is shown in Figure 4.5(d). Its metabolism is mediated in the liver by cytochrome P450-dependent oxidases into 10 major metabolites. After initiation of therapy, steady state concentrations of the active metabolites of Tam are achieved in 4 weeks, suggesting a half-life of 14 days. Metabolites and a small portion of Tam are excreted in the bile as conjugates.

However, breast cancer patients eventually develop hormone resistance (HR) and relapse within five years. There are several mechanisms of drug resistance. Absence of ER expression is the primary mechanism. ER mutations may explain the variations in patients' response to Tam (21–41%) [5]. Alternative mRNA splicing has been identified in normal and malignant breast tissue, with variants lacking one or more exons. There is decreased response in patients with both ER and HER-2, as ER function is strongly influenced by growth factor signaling.

The development of HR is a major clinical problem. Hence, use of ECT is another way to treat ER^+ patients using low dose Tam.

Curcumin (turmeric)

Curcumin (Sigma Aldrich, USA) is the principal curcuminoid, the yellow pigment of the popular Indian spice turmeric. It has been used historically in ancient medicine (ayur vedic – life of science) since about 4000 years ago [35] and in Indian cooking for thousands of years for its various therapeutic properties, including antiseptic, anti-oxidant, anti-inflammatory and anti-tumor characteristics. This is also confirmed by research in the late 20th century. It inhibits production of the free radicals H_2O_2 and NO by macrophages [36, 37] and astrocytes [38] *in vitro*. In experimental autoimmune encephalomyelitis (EAE), curcumin inhibits lymphocyte proliferation, decreases IL-17 production by TH17 cells, and downregulates the expression of Toll-like receptors 4 and 9 [36–42].

Recently, curcumin was utilized to study triple-negative breast cancer cells (TNBC) and the possible molecular mechanisms, and it was shown to inhibit the proliferation of TNBC cells [43]. TNBC, lacking expression of the ER, progesterone receptor (PR) and epidermal growth factor receptor 2 (HER2/EGFR2), is an aggressive breast cancer phenotype with a poor prognosis. Chendil et al. performed an experiment on prostate cancer cells, PC-3, in which they administered curcumin to the cells in tandem with radiation [44]. They found that curcumin had significant positive correlation with the success of the radiation treatment. The curcumin downregulated the Bcl-2 protein and also enhanced radiation-induced apoptosis. In 2011, Sreekanth used curcumin in coordination with paclitaxel. Paclitaxel is highly efficacious against a wide spectrum of human cancers, but drug resistance limits the drug's effectiveness. This study showed how curcumin could effectively downregulate all these survival signals induced by paclitaxel [45]. To study the effects of curcumin in different stages, Kawamori administered curcumin to lab rats both before having colon cancer and after presenting with the disease [46]. The experiment showed that the chemopreventive activity of curcumin is observed when it is administered prior to, during and after carcinogen treatment, as well as when it is given only during the promotion/progression phase (starting late in premalignant stage) of colon carcinogenesis. Figure 4.5(e) shows the structure of curcumin.

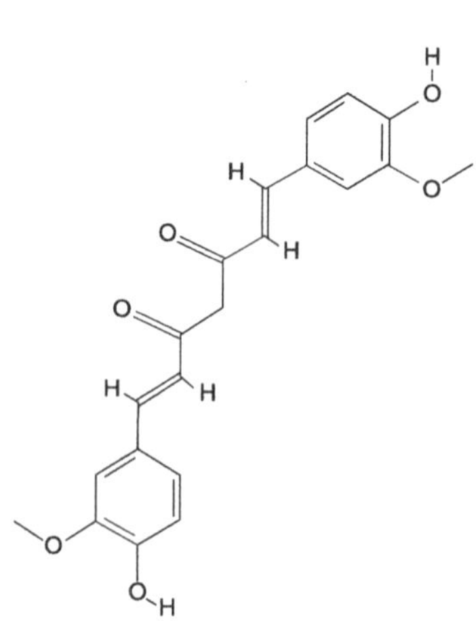

Figure 4.5 (e) Chemical structure of curcumin

4.8.3 Electroporation technique

For efficient electroporation, two conditions have to be satisfied. First, a sufficient amount of chemotherapeutic drug must be present in the targeted tissue/tumor when the pulses are applied. Second, electric pulses have to be of appropriate magnitude and duration to create reversible (transient) pores in the cell plasma membranes. If the pulse intensity is too low, it will not create the pores to uptake the drug molecules. If it is too high, there will be more cell death. Thus, it is critical to choose appropriate electrical parameters to achieve pore formation in the cell membrane without too much cell death.

A pulse generator (pulser) is needed to apply the required pulses. In this study, a BTX ECM830 square wave electroporator (Genetronics, Inc., San Diego, CA) was used (Figure 4.6). This generates square wave pulses under both low voltage and high voltage modes. The voltage changes from 5–500 V, with 1 V resolution under low voltage mode, and from 505–3000 V, with 5 V resolution under high voltage mode. The pulse width ranges from 10–100 ms under low voltage mode, and from 10–600 μs under high voltage mode. A cuvette holder holds the 0.4 mm cuvette with cells and is connected to the generator.

Eight, 1200 V/cm, 100 μs pulses were used in the skin cancer and chest wall breast carcinoma clinical trials [9–13]. In our study, we tested both low voltage (200–500 V/cm), long duration (milliseconds, ms) and high voltage (1200 V/cm), short duration (microseconds, μs) pulses. Eight sequential pulses with a 1 second interval between each pulse (1 Hz frequency) [10] were used in all the cases, to explore the electroporation efficacy of various drug molecules.

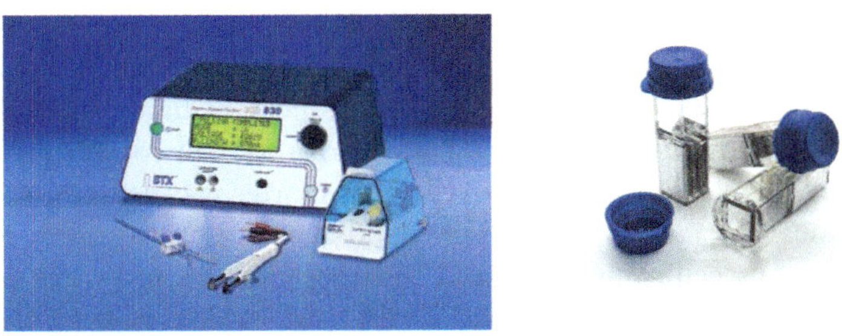

Figure 4.6 BTX ECM 830 square wave electroporator and cuvettes

4.8.4 Characterization assays

The efficacy of electroporation was studied using various assays. These include cell viability and growth assay, fluorescence-activated cell sorter (FACS) assay, Western Blot, fluorescence microscopy and electrical impedance spectroscopy (EIS). For this purpose, the MCF-7 cancer cells were treated with various drugs and a number of electrical and dose parameters. After electroporation, cells were cultured at 37 °C, in a 5% CO_2 incubator for 24 hours before performing the above assays.

Cell viability and growth assay

Electroporated cells were kept in cuvettes for 30 minutes before being transferred to cell culture well plates for incubation. After 24 hours, media and dead, unadhered cells were removed from the wells. Live, adhered cells were dissociated from the incubation flask with 0.25% trypsin/EDTA (ATCC) solution. Live cells were counted and recorded with the Cellometer Auto T4 from Nexcelom Bioscience, LLC, using Trypan Blue assay [47].

Fluorescence microscopy

Since the first use of fluorochromes in biological research in the 1930s, fluorescence microscopy has become a useful tool in biomedical and biological sciences to study specimens, which can be made to fluoresce. The basic idea is based on the phenomenon that certain materials emit energy, detectable as visible light when irradiated with light of a specific wavelength. The intensity of color of the stained cells indicates the nature of the cell. When analyzing fluorescent images, the color intensity of the stained cells is evaluated. A more intensely stained (darker) cell means that the cell is apoptotic, has ceased growing and, in this instance, is undergoing programmed cell death. This function is achieved using two filters (excitation and emission) and one dichromatic mirror. The excitation filter selects the desired regions of incident light. After the desired wavelengths reach the dichromatic mirror, the mirror efficiently passes longer wavelength light while reflecting shorter wavelengths.

In this research, a Nikon Eclipse TE2000-S was used to detect the images of various cell samples. After electroporation, these cancer cell samples were placed into the incubator for 24 hours. Media was removed from the wells containing electroporated cells and they were fixed using 100% ethanol. After 15 minutes, ethanol was removed and PI solution

(25 μL PI per 5 mL of 1 × PBS) was added to stain the cells. Fifteen minutes later, PBS was added after removal and the cells were imaged using a standard set of optical filters for this dye (530 nm excitation/600 nm emission).

Flow Cytometry Assay

To analyze live and dead cells and the nature of death (apoptosis or necrosis), both quantitatively and qualitatively, flow cytometry is a preferred technique as it can count, examine and sort microscopic particles suspended in fluid [48]. The occurrence of apoptosis in a cell depends on the type of cell and the trigger of apoptosis. Fragmentation of the nucleus is the main feature of apoptosis, which can be evaluated by morphology and observed in an increased level of DNA debris in cell samples. Thus, the level of DNA debris can be taken as a method to measure apoptosis, and hence the efficacy of ECT. This method needs cell samples stained using a DNA specific dye, in order to detect the morphologic changes in nucleus. Annexin V is used to mark early stage apoptotic cells. PI is used to mark membrane-permeable – 'dead' – cells, as it penetrates into the groove of the double stranded DNA and then colors the cell nuclei.

Flow cytometry assay was performed using a Cell Lab Quanta™ SC flow cytometer with a 488-nm laser (Beckman Coulter). For this purpose, the electroporated cells were treated using two fluorescence reagents, the Annexin V Incubation Reagent (AVIR), and a binding buffer. The cells were washed twice with 500 μL cold 1 × PBS and were resuspended using AVIR and incubated for 15 minutes in the dark at room temperature. The binding buffer was added and loaded into the flow cytometer.

Western Blot

The biological activity of Taxol is based on its ability to stabilize microtubules, inducing microtubule bundling. Treatment of cells with Taxol disrupts the formation of normal spindles, leading to the arrest of cells in the G2/M phase of the cells cycle and eventually to apoptotic cell death. The effects of Taxol may vary, depending on the cell type and the dosage. Intracellular transmission of extracellular signals is mediated by several groups of sequentially-activated protein kinases, including p38, which belongs to the super family of mitogen-activated protein kinases (MAPK). Taxol-induced apoptosis depends on MAP kinase pathway p38. The activation of p38 MAPK cascades are involved in the Taxol-induced apoptosis, independent of p53 induced G2/M arrest [49–51]. This is studied using the Western Blot, an analytical technique commonly used to identify the specific proteins in the given sample, using electrophoresis [52]. In this study, the total protein concentrations were determined using a p38MAPK protein assay to study the combined effect of electroporation and Taxol chemo drug.

For this purpose, electroporated cells were washed 3× in ice-cold PBS. Whole cell lysate extracts were prepared by lysing cells in 50 μL lysis buffer (20 mmol/L Tris-HCl (pH 7.5), 150 mM/L NaCl, 1 mM/L EGTA, 1% Triton X-100 and complete protease and phosphatase inhibitor cocktails (Sigma)). It was centrifuged at 14 000 × g for 20 minutes at 4 °C to remove the debris and then cell lysate was sonicated (5 × 15 s pulse). The amount of protein in each sample was determined using the BCA protein assay kit (Pierce; Rockford, IL).

Protein lysate (20 μg) was boiled for 5 minutes and subjected to electrophoresis in denaturing 10% SDS-PAGE. Membranes were probed for p38 MAPK (1:1000; Epitomics; Burlingame, California) overnight at 4 °C. Horseradish-peroxidase conjugated anti-rabbit IgG was used as a secondary (1:5000 for 4 h at RT). The membrane was developed using an ECL reagent. The membrane was then stripped using 0.2 M NaOH for 20 minutes at RT. The stripped blot was reprobed with β-actin (1:10 000) as a loading control and developed using horseradish-peroxidase conjugated anti-mouse IgG (1:10 000 for 1 h at RT).

Electrical Impedance Spectroscopy

A biological cell is conventionally modeled as a simple parallel combination of a capacitance (C) and a resistance (R) (Figure 4.7) [53, 54]. The resistance corresponds to the extra-cellular medium and the capacitance represents the cell membrane charge storage capacity and its dielectric nature, which varies with the relative permittivity (εr).

The total impedance, Z is given as

$$Z = R/(1 + j\omega RC) \tag{4.3}$$

where R is the real part (Real Z) and $1/j\omega C$ is the imaginary part (Im Z) corresponding to the resistance and capacitive reactance of the cell. The electrical properties of biological cells and tissues are related to the biological, physiological, morphological and pathological conditions. Thus, the impedance of cancer cells and normal cells could be different

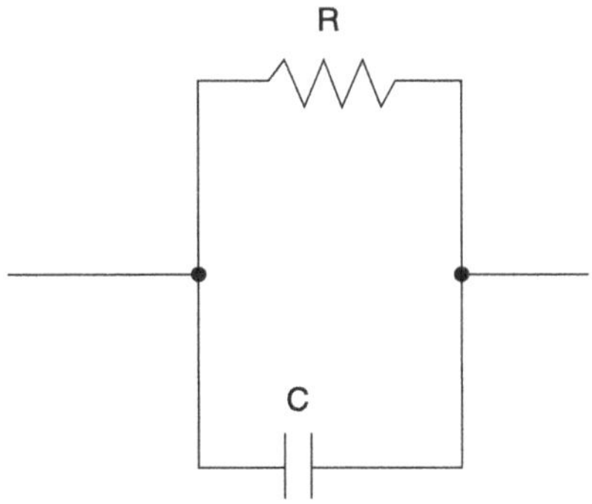

Figure 4.7 Electrical model of biological cell

and hence the electric field required for electroporation also varies from cell to cell for therapeutic electroporation applications [55]. Hence, it is important to study the electrical properties of electroporated cancer cells, to study their permeability changes, both quantitatively and qualitatively. For this purpose, EIS is used.

EIS is an AC measurement technique in which the total impedance, Z (the ratio of voltage and current) is measured over a range of frequencies. It is a non-invasive, non-destructive test technique in which very little energy is dissipated by the system under test, leaving it virtually unaffected. It can be used as a method of identifying and following detectable cellular responses, *ex vivo, in vivo* and *in vitro* [56, 57]. EIS measures the electrical properties, i.e. the conductance (or resistance) and the reactance of any material, as a function of applied voltage frequency, as each polarizable entity within the cell will exhibit its own characteristic response and thus a distribution of relative permittivities will give rise to the impedance spectrograph.

This frequency-dependent relationship between impedance (z), conductivity (σ) and relative permittivity (εr) is given by [58]:

$$Z = Z' + j\omega Z'' = 1 / (\sigma + j\omega\varepsilon_o\varepsilon_r) \qquad [4.4]$$

where Z is the total (complex) impedance, Z' and Z'' are the real and imaginary components of Z respectively, ω is the angular frequency, and ε_o is the permittivity of free space. Both Z' and Z'' were measured, from which the conductivity and relative permittivity can be calculated. The EIS was used to measure the enhanced conduction of the cells under control, drug only and electroporation conditions. For this purpose, the

AC impedance at room temperature was measured using a PGSTAT100 (Autolab) high voltage potentiostat/galvanostat with a compliance voltage of 100 V and a maximum current of 250 mA. It gives 40 data points with a frequency range from 0.5–100 kHz.

To measure the Z for various frequencies, the electroporated cells in the cuvette were connected to the two leads of the potentiostat and a sine voltage of 25 mV amplitude was applied. The samples were scanned at 15 frequency points over the range of 46–10 000 Hz. The data were acquired and impedance plots were plotted for analysis using a MATLAB code. The electrical impedance was displayed as Total (or complex) Z, Real Z (R, the resistive component in Ohm) and Imaginary Z (the capacitive reactance component in negative Ohm) and various plots, such as impedance plots, bode plots, and Cole-Cole plots were plotted for analysis.

4.9 Results and discussion

4.9.1 Cell Viability and dose curves

Doxorubicin

A dose curve was developed to determine the appropriate (the least, but effective) concentration of Doxorubicin for our electroporation experiments (Figure 4.8). The cell viabilities shown indicate the survival rate for cells treated with different concentrations of Doxorubicin only. No electroporation was used.

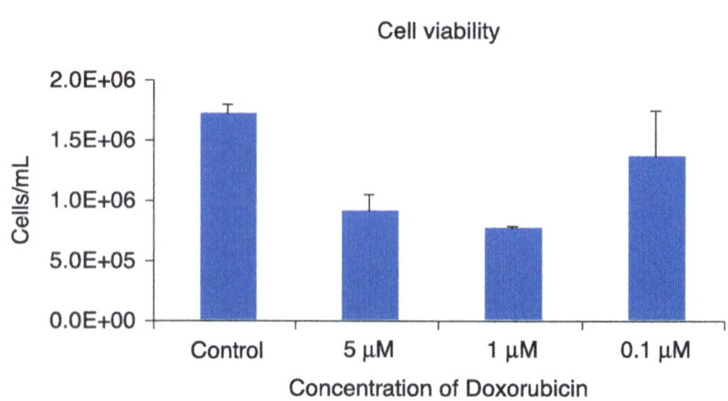

Figure 4.8 Doxorubicin dose curve without electroporation

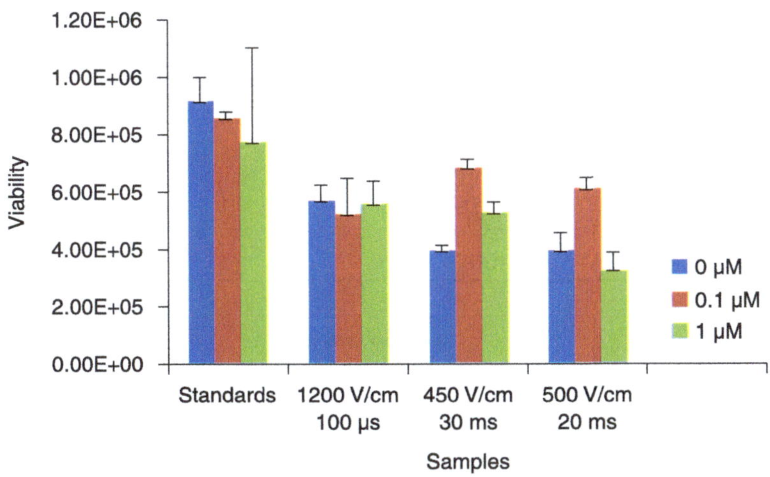

Figure 4.9 **Doxorubicin dose and viability for various pulse parameters**

Figure 4.9 shows the cell viability for various electroporation conditions. The different pulse parameters used include 1200 V/cm, 100 μs; 450 V/cm, 30 ms; and 500 V/cm, 20 ms. Eight pulses at 1 second intervals were used in each case for the doses of 0, 0.1 and 1 μM, each. There is maximum cell death with 0.1 μM. dose at 500 V/cm pulses. The treatment of Dox in combination with the electroporation parameters shows promise for inhibiting the proliferation at a very low dose along with electrical pulses. The cell viability shows a reduction of nearly 90% compared to treatment with Doxorubicin alone, even with a very low dose of 0.1 μM. An increase of dosage to 1 μM showed slightly more cell death.

Taxol

Table 4.6 shows the pulse parameters and dose conditions used. Here 200 μs was used instead of the commonly used 100 μs, because with 100 μs, it was not possible to upload any drug. Hence the pulse duration was increased, keeping the intensity (1200 V/cm) constant. Figure 4.10 shows the cell viability for Taxol at various doses from 100 μM–1 nM for both low intensity, long duration pulses (200 V/cm, 40 ms) and high intensity, low duration pulses (1200 V/cm, 200 μs). The lower pulse intensity was tried in order to study the efficacy of lower intensity, longer duration pulses, compared to the most commonly used 1200 V/cm, 100 μs pulse set. The dose curves indicate that there is slight variation in cell viability from 100 μM–9 nM. Hence, the 9 nM was chosen as the

Table 4.6 Pulse parameters and dosages used for Taxol study

Sample	Electric field intensity (V/cm)	Pulse duration
Control (no treatment)	0	0
Drug control	0	0
EP-200 µs	1200	200 µs
EP-40 ms	200	40 ms

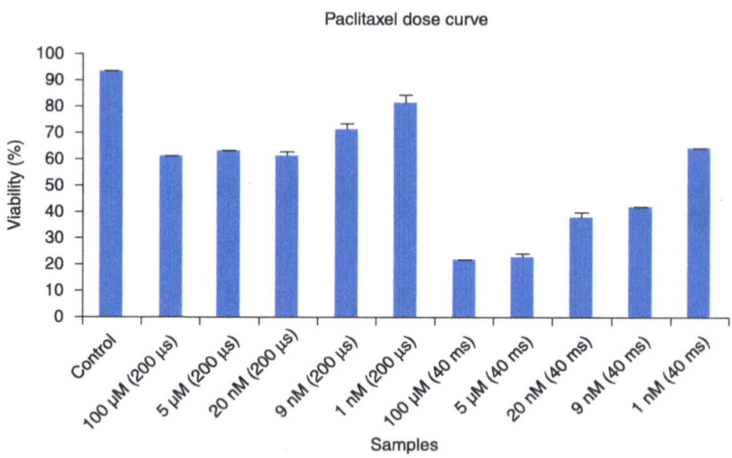

Figure 4.10 Paclitaxel dose curve

standard dose for this study. Also, the results indicate that indeed the lower intensity, longer duration pulses are more effective, owing to their higher energy content due to longer durations. We used 8 pulses at 1 second intervals in these tests. Reducing the number of pulses from 8 down to 4 or even 2 will be interesting to study the efficacy of drug treatment on the cells. The energy content is relatively higher with the 8 ms pulses that have durations in the tens of milliseconds. Reducing the pulse number will effectively reduce the energy content and may possibly be gentler on cells, but we need to determine if this will work as well.

Bleomycin and Taxol

Figure 4.11 shows the results for delivering Bleomycin and Paclitaxel, for the control (no treatment), drug only, and the two electroporated conditions (1200 V/cm, 200 µs and 200 V/cm, 40 ms). The dose used was

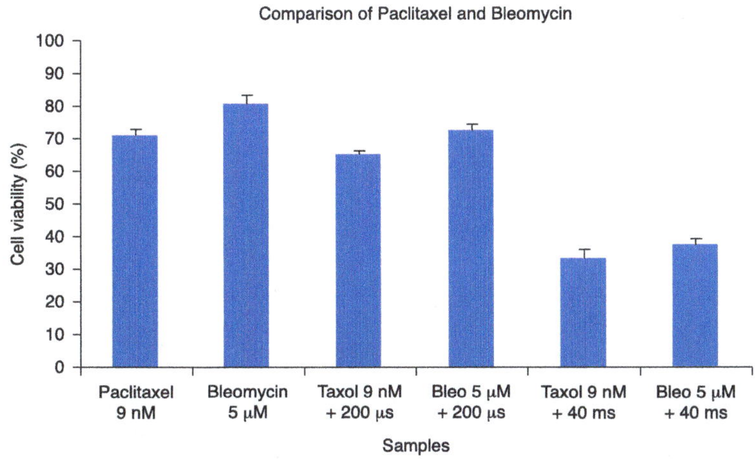

Figure 4.11 **Comparison of Paclitaxel and Bleomycin under various conditions**

9 nM for Taxol and 5 μM for Bleo. The viability of cells with Taxol only is reduced approximately 20% compared to the control. Cell viability of 9 nM Paclitaxel and 200 μs duration electrical pulse decreased to 65%. The same concentration of Paclitaxel treated cells with 40 ms electroporated had a further 30% reduction compared to the 200 μs electroporated cells. Cell viability with Paclitaxel and 40 ms electroporation showed a reduction of more than 60% compared to the control and 40% compared to the Paclitaxel only. These results indicate the enhanced efficacy of the combination therapy with electrical pulses and the chemo drug at concentrations as low as 9 nM. Similar results were obtained for Bleo, but at a larger dose of 5 μM. This indicates that the highly toxic Taxol is an ideal candidate for ECT using reduced dosages.

Tamoxifen and Bleomycin

Figure 4.12 shows the results of the Tam study where the live cell numbers, and hence the viability of the electroporated MCF-7 cells, at 5 μM are calculated. In this case, the pulses applied were 200 V/cm, at 10, 20, 30 and 40 ms durations, denoted by LP1, LP2, LP3 and LP4 respectively. NoEPorT represents the control samples with no treatment and Tam only denotes drug only samples, with no electroporation. As expected, with longer pulses, more energy is being applied and hence

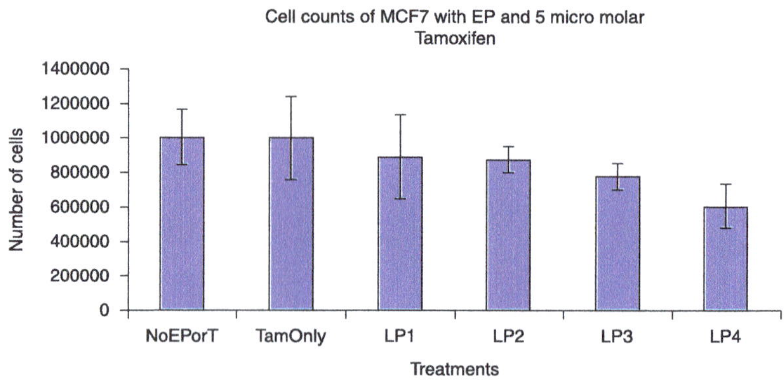

Figure 4.12 Cell viability due to millisecond pulses using Tam (LP1 = 10 ms, LP2 = 20 ms, LP3 = 30 ms and LP4 = 40 ms)

more cell death compared to the control or the drug only, or other lower pulse duration samples.

Figure 4.13 shows a comparison of Tam versus Bleo under the same conditions. They have very similar cell death values for these conditions. The 40 ms samples have almost 50% of the cells as those of the drug only samples. However, in real life, Tam is much gentler on humans than Bleo, but its response rate is only 21–41% [6]. Hence, use of EP is strongly recommended for targeted delivery. In addition, EP could be used for the

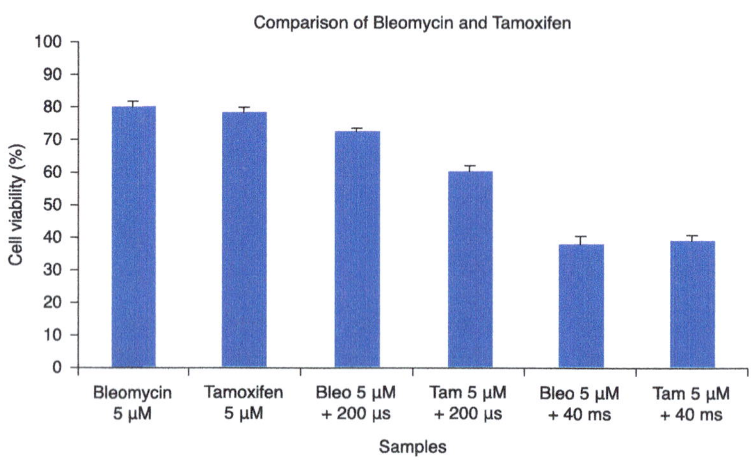

Figure 4.13 Comparison of Bleomycin and Tamoxifen under various conditions

marginally ER negative patients; it depends on how they are determined; some pathologists might decide the border cases as ER negative [30], whereas they could be slightly ER positive. Hence, it is preferred to use EP for delivering lower doses of the chemo and hormone drugs. More studies using both micro- and millisecond pulses were also done using Tamoxifen [47].

Curcumin

The results of the experiments at 50 μM and at pulse conditions of 1200 V/cm, 100 μs and 600 V/cm, 100 μs are shown in Figure 4.14. While curcumin alone could cause cell death, there was more cell death, 60% viability compared to 100% viability for control at 50 μM, when curcumin was used in coordination with electroporation. The results show the inverse linear relationship between the voltage used and the viability of the cells.

4.9.2 Fluorescence Microscopy

Figures 4.15(a–d) show fluorescence microscopy images using Bleomycin at 1200 V/cm and 200 V/cm. The dead cells are stained darkly. Figure 4.15(a) shows the control sample with no treatments, where there are more live cells in this than the others. Figure 4.15(b) shows the Bleomycin only sample, in

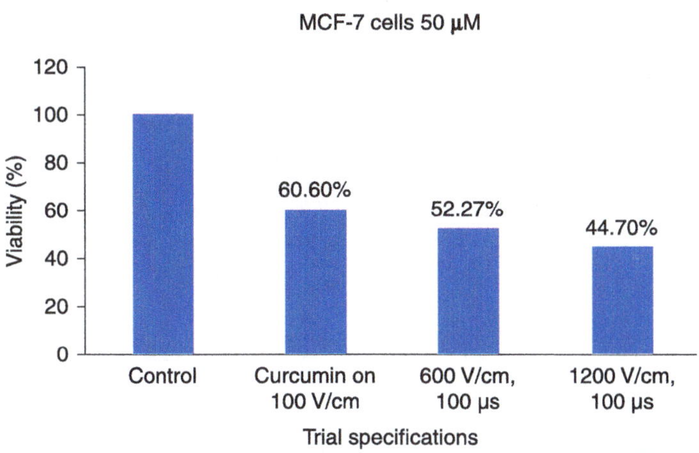

Figure 4.14 Effect of Curcumin (turmeric) on MCF-7 breast cancer cells at 50 μM concentration

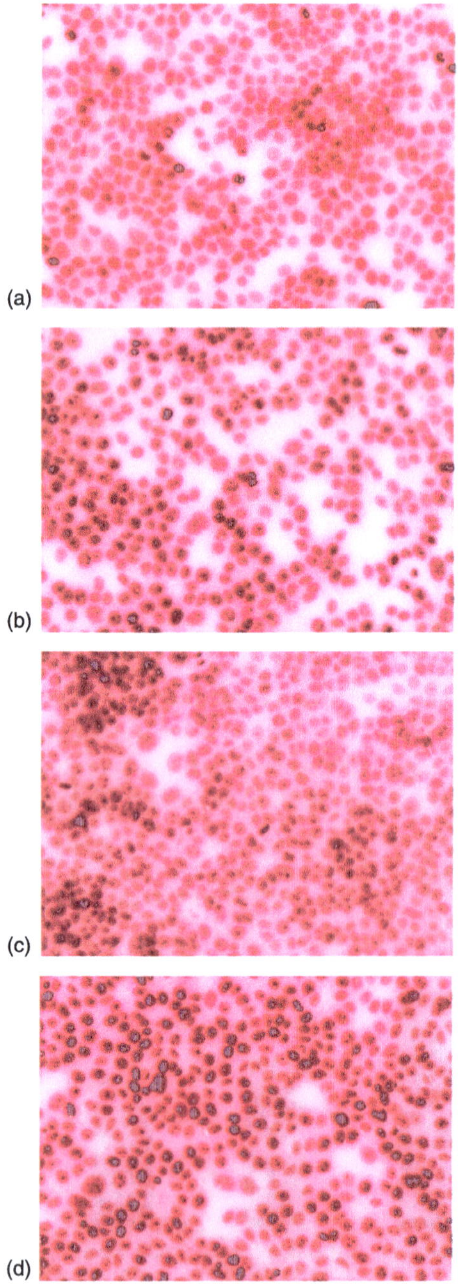

Figure 4.15 Fluorescence microscopy images of MCF-7 breast cancer cell samples (a) control (b) Bleomycin (c) Bleomycin + 200 μs (d) Bleomycin + 40 ms

which there are more dead cells than the control, but less than the electroporated samples. Figures 4.15(c) and (d) show the electroporated samples with Bleomycin using 1200 V/cm, 200 μs and 200 V/cm, 40 ms samples. These two samples have more cell death than the other two samples. Of all the samples, the sample treated with 200 V/cm, 40 ms electroporation at 5 μM concentration Bleomycin have more dead cells than the other samples, while the control has the least dead cells. These results also illustrate that due to the synergy of electroporation and chemo drug, there are more dead cells compared to drug only, and this shows the efficacy of the combination therapy of chemodrug and electrical pulses.

4.9.3 Flow cytometry

Figures 4.16(a–d) display results obtained using flow cytometry for Taxol. There are four areas in each image showing the status of the cells. These are live (blue), early apoptotic (purple), apoptotic (brown) and dead (red). It also indicates the various conditions of the cells by using two parameters, FL1 and FL3. FL1 displays the live cell data from Annexin V, while FL3 displays the dead cell data from PI. The peak emission wavelength for Annexin V-FITC is 528 nm, while the peak emission wavelength for PI is 620 nm. Figure 4.16(a) shows the control sample. In this case, a large number of cancer cells are located in the live area. When these cells were treated with 9 nM Taxol only, more cancer cells were moved from the live area to the apoptotic and dead areas, as illustrated in Figure 4.16(b). With electroporation, still more cells were dead compared to the control and drug only samples. They moved from live and apoptotic areas to the dead area. Especially when comparing 40 ms and 200 μs electroporated samples with Taxol, more cells moved from the apoptotic area to the dead area, as illustrated in Figures 4.16(c and d). Similar results were obtained for Bleo with 5 μM concentration. Table 4.7 shows a comparison of the results for Taxol and Bleo.

There is a good correlation between our results and the previous research findings that a number of anticancer drugs induce apoptosis in cancer cells [59]. Apoptosis occurs in most solid cancers. It is preferred to have apoptotic cell death, as opposed to necrosis, which involves inflammation, swelling of organelles, clumping of chromatic cells, and membrane disintegration. In contrast, apoptotic cells contain chromatin compacted into sharply delineated masses, and condensation of the cytoplasm occurs. No inflammatory reaction is noted due to apoptotic cell death. As seen in Figure 4.16, most MCF-7 cancer cells were located

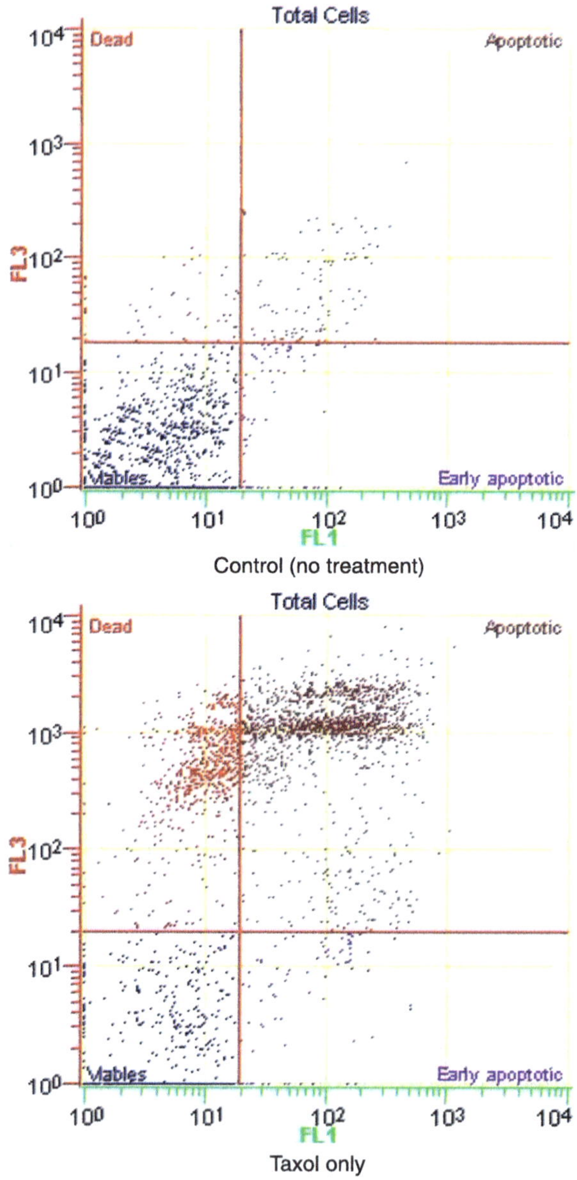

(a) Control (no treatment)

(b) Taxol only

Figure 4.16 Flow cytometry histograms illustrating the status of cells for control, drug only and electroporated samples

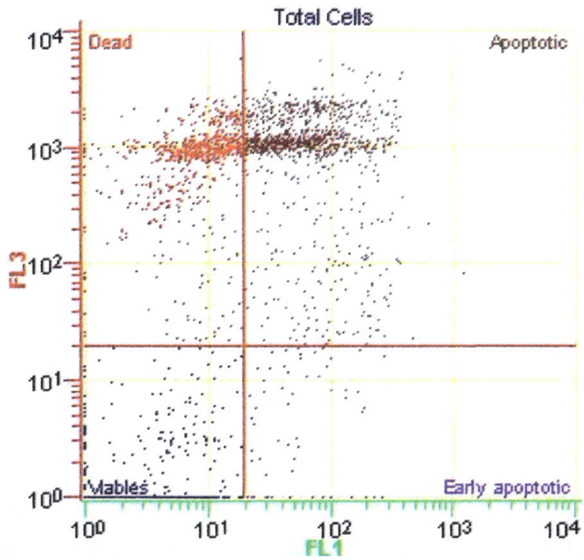

(c) Taxol + 200 μs

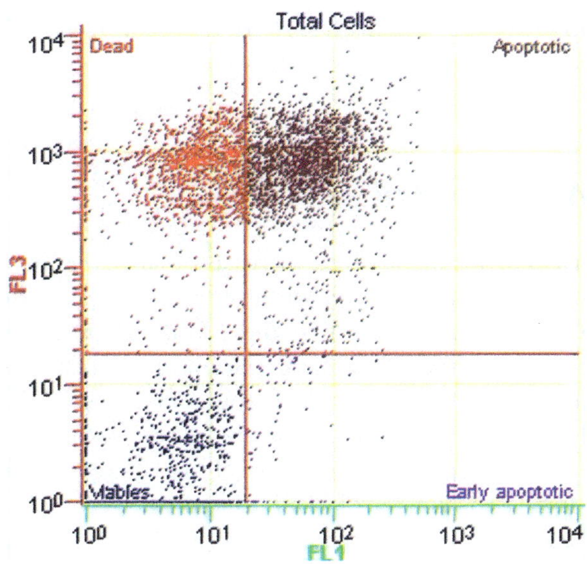

(d) Taxol + 40 ms

Figure 4.16 Continued

Table 4.7 Various statuses of the cells for control, drug only and electroporated samples [49]

Item	% Status			
	Dead	Early Apoptotic	Apoptotic	Viable
Control	3.97	2.19	4.03	89.81
Taxol only	28.14	5.71	43.79	22.36
Taxol + 200 μs	39.48	2.67	42.15	15.70
Taxol + 40 ms	67.92	0.53	24.96	6.59
Bleomycin only	18.74	4.85	36.73	39.68
Bleomycin + 200 μs	28.93	1.07	43.38	26.62
Bleomycin + 40 ms	64.57	4.25	23.62	9.56

in the dead area after using electrical pulses to deliver the anticancer drug. The results indicate that electrical pulses-mediated anticancer drug delivery enhanced the efficacy of curing breast cancer, because this technique can improve the permeability of the cell membrane in MCF-7 breast cancer cells.

4.9.4 Western Blot

The Western Blot results showed increased p38MAPK expression with Taxol (Figure 4.17). Significant elevation of p38MAPK activities in Taxol-treated MCF-7 cells suggested that these kinases could mediate Taxol-induced apoptosis. There is good correlation of our results with those of other researchers [49]. Compared to the control, MCF-7 cells treated with Taxol only and Taxol + EP show increased expressions of p38MAPK. Taxol only shows a higher expression level than Taxol + EP. This could be due to the increased number of dead cells in the electroporated samples being washed away during preparation of the assay.

4.9.5 Impedance spectroscopy

Bode plots

The plots of total impedance, Z versus frequency and phase angle, φ versus Frequency, are known as Bode plots. The magnitude is the ratio of

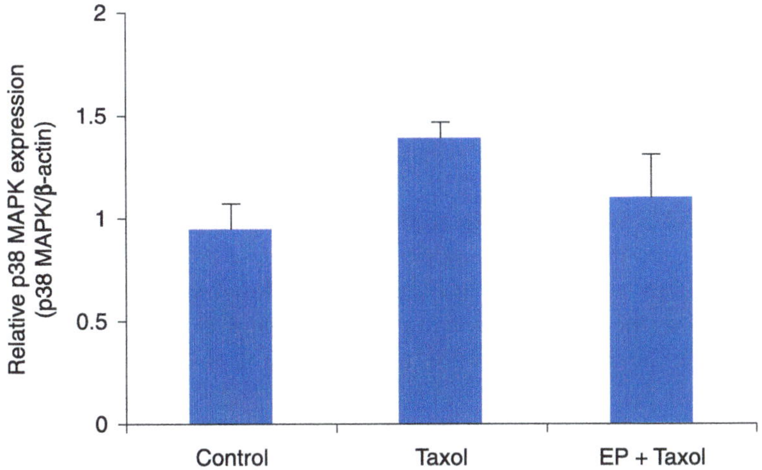

Figure 4.17 Western Blot results illustrating higher protein level for EP+Taxol than control sample

the voltage applied and the resultant current. In addition to the magnitude, there is also a relative time delay between the voltage and the current, given by the phase. A resistance has a zero phase angle, meaning the voltage and the current are in phase, so there is no time delay. However, there is a 90-degree phase shift or delay when a capacitance only is in the circuit (ideal case for a dielectric). In this case, the current leads the voltage and it is represented by negative 90 degrees, corresponding to the capacitive reactance of $1/j\omega c$. In these Bode plots, magnitudes of Z decrease with increase in frequency and approach R at high frequencies [53,54]. At lower frequencies, phase φ is equal to 90 degrees as for a pure dielectric, but at higher frequencies it decreases approaching the phase angle of resistance (zero) or the combination of R and C.

Figures 4.18 to 4.21 show the Bode plots obtained in this study. Figures 4.18(a) and (b) show the Bode plots for the control sample. Figures 4.19(a) and (b) show that of the drug only, in this case using Taxol at 9 nM. Figures 4.20 and 4.21 show the comparison of the variation of the two electroporated samples with the two drugs, Taxol and Bleo. Figure 4.20 shows the comparison with 1200 V/cm, 200 μs pulses and Figure 4.21 shows that of 200 V/cm, 40 ms pulses.

Figure 4.18(a) shows the magnitude Bode plot of the control sample. Here, the impedance is in the order of a few $10^7\,\Omega$ at low frequencies. It reduced to a few hundred ohms at higher frequencies. When comparing this with the drug only sample values (Figure 4.19(a)), a similar trend is seen. However, with the electroporated samples, the maximum magnitudes of the total impedances vary by up to 1200 Ω only (Figures 4.20(a) and 4.21(a)). There is up to four orders of magnitude reduction due to electroporation. This indicates the enhanced conduction or permeability of the samples electroporated with two different pulse parameters.

This trend of variation between the control and electroporated samples is also reflected in the phase angle plots. The phase angle increases by up to 90° at high frequencies from about 40° at low frequencies, for both control and the drug only samples (Figures 4.18(b) and 4.19(b)). However, in the case of electroporated samples, the phase angle is reduced

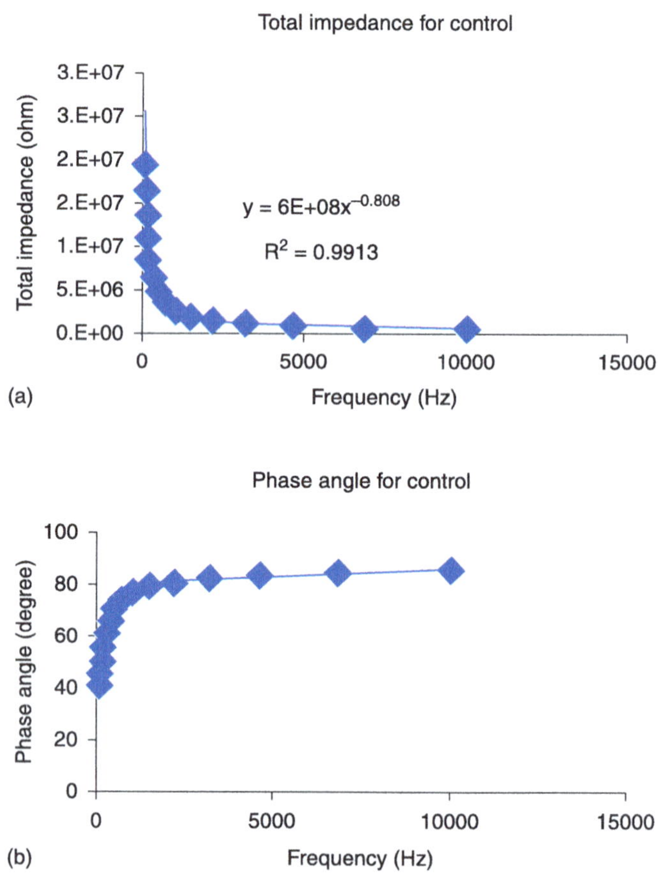

Figure 4.18 Total impedance and phase angle (Bode) plots of the control sample

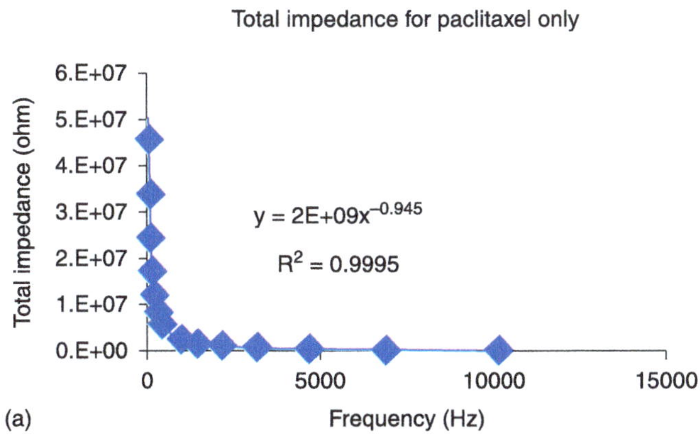

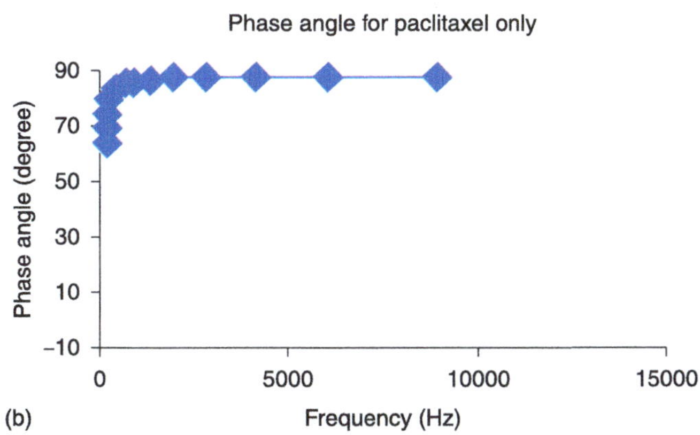

Figure 4.19 **Total impedance and phase angle plots of the Taxol only sample**

from 78° at low frequencies to 30° at higher frequencies (Figures 4.20(b) and 4.21(b)). Both the trend and the magnitude of these electroporated samples varied from the control and the drug only samples. This could be due to the variation of the cell characteristics with electroporation using the drugs.

While the total impedance versus frequency plots follow the typical similar style in all the samples, whether electroporated or not, the phase angle versus frequency of the control and the drug only samples have a trend opposite to those of electroporated samples. There is an increase in the phase angle with frequency in control and drug only samples, while the electroporated samples with the drug follows the expected trend, namely decrease in phase angle with increase in frequency from 78–30°.

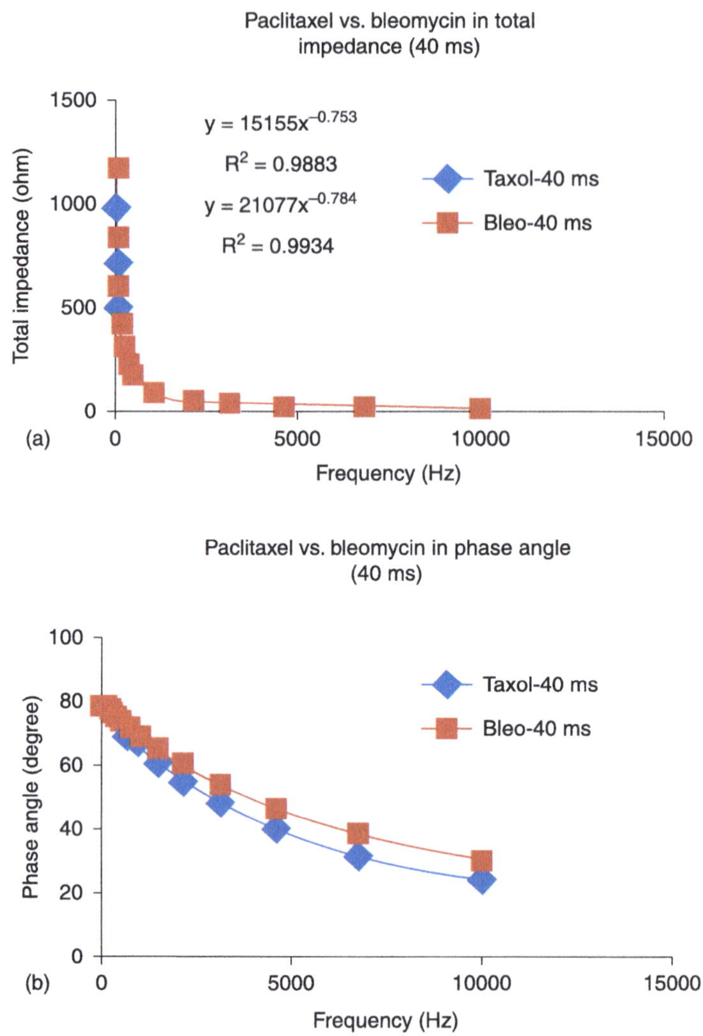

Figure 4.20 Comparison of Bode plots of the electroporated samples with eight 200 V/cm, 40 ms pulses

These results correlate well with the findings of other researchers [53, 54]. The increase in conductivity due to electroporation was observed in the rat liver by Ivorra and Rubinsky, using frequencies from 1–65 kHz (8 points) [53].

All the impedance plots follow the fractal power law with a negative exponent (i.e. inverse power law) typical of bio-dielectrics [55, 56]. The

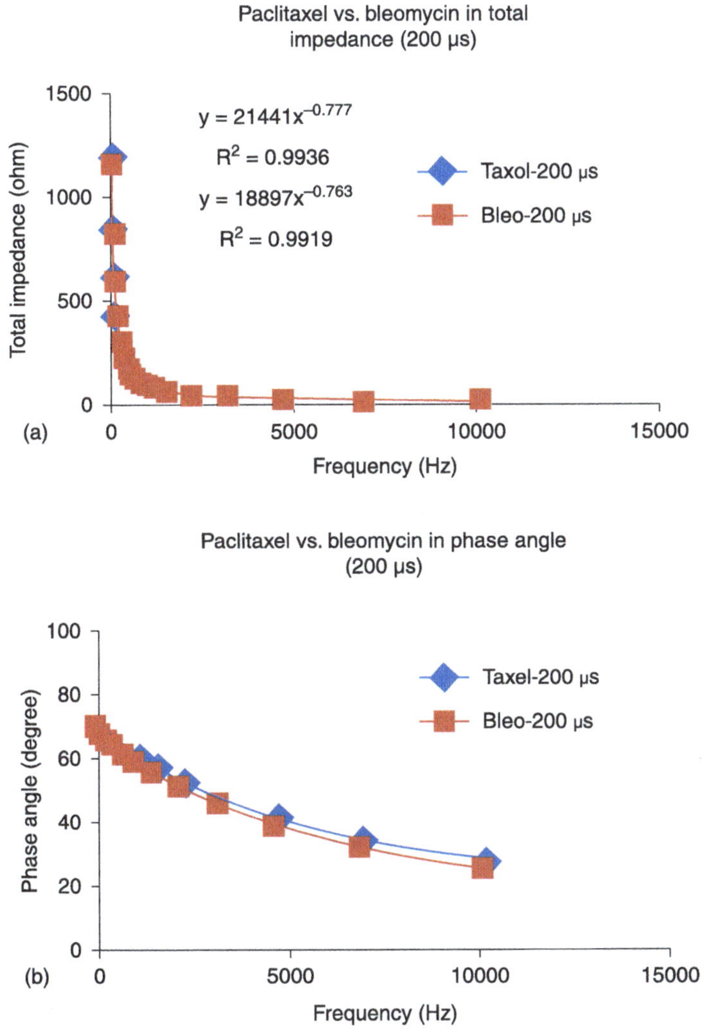

Figure 4.21 Comparison of Bode plots of the electroporated samples with eight 1200 V/cm, 200 μs pulses

variations in the negative power indices reflect the differences between electroporated, and the control and drug only samples. The electroporated samples have a lower index of around –0.5, while the control and drug only samples have indices of more than 0.5. It is –1 for a perfect dielectric or insulator. These results agree very well with the previous results [55].

Cole-Cole plots

The plot of Real Z (x-axis) versus Img Z for each frequency is known as a Cole-Cole plot [53,60]. It is used to represent the variation of the conductive part versus the membrane in the case of cancer cell/tissue studies for the frequency range studied. The low frequency data are on the right side of the plot, having higher magnitudes than the high frequency data, which are on the left side, as the impedance falls with increase in frequency, as seen in Figures 4.18–4.21. For a truly complex impedance, for a wide range of frequencies, the Cole-Cole plot is a half circle [60]. For a narrow range of frequencies, as in this study, they will form an arc or part of the semicircle. Figures 4.22 and 4.23 show the Cole-Cole impedance plots of this study.

Figure 4.22 shows the Bode plots of the control and the drug only samples. Figure 4.23 the electroporated sample with 200 V/cm, 40 ms pulses. The drastic reduction in the impedance values between the electroporated and the non-electroporated samples can be observed in these two figures. A fit of the trend line shows the trend of these graphs with a second-order polynomial (similar to that of a circle) semicircle with $R^2 = 1$, illustrating that they follow the semicircle trend.

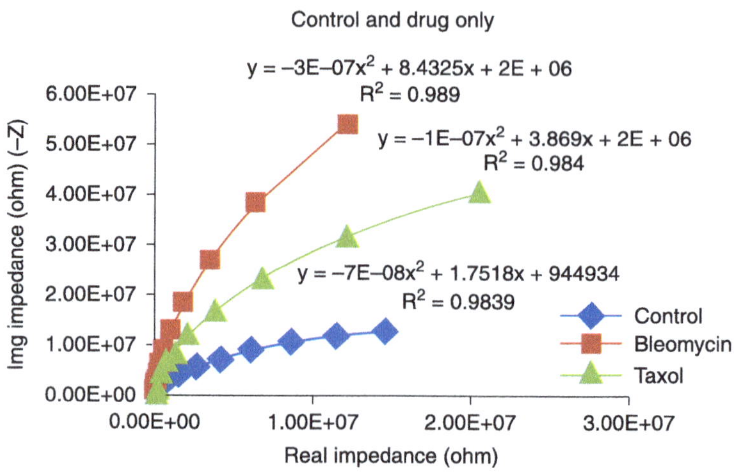

Figure 4.22 Cole-Cole Impedance plots indicating the high magnitudes of R and Xc for control and drug only samples

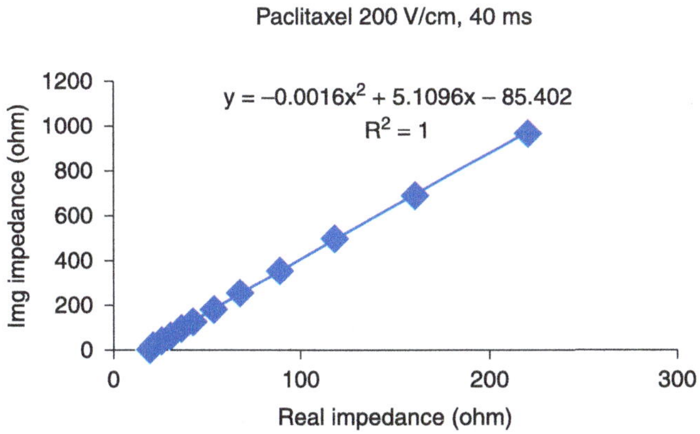

Figure 4.23 Cole-Cole Impedance plots indicating the low magnitudes of R and Xc for the electroporated sample

4.10 Conclusions

Our study assessed the prospects of using electrical pulses to enhance the effect of minute doses of chemotherapeutics, hormone drug and curcumin for MCF-7 breast cancer cells. We delivered both low voltage, millisecond and high voltage, microsecond pulses, viz. 200 V/cm, 40 ms and 1200 V/cm, 200 μs pulses, at 1 Hz to enhance the uptake of chemodrugs, Dox, Taxol and Bleomycin, hormone drug Tamoxifen and natural herb curcumin into MCF-7 breast cancer cells.

The flow cytometry spectra indicate death by apoptosis, the desired phenomena in cell death. The Western Blot results indicate the MAPK38 protein to be higher than the control for electroporated samples, indicating the effectiveness of the technique. The impedance spectroscopy results compared with non-electroporated samples, such as control and drug only samples, indicate the enhanced conduction of the electroporated samples. There is a four to five orders of magnitude enhanced conduction between the electroporated and the non-electroporated samples, indicating the efficacy of the technique in opening up pores allowing the transport of xenomolecules of chemo drugs across the cell plasma membrane, which otherwise are impermeable to the drugs.

These results indicate that the MCF-7 cell line is receptive to electroporation. It is also demonstrated that the combination therapy, electroporating cells along with a low dose of the chemo drug is more effective than treatment with the drug alone. This supports the hypothesis that electroporation and drug combination therapy will increase the cell's response to low dose chemo drugs including Dox, Taxol, Bleomycin, Tamoxifen and Curcumin, and enhance its cancer curing capabilities. Overall, in the majority of the samples, the viable cell count decreased in the electroporated sample, indicating that the proliferation control was achieved in these cells due to apoptosis. The Western Blot, impedance spectroscopy, fluorescence microscopy and the flow cytometry results corroborated the cell viability data and showed both quantitatively and qualitatively the proliferation control.

Our findings indicate that pulses as low as 200 V/cm, but of millisecond duration, could be used to trigger cell death in the MCF-7 cancer cells, compared to the conventional 1200 V/cm, 100 μs pulses and very low dose chemo drugs. Considering that only a 5 μM dose in the case of Bleo and 9 nM in the case of Taxol were used, these results are encouraging for the transfer of this technique to clinics for patients to have another treatment modality with lower cost and fewer side effects.

This is especially important in the US, where oncology is one of the most profitable sectors in medicine. Cancer drugs cost from hundreds to thousands of dollars per month. For example, Avastin (Genentech) costs up to $100 000 a year and Erbitux, another cancer drug, costs around $10 000 a month and pulled in revenues of $1.3 billion in 2007 for the manufacturer, ImClone [61]. Fifty percent of all personal bankruptcies and 50% of foreclosures in the US were due to medical expenses, especially cancer-related [62]. 'Nobody talks about the financial impact of cancer, and it's huge', as indicated in a September 2008 article [63]. Single mothers with one or two children, and breast cancer, are too ill to work and nearly bankrupted by chemotherapy. Some lose their jobs, and even starve. They struggle with astronomical co-payments and deductibles, at a time when they are too sick to work. It is critical to develop alternative, efficient and affordable cancer therapies. Towards this, the goal of our research is to develop safe and efficient novel therapies to treat breast tumors, as conventional treatments, such as surgery, radiation therapy and chemotherapy does not serve all patients, have serious side effects, are costly and do not work for all patients. To enhance drug uptake for breast cancer, we propose electrical pulse-mediated therapy that is less costly, has fewer side effects, and is simple (could be performed as an outpatient treatment [10]), requiring no costly outfits, such as a radiation facility.

4.11 Acknowledgments

We are extremely grateful to Dr. Jill Hutchcroft with the Purdue University Cancer Center Flow Cytometry Facility located in the Bindley Bioscience Center for all her training, help and support in doing the various experiments. We also thank Nick Lenarduzzi for the fluorescence microscopy and Caleb Larsen for curcumin experiments, Ryan Muir for guidance and training on impedance microscopy, and Brandon Bergman for editing.

4.12 References

1. Haagensen C D (1986), *Diseases of the Breast*, 3rd edition, PA: WB Saunders Co.
2. Ferla J H, Shin R, Bray F, Forman D, Mathers C and Parkin D M (2010), 'Estimates of worldwide burden of cancer 2008: GLOBOCAN 2008', *Int J Canc*, **127(12)**: 2893–917.
3. Andolina V F and Lille S L (2011), *Mammographic Imaging: A Practical Guide*, 3rd edition. PA: Wolters Kluwer Health-Lippincott Williams & Wilkins.
4. Kingsbury K (2007), 'The changing face of breast cancer', *Time Magazine*, Oct 15.
5. Lambert L and Keyomarsi K (2007), 'Cell cycle deregulation in breast cancer; insurmountable chemoresistance or Achilles heel', in: *Breast Cancer Chemosensitivity*, Yu D and Hung M-C (eds), New York: Springer Science Business Media, 52–69.
6. Gonzalez-Augulo A M, Morales-Vasquez F and Hortobagyi G N (2007), 'Overview of resistance to systemic therapy in patients with breast cancer', in: *Breast Cancer Chemosensitivity*, Yu D and Hung M-C (eds), New York: Springer Science Business Media, 1–22.
7. Tamkins T (2009), 'Medical bills prompt more than 60 percent of US Bankruptcies', Available at: *CNNhealth.com*, June 5.
8. Haltiwanger S (2010), *The Electrical Properties of Cancer Cells*. Available at: *http://www.royalrife.com/haltiwanger1.pdf*, July.
9. Heller R, Jaroszeski M J, Glass L F, Messina J L, Rapaport D P, et al. (1996), 'Phase I/II trial for the treatment of cutaneous and subcutaneous tumors using electrochemotherapy', *Cancer*, **77**: 964–71.
10. Gehl J and Geertsen P F (2000), 'Efficient palliation of haemorrhaging malignant melanoma skin metastases by electrochemotherapy', *Melanoma Res*, **10**: 1–5.
11. Larkin J O, et al. (2007), 'Electrochemotherapy: aspects of practical development and early clinical experience', *Ann Surg*, **245(3)**: 469–79.
12. Campana L G, Mocellin S, Basso M, Puccetti O, De Salvo G L, et al. (2009), 'Bleomycin-based electrochemotherapy: clinical outcome from a single Institution's experience with 52 patients', *Ann Surg Oncol*, **16**: 191–9.

13. Li S, Zhang X, and Xia X (2002), 'Regression of tumor growth and induction of long-term antitumor memory by interlukin 12 electrogenetherapy', *J Natl Canc Inst*, **94**: 762–8.
14. *http://www.who.int/mediacentre/news/releases/2003/pr27/en/*
15. Udayakumar P (2011), 'Introduction and routes of drug administration', *Medical Pharmacology*, New Delhi: CBS Publishers & Distributors.
16. King R J B and Robins M W (2006), *Cancer Biology*, 3rd edition. New York: Pearson Prentice Hall.
17. Udayakumar P (2011), 'Cancer chemotherapy', in: *Medical Pharmacology*. New Delhi: CBS Publishers & Distributors.
18. Osman A M, Nemnem M M, Abou-Bakr A A, Nassier O A and Khayyal M (2009), 'Effect of methimazole treatment on doxorubicin-induced cardiotoxicity in mice', *Food Chem Toxicol*, **479(10)**: 2425–30.
19. Koppelmans V, Breteler M, Boogerd W, et al. (2012), 'Neuropsychological performance in survivors of breast cancer more than 20 years after adjuvant chemotherapy', *J Canc Oncol*, 1 **30**: 1080–6.
20. Dev S B, et al. (2000), 'Medical applications of Electroporation', *IEEE Trans Plasma Sci*, **28(1)**: 206–23.
21. Leontiadou H, Mark A E, and Marrink S J (2004), 'Molecular dynamics simulations of hydrophilic pores in lipid bilayers', *Biophysical J*, **86(4)**: 2156–64.
22. Schwan H P (1985), 'Dielectric properties of cells and tissues', in: *Interactions between Electromagnetic Fields and Cells*, Chiabrera A, Nicolin C and Schwan H P (eds), New York: Plenum Press.
23. O'Hare M J, et al. (1989), 'Electropermeabilization and electrosensitivity of different types of mammalian cells', in: *Electroporation and Electrofusion in Cell Biology*, Nuemann E, Sowers A E, and Jordan C A (eds). New York: Plenum Press, 319–30.
24. Hui S W and Li L H (2000), '*Vitro* and *Ex vivo* gene delivery to cells by electroporation', in: Jaroszeski M, Heller R and Gilbert R (eds), *Electrochemotherapy, Electrogenetherapy, and Transdermal Delivery*, New Jersey: Humana Press, 157–71.
25. Davalos R and Rubinsky B (2013), 'Tissue ablation with irreversible electroporation', US Patent 8282631, April.
26. Horwitz K B, Costlow M E and McGuire W L (1975), 'MCF-7: a human breast cancer cell line with estrogen, progesterone, and glucocorticoid receptors', *Steroids*, **26(6)**: 785–95.
27. Anait S and Jordan V C (1997), 'MCF-7: the first hormone-responsive breast cancer cell line', *Canc Res*, **57**: 3 071–8.
28. *www.mcf7.com*, April 2013.
29. *www.ignagen.com*, April 2013.
30. Love S M with Lindsey K (2000), *Dr Susan Love's Breast Book*, 3rd edition. New York: Capo Press.
31. Doxorubicin (Intravenous Route) (2009), Available at: *Mayoclinic.com*, 1 November and *http://www.mayoclinic.com/health/drug-information/DR600581*, 4 December
32. Hamilton S (2005), Doxorubicin, Cleveland Clinic. Available at: *Chemocare.com* and *http://www.chemocare.com/bio/doxorubicin.asp*, 4 December.

33. Alberts D S, Chen H-S G, Mayersohn M, Perrier D, Moon T E and Gross J F (1979), 'Bleomycin pharmacokinetics in man', *Canc Chemother Pharmacol*, **2**: 127–32.
34. Baum M, Dossett J A, Patterson J S, Smiddy F G, Wilson A, et al. (1983), 'Improved survival among patients treated with adjuvant tamoxifen after mastectomy for early breast cancer', *Lancet*, **2(8347)**: 450.
35. Agarwal B B, Sundaram C, Malani N and Ichkawa H (2012), 'Curcumin: the Indian solid gold', Available at: *www.medify.com*, 10 July.
36. Joe B and Lokesh B R (1994), 'Role of capsaicin, curcumin and dietary n-3 fatty acids in lowering the generation of reactive oxygen species in rat peritoneal macrophages', *Biochim Biophys Acta*, **1224**: 255–63.
37. van Meeteren M E, Hendriks J J, Dijkstra C D and van Tol E A (2004), 'Dietary compounds prevent oxidative damage and nitric oxide production by cells involved in demyelinating disease', *Biochem Pharmacol*, **67**: 967–75.
38. Soliman K F and Mazzio E A (1998), '*In vitro* attenuation of nitric oxide production in C6 astrocyte cell culture by various dietary compounds', *Proc Soc Exp Biol Med*, **218**: 390–7.
39. Xie L, et al. (1994), 'Amelioration of experimental autoimmune encephalomyelitis by curcumin treatment through inhibition of IL-17 production', *Int Immunopharmacol*, **9**: 575–81.
40. Riccio P (2011), 'The molecular basis of nutritional intervention in multiple sclerosis: a narrative review', *Complement Therap Med*, **19**: 228–37.
41. Kanakasabai S, et al. (2011), 'Differential regulation of $CD4^+$ T helper cell responses by curcumin in experimental autoimmune encephalomyelitis', *J Nutr Biochem*. Available at: *http://dx.doi.org.laneproxy.stanford.edu/10.1016/j.jnutbio.2011.10.002*
42. Chearwae W and Bright J J (2008), '15-deoxy-Δ-prostaglandin J_2 and curcumin modulate the expression of Toll-like receptors 4 and 9 in autoimmune T lymphocyte', *J Clin Immunol*, **28**: 558–70.
43. Xie L, Li X K and Takahara S (2011), 'Curcumin has bright prospects for the treatment of multiple sclerosis', *Int Immunopharmacol*, **11**: 323–30.
44. Chendil D (2004), 'Curcumin confers radiosensitizing effect in prostate cancer cell line PC-3', *Oncogene*, **23(8)**: 1599–607.
45. Sreekanth C N, Bava S V, Sreekumar E, and Anto R J (2011), 'Molecular evidences for the chemosensitizing efficacy of liposomal curcumin in paclitaxel chemotherapy in mouse models of cervical cancer', *Oncogene*, **30(28)**: 3159–52.
46. Toshihiko K (1999), 'Chemopreventive effect of curcumin, a naturally occurring anti-inflammatory agent, during the promotion/progression stages of colon cancer 1', *Canc Res*, **59(3)**: 597–601.
47. Sundararajan R, et al. (2012), 'Effective proliferation control of human cancer cells using electrical pulses', *IEEE Trans Dielect Elect Insul*, **19(6)**: 2225–36.
48. Ibey B L, Pakhomov A G, Gregory B W, Khorokhorina V A, Roth C C, et al. (2010), 'Selective cytotoxicity of intense nanosecond-duration electric pulses in mammalian cell', *Biochim Biophys Acta*, **1800(11)**: 1210–19.

49. Bacus S S, et al. (2001), 'Taxol-induced apoptosis depends on MAP kinase pathways (ERK and p38) and is independent of p53', *Oncogene*, **20**: 147–55.
50. Reshkin S K, et al. (2003), 'Paclitaxel induces apoptosis via protein kinase A- and p38 mitogen-activated protein-dependent inhibition of the Na^+/H^+ exchanger (NHE) NHE isoform 1 in human breast cancer cells', *Clin Canc Res*, **9**: 2366–73.
51. Kuo H-C, et al. (2006), 'Enhancement of esculetin on taxol-induced apoptosis in human hepatoma HepG2 cells', *Toxicol Appl Pharmacol*, **210**: 55–62.
52. Huang T-S, Shu, C-H, Chao, Y, Chen S-N and Chen L-L (2000), 'Activation of MAD 2 ceheckprotein and persistence of cyclin B1/CDC 2 activity associate with paclitaxel-induced apoptosis in human masopharyngeal carcinoma cells', *Apoptosis*, 5, 235–41.
53. Ivorra A and Rubinsky B (2007), '*In vivo* electrical impedance measurements during and after electroporation of rat liver', *Bioelectrochemistry*, **70**: 287–95.
54. Pliquett U and Prausnitz M (2000), 'Impedance spectroscopy for rapid and noninvasive analysis of skin electroporation', in: Jaroszeski M, Heller R and Gilbet R (eds), *Electrochemotherapy, Electrogenetherapy and Transdermal Delivery*. New Jersey: Humana Press, 377–406.
55. Dean D A, Ramanathan T, Machado D and Sundararajan R (2008), 'Electrical Impedance Spectroscopy study of biological tissues', *J Electrostat*, 66: 165–77.
56. Wang R Y, Zhang T, Bao Q and Rawson D M (2006), 'Study on fish embryo responses to the treatment of cryoprotective chemicals using impedance spectroscopy', *Eur Biophys J*, **35**: 224–30.
57. Soley A, Lecina M, Gamez X, Cairo J J, Riu P, et al. (2005), 'On-line monitoring of yeast cell growth by impedance spectroscopy', *J Biotechnol*, **118**: 398–405.
58. Osterman K S (2004), 'Non-invasive assessment of radiation injury with electrical impedance spectroscopy', *Phys Med Biol*, **49**: 665–83.
59. Holzapfel C, Vienken J and Zimmermann U (1982), 'Rotation of cells in an alternating electric field: theory and experimental proof', *J Memb Biol*, **67(1)**: 13–26.
60. Bauchot A D, Harker F R and Arnold W M (2000), 'The use of electrical impedance spectroscopy to assess the physiological condition of kiwi fruit', *Postharvest Biotechnol Technol*, **18**: 9–18.
61. Arnst C and Capell K (2008), 'Lilly's risky rush into cancer drugs', *Businessweek*, October 20, p 72.
62. Feder J (2009), 'Federal action is required', *US News & World Report*, February, p 6.
63. Phillips B J (2008), 'Up from under', *Reader's Digest*, September, pp 17–18.

5

Why electroporation is a useful technique for cancer treatments

Steve Haltiwanger

DOI: 10.1533/9781908818294.103

Abstract: The electrical properties of cancer cells differ from normal cells in their composition, surface charge, membrane potential and resistivity and permittivity. This could be utilized to effectively electroporate the cell membranes of these cells for enhanced uptake of normally impermeable molecules including chemo therapeutic drugs. Described in this chapter are the detailed electrical properties of cancer cells and how they could be utilized for electrochemotherapy.

Key words: electroporation, cancer cells, electrical properties, plasma membrane composition, lipid peroxidation, electrochemotherapy, irreversible electroporation, nanosecond pulses.

5.1 Introduction

The purpose of this chapter is to describe cancer as a living system existing inside another living system, which is the body. If it is accepted that all systems have boundaries, then an exploration of some of the mechanisms that cancer uses to establish and maintain its boundaries within the body is worth exploring. The boundary conditions established by cancer enable it, as a living system, to maintain its integrity by separating its internal environment from the outside environment. This chapter will explain and review the electrical properties of cancer and how cancer cells have abnormal boundary conditions that affect their response to therapeutic interventions.

It is a well-accepted biological fact that the plasma membrane of a cell is a boundary that separates the internal milieu of the cell from the external environment. From the perspective of analyzing boundary conditions, it is important to note that cancer cells have membranes that are different from normal cells. It has been determined that cholesterol accumulates in the membranes of solid tumors and cancer cells activate *de novo* lipogenesis, which promotes the production and accumulation of saturated lipids in cancer cell membranes [1, 2]. Because the membrane composition of cancer cells differs from normal cells in lipid and sterol content, and the expression of membrane proteins is different in cancer, this has a direct influence on the fluidity, permeability and conductivity of the membranes [3, 4].

Production of reactive oxygen species occurs in excess in many different types of cancers, creating lipid peroxidation in the cell membranes. This is one mechanism that can contribute to membrane degeneration and increased permeability in cancer cells [5, 6]. Cancer cells compensate for the effects of lipid peroxidation by increasing the content of saturated acids and reducing the content of unsaturated fatty acids in their cell membranes. This provides increased protection against oxidative-induced lipid peroxidation [2].

The cell membrane in electrical terms is a leaky dielectric, which is a somewhat porous electrical insulator. The dielectric characteristics of a material include both conductive and capacitive properties [7]. One of the most important boundary functions of a cell membrane is the regulation of different mineral ion concentrations on each side of the cell membrane [8]. In electrical terms, mineral ions are conductors. The combination of a dielectric lipid bilayer coated on either side by layers of conducting mineral ions is an electrical capacitor.

The passage of electrically charged ions through a membrane will create a flow of electric currents through that membrane. These ions in turn will affect the metabolism of the cell and the potential of the cell membrane. Because the distribution of electrical charges is different on either side of a cell membrane, this creates a cell membrane potential (a voltage difference across the membrane) and produces an electrical field around the cell membrane [9].

The physical properties of cell membranes limit the movement of charged ions across the membranes to specialized transmembrane ion channels. Ion channels are transmembrane protein molecules that contain aqueous pores connecting the inside of the cell to the extracellular space. These channels can open and shut in response to a variety of signals. The passage of charged molecules through ion channels in the cell membrane endows the membrane with an electrical conductive property, allowing inward and outward current flows [10]. This is one factor that establishes electric circuits in biological tissues.

Normally the bi-layer lipid structure of cell membranes makes them relatively impermeable to charged molecules. Because of altered membrane composition and structure, cancer cells are more permeable than normal cells. This increase in membrane permeability results in potassium, magnesium and calcium moving out of cancer cells and there is a corresponding inflow of sodium and water [11, 12]. Movement of potassium out of the cell interior is also regulated by acidity of the cell interior and chemical and electrical gradients to the potassium ions [13].

Because cancer cell membranes are more permeable than healthy cells, they have greater conductivity. The electrical conductivity and permittivity of cancerous tissue have been found to be greater than the electrical conductivity and permittivity of normal tissues [14]. Because cancerous cells demonstrate greater permittivity, which is the ability to resist the formation of an electrical field, they have weaker fields than normal cells.

Cancer cells exhibit different electrochemical properties and a different distribution of electrical charges than normal tissues [15, 16], and contain an excess amount of negative charges on their surface. Cure has proposed that the accumulation of an excessive amount of negative charges on the exterior surface of cancer cells will depolarize the cancer cell membranes. He postulates that the depolarization (fall in membrane potential) of the cancer cell membrane due to the accumulation of excess negative surface charges may precede and create the reduction in intracellular potassium and the rise in the intracellular sodium launching the cell into a carcinogenic state [15]. Changes in the surface electrical charge to a highly negative state compared to normal cells, along with alterations in intracellular mineral concentrations, causes cancer cells to have lower membrane potential and lower electrical impedance [9, 17].

The result of these mineral movements, membrane composition changes, energy abnormalities and membrane charge distribution abnormalities is a drop in the normal membrane potential and a change in membrane capacitance (13, 18–21]. By comparing the electrical nature of cancer cells and their response to applied electrical stimulation, insight into how electrochemotherapy (ECT) works may be gained. This is the basic goal of this chapter.

5.2 What is electroporation (EP)?

Electroporation describes a process whereby externally applied electric fields are able to destabilize membranes and induce formation of pores [22]. The process is based on the polarization of material in electric fields [23]. When an electric field exerts force (pressure) on charges in biological tissue, the charges are either conducted and they move or if they are restricted in their movement, they accumulate. The accumulation and redistribution of charges under the influence of an electrical field is characterized by the polarizibility of the material. Exposing cells to an external electric field will result in membrane polarization, as charges accumulate at the membrane surfaces. This accumulation of charges produces an electric field inside the membrane, which in turn begins to deform the structure of the membrane. When the membrane is subjected to a sufficiently high field strength, pores begin to form in the membrane and ionic currents are produced as the outer surface charges flow through the membrane [23].

Electroporation occurs when an externally applied electric field of sufficient strength causes the externally induced transmembrane potential, V_m, to exceed a critical value, V_c, which is usually about 1 V in most cells [24]. When the field within the cell membrane exceeds the V_c of the V_m, dielectric breakdown occurs in the lipid bilayer, creating a phase transition that causes pores to form in the membrane [25, 26]. The pores that are created produce a dramatic increase in membrane permeability, along with a corresponding increase in membrane conductance [27]. Under normal physiological conditions, the electrical conductivity of the plasma membrane (PM) is much lower than the electrical conductivity of the extracellular medium and the cytoplasm [28, 29].

The creation of transient membrane pores during electroporation increases membrane conductivity greater than the conductivity produced by the ion channels [30]. This high-conductivity state results from structural rearrangement of the membrane phospholipids [31]. The rapid change in membrane conductance results in an elevation in the intracellular field, which may produce biochemical changes inside of the cells. The high-conductivity state is determined by the pulse amplitude and duration, which controls the number and size of pores produced [30].

The electrical charges associated with semiconducting proteins and extracellular matrix (ECM) proteoglycans also contribute to the innate conductivity of a tissue. So the electrical properties of tissues also relates to electron availability, which can be affected by such factors as the degree of tissue acidity, the degree of tissue hypoxia, and the availability of electron donors such as antioxidants, and the presence of electrophilic compounds on the cell membrane and in the ECM. Since all of these properties are altered in cancer, they may also contribute to the increased conductivity of cancer cells.

Overall, the goal of electroporation is to provide a short high-voltage pulse of sufficient field strength and pulse duration strong enough to create transient membrane pores (reversible dielectric breakdown) to allow passage of drugs, nutrients and genes into the cell, but not so strong as to produce irreversible dielectric breakdown of the cell membrane [32]. Electroporation makes it possible to go beyond the limitation of only lipophilic substances being able to pass the skin barrier. Use of electroporation can facilitate reversible electrical breakdown of cell membranes, creating transient structural rearrangement of cell membranes, which promotes rapid molecular transport into cells of non-lipophilic substances. Molecular transport occurs because the creation of transient (electrical pores) in cell membranes promotes both diffusion and electrophoretic movement of molecules into cells [33].

5.3 Irreversible electroporation (IRE)

The effects of high voltage impulses on living cells will create cell destruction by producing a dielectric breakdown of the cell membrane when the field strength and pulse duration are above V_c [34]. IRE uses electrical pulses alone (no chemo drugs) to kill cells in tumors by irrecoverable disruption of the cell membrane. The mechanism is non-thermal [35]. IRE characteristically uses pulses of a long duration and/or of high voltage to produce an irreversible membrane destabilization [36]. IRE can promote cell death by necrosis. In a sense, it is a non-thermal means of electrocution, where the cell membrane is radically disrupted, creating 'lethal biochemical imbalances' [30].

5.4 Electrochemotherapy (ECT)

Electrochemotherapy is the name of the procedure when EP is combined with the injection of cytotoxic drugs. It enhances the diffusion of chemotherapeutic drugs into the interior of tumor cells. ECT of solid tumors provides the ability to selectively kill tumor cells without harming normal surrounding tissue [37]. It is different from IRE, by causing cell death predominantly by the mechanism of drug-induced apoptosis, although some necrosis also occurs [30] and it is reversible. In ECT, an electric pulse is used to create membrane permeability, which allows large molecules to pass through the membrane producing intracellular cytotoxicity. IRE uses a different mechanism to destroy cancer cells by creating unrecoverable membrane damage leading to cell necrosis. Electroporation of solid tumors is a local treatment that when combined with intratumoral administration of cytotoxic drugs, such as bleomycin or cisplatin, can result in an exponential increase in cytotoxicity of up to a 1000-fold [38, 39].

5.5 Example of a hydrophilic agent used with electrochemotherapy

Bleomycin is a cytotoxic agent that is hydrophilic, so it does not easily cross the cell membranes. The use of ECT creates transient pores that temporarily increase the permeability of cell membranes, permitting passage of the drug into the cell interior [40]. Clinically, bleomycin combined with ECT has been reported to be successful in treatment of melanomas, squamous cell carcinomas, basal cell carcinomas, chondrosarcoma, breast cancer metastasis, Kaposi's sarcoma and soft tissue sarcomas [41–45].

5.6 Local delivery by intratumoral injection versus systemic administration in EP

Local tumor infiltration of a solid tumor by a chemotherapy drug followed by EP is superior to systemic administration of the drug for a number of reasons:

1. Lower amounts of the drug can be used, thus reducing systemic blood levels and side-effects.
2. Direct injection produces higher therapeutic concentrations at the target site.

3. An intratumoral injection allows delivery that bypasses the tumor vascular system, which is generally an obstacle to achieving effective concentrations when using systemic chemotherapy [46–48].
4. Increasing the uptake of chemotherapy by local delivery leads to both necrosis and apoptotic death of tumor cells, thus resulting in a high rate of local control and extended survival [49].

5.7 Prerequisites for effective ECT

1. Effective ECT requires the delivery of an effective electrical field with a selection of optimal pulse parameters and pulse delivery. Many studies have used 8 electric pulses, amplitude over distance 1300 V/cm, duration 100 μs and frequency 1–5 Hz [50].
2. In order for ECT to be effective, all of the tumor cells must be subjected to a high enough (i.e. over threshold) electric field. This means several applications have to performed, so that the entire tumor is exposed to the electrical field during treatment [29, 51].
3. Parallel plate electrodes are used for superficial tumors and needle electrode arrays are used for deeper tumors. The clinician must choose the appropriate electrodes, which when correctly placed will deliver electric pulses of sufficient amplitude [50].
4. The selected drug can be delivered either intravenously or intratumorally. However, clinical response requires that the drug must achieve sufficient distribution and concentration inside of the tumor.
5. The application of the electric pulses is time dependent after administering the selected cytotoxic drug. After intratumoral injection, timing can be 1–10 minutes and 8–28 minutes after intravenous injection [50–53].

5.8 ECT can overcome multidrug resistance

Cancer cells that over-express membrane glycoproteins, known as P-glycoproteins, can exhibit the phenomenon of multidrug resistance (MDR). MDR proteins are membrane transporters that lack substrate specificity, so they possess the ability to bind and expel a wide variety of diverse chemical compounds. Tumors that over-express MDR transporters are typically chemo-resistant, since they prevent cytotoxic agents from achieving effective intracellular concentrations (54, 55]. ECT can overcome the problem of MDR in several ways. First, ECT is capable of increasing the concentrations of cytotoxic drugs inside of tumor cells [56]. In addition, electrical stimulation can impair the translocation of MDR proteins to their functional location at the plasma membrane [54].

5.9 Intense nanosecond pulsed electric fields (nsPEFs)

Nanosecond pulsed electric fields use high intensity electric fields of 10–100s of kV/cm that are applied in nanosecond (1–300) durations. Although the energy density of nsPEFs is very low, because of the short duration of the pulses, the power is extremely high [57]. Intense nsPEFs produce both cellular and subcellular responses in mammalian cells [58]. Due to the high frequency of the applied fields, there is penetration of the electrical field into the interior of cells. The effects produced by nsPEFs include translocation of phosphatidylserine, formation of nanopores in the plasma and organelle membranes, calcium release, nuclear DNA damage, apoptosis, and a profound increase in inward electrical conductance at negative membrane potentials, which is thought to result from nanosecond electroporation of the cell membrane and the movement of ions into the cell [59–64]. The penetration of an nsPEF into the interior of cells can occur as long as the pulse rise time is faster than the charging time of the cellular membrane, which is about 0.1–1 μs. When pulse time is fast enough, polarization of the interior charges will not have sufficient time to redistribute to counteract the imposed field [65].

Preliminary work, using paired tumor and normal skin cell lines, has found that tumor cells have a stronger response to nsPEFs than do normal cells. If this work is replicated, it indicates the potential for benefit of the use of nsPEFs in cancer treatment [66]. One of the facts that may account for the observation that tumor cells have stronger responses to nsPEFs, is that tumor cells have different dielectric properties, and these dielectric properties of the cell membrane have a strong impact on the response to electroporation [67]. This finding supports the hypothesis that the different physiochemical properties of tumor cells are involved in the greater sensitivity to electroporation therapies than normal cells.

Nanosecond electric field pulses have been reported to produce significant changes in tissue conductivity and strong local (intracellular) electric fields, but only small changes in membrane permeability [58, 68]. Although the pores created by nsPEFs are only nanometers in size, the number of pores created in the plasma membrane is very high, so that average tissue conductance is even greater than for longer IRE pulses [63]. Nanosecond pulsed electric fields greater than 20 kV/cm, with rise times of 30 ns and durations of 300 ns, have been found to penetrate into the interior of tumor cells of fibrosarcomas and melanomas. These fields have been found to create only a 3 °C rise in temperature. The interesting effect is not on the plasma membrane, but instead it appears that these nanosecond fields essentially electrocute the cell nuclei, causing tumor cell nuclei to rapidly shrink and tumor blood flow to stop. The tumor blood flow is disrupted by the treatment, because the tumor blood vessels have become leaky [68].

The evasion of apoptosis is a characteristic and possibly a ubiquitous cancer hallmark. Yet nsPEF studies demonstrate that *promotion of apoptosis* in cancer is a consistent finding in both *in vitro* and *in vivo* studies [65, 69]. The nsPEFs can also be used for IRE, without the use of drugs [70]. However, with reversible nsPEFs, cell death occurs by apoptosis and in IRE, cell death occurs by necrosis [71, 72]. The nsPEFs penetrate the cell membrane to target intracellular structures such as the cytoskeleton, nuclear membrane and DNA, while producing only transient permeability of the plasma membrane. The action of nsPEFs is primarily intracellular. IRE, on the other hand, produces unsurvivable disruption of plasma membranes [37, 73]. Older literature contends that the PM is not affected by nsPEFs. However, research performed by Esser's group demonstrates that the PM does respond to nsPEFs and in fact nsPEFs create pore densities that are much higher than even those produced by longer conventional EP pulses. Esser terms this a supra-EP response. The difference in PM permeability produced by nsPEFs is that all of the pores in the PM are only of nanometer size, which '. . . leads to utterly different electropermeability values than for conventional EP, including IRE pulses' [63].

5.10 Electroporation therapies can produce a reduction in blood flow to tumors

Both EP and ECT can produce a profound anti-angiogenic effect in tumors, by reducing tumor blood flow. Both create a vasoconstrictive response starting within seconds. However, with EP tumor, blood flow gradually reverses over 24 hours, whereas ECT leads to an irreversible reduction in blood flow as bleomycin influx damages endothelial cells in tumor vasculature causing capillary leakage [74, 75]. Also, nsPEFs will all cause a reduction in blood flow to tumors [68]. The following section reviews some of the physical properties of tumors as they relate to their electrical properties and response to electric fields.

5.11 Properties of solid tumors

1. irregular vascularization;
2. heterogeneous concentrations of oxygen;
3. pH variations.

The most important physicochemical alterations found in tumors include a shift of the microenvironmental extracellular *pH towards acidic* values, a *reduced oxygen supply* and *non-uniform nutrient delivery*. These are a consequence of chaotic vascularization, resulting in an inadequate blood supply [76]. Solid cancerous tumors present several barriers to effective drug delivery. Typically solid tumors have irregular vascularization that prevents chemotherapeutic agents from being able to fully penetrate and reach all the cancer cells at therapeutic levels [77]. In addition, solid tumors have concentrations of oxygen and pH that vary extensively in different areas of the tumor [78]. The pH level can fluctuate extensively in range, with some areas being rather acidic (<6.5) to other areas where pH values are more basic (7.5). The same heterogeneity occurs with oxygen. Tumor areas near blood vessels can be well oxygenated, whereas poorly vascularized tumor areas can be very hypoxic [79, 80]. The extensive variations in oxygen and pH in the same tumor will promote the proliferation of heterogeneous populations of cells that utilize different metabolic pathways, which can confer increased drug resistance [79, 81]. In the hypoxic areas, post-translational and transcriptional changes occur that promote cell survival, metastasis and resistance to chemotherapy [82, 83].

Fortunately, electroporation strategies can provide a solution, by utilizing a therapeutic technique based on an electromagnetic mechanism that can reach all the tumor cells in a localized area. Use of electroporation, especially when combined with intratumoral injection, provides a means of delivering oncological agents that circumvents the limitation of vascular delivery to poorly vascularized areas.

5.12 Why do tumors have increased susceptibility to (EP) permeabilizing pulses, compared to that of normal tissue?

Membrane composition of cancer cells differs from normal cells in the amount of negative surface charges present. Most cancers possess an excess of fixed electronegative charges on their surfaces [84]. During the course of cancerization, cells become more electronegative and the cell membranes become more permeable [12]. Negatively charged molecules such as phosphatidylserine, heparin sulfate and sialic acid are more abundant on cancer cell membrane surfaces compared to normal cells. The negative charge is also augmented, because cancer cells have more micro-villi, creating a larger surface area than normal cells. In normal cells, phosphatidylserine is normally located on the inner leaflet of the plasma membrane. However, the expression of phosphatidylserine on the surface of cancer cells may prove to be potential uniform marker of tumors [85, 86].

Sialic acid is another source of negative charges found in excess on cancer cell membranes. Sialic acid residues are linked to glycolipids and glycoproteins on the cell membrane surface. The large numbers of negative charges on tumor cell surfaces provide a protective barrier against the immune system and facilitate migration and invasion [85]. Human chorionic gonadotropin (hCG), a hormone usually associated with pregnancy, can be found on the surface of cancer cells. Dr Acevedo has proposed that the presence of hCG on the surface of cancer cells may be a universal marker for cancer [87, 88]. It remains to be seen whether it can be confirmed that all cancer cells have this marker. However, the hCG contains large amounts of sialic acid; this is another factor that contributes to *cancer cells having a stronger cell surface negative charge than normal cells* [89]. Since the negatively charged electric field around cells are created by sialic acid residues, any factor that increases or decreases the number of sialic acid residues will change the degree of surface negativity a cell exhibits.

Strong negative surface charges on cancer cells are a defensive measure against immune system attack. Because immune defense cells such as NK cells and macrophages also have a negative surface charge, these cells are repulsed by the strong negative electrical field of cancer cells when they try to approach and terminate cancer cells [16, 89, 90]. Negatively charged hCG is present on the cell membranes of embryonic and fetal cells, sperm cells and cancer cells, irrespective of type or origin [87, 88]. Membrane-associated hCG makes all these cells immunologically inert.

The very high surface negativity of cancer cells creates an electrical defense shield. This electrical barrier may be relevant for immune resistance, but it is non-significant when electroporation therapies are used.

5.13 Membrane composition and mineral concentrations of cancer cells affect the electrical properties

Membrane composition and fluidity is different in cancer compared to normal cells. Fluidity can be either elevated or reduced, depending on cancer cell type [85]. Cells with an increased percentage of membrane cholesterol are less fluid than cells with a lower cholesterol/phospholipid ratio [91]. Cancer cells have different lipid and sterol content than normal cells. Because of the lipid/sterol composition, water and sodium content of cancer cells differ from normal cells; this will affect the V_m as well as the conductivity, dielectric and permittivity properties of cancer cell membranes [14, 92]. Tumor tissue exhibits a larger permittivity and greater conductivity than normal tissues [14]. The electrical conductivity of a tissue depends on both the physico-chemical bulk properties, such as properties of tissue fluids and solids and the microstructural properties (i.e. the geometry of microscopic compartments). In turn, the electrical conductivity and permittivity of biological materials will vary characteristically, depending on the frequency applied. Because cancerous cells demonstrate greater permittivity, which is the ability to resist the formation of an electrical field, they will respond to external electrical fields differently from normal cells.

Tumor cells characteristically have a higher water content and sodium concentration than normal cells, 'as well as different electrochemical properties of their cell membranes' [17]. A change in mineral content of the cell, particularly an increase in the intracellular concentration of positively charged sodium ions and an increase in negative charges on the cell coat (glycocalyx), are two of the major factors causing cancerous cells to have lower membrane potential and lower electrical impedance than healthy cells [9, 16, 17, 93].

Alterations in the plasma *membrane composition* of cancer cells, along with the *relative permeability* to diffusible ions and the concentration gradient of these ions across the membrane, produces a lower V_m in cancer cells compared to normal cells [94]. A healthy cell will maintain a high intracellular concentration of potassium and a low concentration of sodium. When increased membrane permeability occurs in injured and cancerous cells, sodium and water will flow into the cells, while potassium, magnesium, calcium and zinc are lost from the cell interior. Collectively, the changes in membrane surface negativity, composition, structure and permeability, along with the changes in mineral concentrations across the membrane, result in a fall in cell membrane potential [9, 18–20].

By way of comparison, healthy cells have a membrane potential of about -60 to –100 mV [15]. When cancer cells begin cell division and DNA synthesis, the membrane potential falls to around –15 mv [16]. Since the membrane potential in a cancer cell is consistently weaker than the membrane potential of a healthy cell, the electrical field across the membrane of a cancer cell will be reduced. The reduction in membrane electrical field strength will in turn cause alterations in the metabolic functions of the cell.

5.14 Oxygen levels vary in solid tumors

Tumors have gradients of O_2 levels, based on the distance of tumor cells from a functional blood vessel. Tumor cells surrounding a blood vessel are well oxygenated, whereas tumor cells distant from vessels are poorly oxygenated and express high levels of transcription factors including HIF-1 [95]. *Hypoxic conditions* within tumors activate signaling pathways (hypoxia-inducible factor (HIF-1), vascular endothelial growth factor (VEGF) and others), which initiate tumor cell invasion, migration, adhesion and angiogenesis [96]. HIF-1 induces the expression genes that code for proteins that increase: uptake of glucose (e.g. glucose transporter-1; conversion of glucose to pyruvate (glycolytic enzymes); generation of lactate and H^+; and the efflux of lactate and H^+ ions out of the cells into the extra cellular matrix (ECM) [97, 98]). The severity of hypoxia and acidosis in tumors can affect tumor cell invasiveness, metastasis, the risk of recurrence, and resistance to chemotherapy and radiation therapy [99].

A decrease of oxygen with increasing distance from blood vessels leads to a metabolic switch of hypoxic tumor cells from respiration to anaerobic glycolysis with extracellular accumulation of lactic acid. Cancer cells can also use glycolysis for energy production, even under normal oxygen conditions [100]. Other phenotypes of tumor cells in the same tumor can use oxidative phosphorylation under aerobic conditions [101]. Hence, both well oxygenated and hypoxic areas exist in the same tumor containing a heterogeneous population of cells with different metabolic phenotypes [97].

Tumors with areas of mixed oxygenation will often contain heterogeneous groups of cells that exhibit biochemical diversity. The same tumor will have some cells that are utilizing different metabolic reactions to create energy than other groups of cells in the same tumor. This is one reason why different cell populations in the same tumor will respond differently to treatment measures. Some cells will be killed by some treatments, while other cells will survive and in a sense be selected for further growth. Tumor cells will express larger amounts of glucose receptors/transporters on their cell surface, in order to increase their sugar uptake. In fact, hypoxia stimulates the transcription of numerous genes, including genes that code for enzymes of the glycolytic pathway and cell membrane glucose transport proteins GLUT-1 and GLUT-3 [102].

5.15 Glycolysis and pH

Cells that are transformed into cancer cells have a highly altered energy metabolism, which includes increased reliance on glycolysis and a shift to the use of glutamine in the tricarboxylic acid (TCA) cycle [103]. In normal cells, glycolysis is inhibited by the presence of oxygen, which allows mitochondria to oxidize pyruvate to CO_2 and H_2O; this inhibition is termed the 'Pasteur effect'. Conversion of glucose to lactic acid in the presence of oxygen is known as aerobic glycolysis, or the 'Warburg effect'. Increased aerobic glycolysis is almost universally observed in cancers [100]. Persistent metabolism of glucose to lactate, even in aerobic conditions, is an adaptation to intermittent hypoxia.

Increasing glycolysis could create intracellular acidification and a reversal of the metabolic switch back to oxidative-phosphorylation when aerobic conditions are present. Tumors can sustain aerobic glycolysis by avoiding cytosolic acidification by activating transporters that pump out H^+ ions and lactate to maintain an alkaline interior [98]. Up-regulation of glycolysis requires control of intracellular H^+ ions (acidity) through up-regulation of multiple families of H^+ membrane transporters to maintain intracellular pH [95]. The $Na(^+)/H(^+)$ exchanger and the $H(^+)$/lactate co-transporter are two of the important membrane transporters that are activated in cancer cells [104].

Tumor cells and normal proliferating cells can activate glycolysis, under both aerobic and hypoxic conditions, by increasing intracellular alkalinization [105,106]. Cancer cells, which have high rates of aerobic glycolysis, can increase their intracellular pH to 7.12–7.65 compared with 6.99–7.20 in normal cells [104]. Up-regulation of the Na^+/H^+ exchanger is an essential condition for growth and proliferation of tumor cells [107]. This transporter brings in Na^+ and exports H^+ ions out of the tumor cells. This contributes to an elevation of intracellular sodium levels and reduction in potassium and magnesium levels and alkalinization of the interior of cancer cells [108]. High Na^+ permeability and K^+ loss may be characteristic of cells with a high aerobic glycolytic activity. The sodium pump (membrane Na/K ATPase) is impaired in cancer, leading to high intracellular Na^+ and low intracellular K^+. Malignant tumors have a much higher Na:K ratio [109]. In general, intracellular pH elevation occurs when proton transport (export) mechanisms are up-regulated. Increased H^+ ion export from cancer cells is accompanied by elevations in sodium and depletion of potassium inside cancer cells [76].

Aerobic glycolysis increases lactate secretion, which acidifies the surrounding ECM and facilitates tumor metastasis and immune escape. Immune system function is compromised by excessive acidity [110]. Acidification of the microenvironment also enables tumor invasion through destruction of adjacent normal populations, acid-induced degradation of the ECM, promotion of angiogenesis, selection of aggressive tumor cells, suppression of apoptosis, induction of pro-invasive mediators and stimulation of proteolytic enzyme secretion [76,80]. Acidic regions are frequently found within solid tumors ($pH < 6.8$), as well in the tissue surrounding the tumor [80]. Up-regulation of glycolysis leads to surrounding tissue acidosis, requiring evolution to phenotypes resistant to acid-induced cell toxicity. Subsequent cell populations with up-regulated glycolysis and acid resistance have a powerful growth advantage, which promotes unconstrained proliferation and invasion [95]. Acidity has a role in resistance to chemotherapy, proliferation and metastatic behavior by the alteration of the tumor microenvironment through changes of the pH gradient between the extracellular environment and cell cytoplasm [111].

The body buffers acids with minerals, in particular calcium, magnesium and potassium, and through the bicarbonate stores in the body. Systemic alkalinization is effective as a cancer treatment in some studies [112]. Serum concentrations of sodium bicarbonate can be increased by oral ingestion. 'Consequent reduction of tumor acid concentrations significantly reduces tumor growth and invasion without altering the pH of blood or normal tissues' [113]. In consideration of the pH abnormalities of solid tumors, the use of pH buffers combined with electroporation may be warranted, based on the physiochemistry of cancer.

5.16 Conclusions

Cancer cells have a number of properties that produce unique electrical characteristics. They have a different membrane composition and mineral content than normal cells. These alterations directly influence the electrical properties of cancer cells. The electrical field surrounding cancer cells plays an important role as a boundary separating cancer from the immune system. The high surface negativity of cancer cells creates an electrical defense shield and is a factor in immune resistance. The physiochemical properties that make cancer resistant to radiotherapy and chemotherapy can be overcome by the application of external electrical fields. Electroporation combined with directed chemotherapy can circumvent the obstacles of a chaotic vasculature system, hypoxia, acidosis and multidrug resistant transporters that make systemic chemotherapy less effectual in controlling solid tumors. Although ECT has mainly been used on surface tumors, recent advances with needle array applications are demonstrating the benefit in treating deeper tumors.

5.17 References

1. Freeman M R and Solomon K R (2004), 'Cholesterol and prostate cancer', *J Cell Biochem*, **91(1)**: 54–69.
2. Rysman E, Brusselmans K, Scheys K, et al. (2010), '*De novo* lipogenesis protects cancer cells from free radicals and chemotherapeutics by promoting membrane lipid saturation', *Cancer Res*, 70(20): 8117–26.
3. Dianzani M U (1989), 'Lipid peroxidation and cancer: a critical reconsideration', *Tumori*, **75**(4): 351–7.
4. Li M and Xiong Z G (2011), 'Ion channels as targets for cancer therapy', *Int J Physiol Pathophysiol Pharmacol*, **3**(2): 156–66.
5. Skrzydlewska E, Sulkowski S, Koda M, et al. (2005), 'Lipid peroxidation and antioxidant status in colorectal cancer', *World J Gastroenterol*, 2005 Jan **11**(3): 403–6.
6. Dargel R (1992), 'Lipid peroxidation – a common pathogenetic mechanism?', *Exp Toxicol Pathol*, **44**(4): 169–81.
7. Reilly J P (1998), *Applied Bioelectricity: From Electrical Stimulation to Electropathology*. New York: Springer.

8. Cho Y S, Resasco D, Schaff J and Slepchenko B (1999), 'Electrodiffusion of ions inside living cell', *IMA J Appl Math*, **62(3)**: 207–26.
9. Cone C D Jr (1985), *Transmembrane Potentials and Characteristics of Immune and Tumor Cells*. Boca Raton, FL: CRC Press.
10. Aidley D J and Stanfield P R (1996), *Ion Channels: Molecules in Action*. Cambridge: Cambridge University Press.
11. Nieper H A (1985), *Dr Nieper's Revolution in Technology, Medicine and Society*. Oldenburg, Germany: MIT Verlag.
12. Seeger P G and Wolz S (1990), *Successful Biological Control of Cancer: By Combat against the Causes*. Gesamtherstellung: Neuwieder Verlagsgesellschaft mbH.
13. Haltiwanger S G (2003), 'The electrical properties of cancer cells', paper presented at: *Rife 2003 International Health Conference*; September 26–28, Seattle, WA. Available at: *http://www.royalrife.com/haltiwanger1*
14. Foster K R and Schepps J L (1981), 'Dielectric properties of tumor and normal tissues at radio through microwave frequencies', *J Microwave Power*, **16**: 107–19.
15. Cure J C (1991), 'Cancer an electrical phenomenon', *Resonant*, **1(1)**: 937–51.
16. Cure J C (1995), 'On the electrical characteristics of cancer', paper presented at: *The Second International Congress of Electrochemical Treatment of Cancer*; October, Jupiter, FL.
17. Blad B and Baldetorp B (1996), 'Impedance spectra of tumour tissue in comparison with normal tissue; a possible clinical application for electrical impedance tomography', *Physiol Meas*, **Suppl 4A**: A105–15.
18. Cone C D Jr (1970), 'Variation of the transmembrane potential level as a basic mechanism of mitosis control', *Oncology*, **24(6)**: 438–70.
19. Cone C D Jr (1974), 'The role of surface electrical transmembrane potential in normal and malignant mitogenesis', *Ann NY Acad Sci*, **238**: 420–35.
20. Cope F W (1978), 'A medical application of the Ling Association-Induction Hypothesis: The high potassium, low sodium diet of the Gerson cancer therapy', *Physiol Chem Phys*, **10(5)**: 465–8.
21. Haltiwanger S G (1998), 'Clinical use of mineral transporters and their effects on cell membrane capacitance', paper presented at: *Second International Congress of BioEnergetic Medicine; Institute of Quantum and Molecular Medicine*; February 20–22, Melbourne, FL.
22. Chang D C, Chassy B M, Saunders J A and Sowers A E (1992), *Guide to Electroporation and Electrofusion*. San Diego: Academic Press.
23. Dimitrov D S (1995), 'Electroporation and electrofusion of membranes', in: Lipowsky R and Sackmann E (eds), *Handbook of Biological Physics*, vol. 1. Elsevier Science BV, 851–900.
24. Kinosita K Jr, Ashikawa I, Saita N, Yoshimura H, Itoh H, et al. (1988), 'Electroporation of cell membrane visualized under a pulsed-laser fluorescence microscope', *Biophys J*, **53(6)**: 1015–9.
25. Gaynor P T and Bodger P S (1994), 'Ionisation of dielectric spherical membranes: a balloon model of electroporation of biological cells', *IEEE Proc Sci Meas Technol*, **141(3)**: 190–6.

26. Partenskii M B and Jordan P C (2005), 'Negative capacitance and instability at electrified interfaces: Lessons from the study of membrane capacitors', *Condens Matter Phys*, **8(2)**: 397–412.
27. Prausnitz M R, Bose V G, Langer R and Weaver J C (1993), 'Electroporation of mammalian skin: a mechanism to enhance transdermal drug delivery', *Proc Natl Acad Sci USA*, **90(22)**: 10504–8.
28. Kotnik T, Bobanovic F and Miklavćić D (1997), 'Sensitivity of transmembrane voltage induced by applied electric fields – a theoretical analysis', *Bioelectrochem Bioenerg*, **143**: 285–91.
29. Miklavčič D, Beravs K, Semrov D, et al. (1998), 'The importance of electric field distribution for effective *in vivo* electroporation of tissue', *Biophys J*, **74(5)**: 2152–8.
30. Esser A T, Smith K C, Gowrishankar T R, et al. (2007), 'Towards solid tumor treatment by irreversible electroporation: intrinsic redistribution of fields and currents in tissue', *Technol Cancer Res Treat*, **6(4)**: 261–74.
31. Weaver J C (2003), 'Electroporation of biological membranes from multicellular to nano scales', *IEEE Trans Dielectr Electr Insulat*, **10**: 754–68.
32. Weaver J C and Chizmadzhev Y A (1996), 'Electroporation', in: Polk C and Postow E (eds), *Handbook of Biological Effects of Electromagnetic Fields*. Boca Raton, FL: CRC Press, 247–74.
33. Prausnitz M R (1999), 'A practical assessment of transdermal drug delivery by skin electroporation', *Adv Drug Deliv Rev*, **35(1)**: 61–76.
34. Johnstone P T and Bodger P S (1997), 'High voltage disinfection of liquids', *IPENZ Trans*, **24(1)**: 30–35.
35. Neal I R, Garcia P, Robertson J, et al. (2012), 'Experimental characterization and numerical modeling of tissue electrical conductivity during pulsed electric fields for irreversible electroporation treatment planning', *IEEE Trans Biomed Eng*, **59(4)**: 1076–85.
36. Corović S, Pavlin M and Miklavćić D (2007), 'Analytical and numerical quantification and comparison of the local electric field in the tissue for different electrode configurations', *Biomed Eng Online*, **15(6)**: 37.
37. Breton M and Mir L M (2011), 'Microsecond and nanosecond electric pulses in cancer treatments', *Bioelectromagnetics*, on line: February 2012. 106–23.
38. Mir L M and Orlowski S (1999), 'Mechanisms of electrochemotherapy', *Adv Drug Deliv Rev*, **35(1)**: 107–18.
39. Edhemovic I, Gadzijev E M, Brecelj E, et al. (2011), 'Electrochemotherapy: a new technological approach in treatment of metastases in the liver', *Technol Cancer Res Treat*, **10(50)**: 475–85.
40. Glass L F, Pepine M L, Fenske N A, et al. (1996), 'Bleomycin-mediated electrochemotherapy of metastatic melanoma', *Arch Dermatol*, **132(11)**: 1353–7.
41. de Bree R, Tijink B M, van Groeningen C J, et al. (2006), 'Electroporation therapy in soft tissue sarcoma: a potentially effective novel treatment', *Sarcoma*, 2006: 85234.
42. Spugnini E P, Baldi A, Vincenzi B, et al. (2007), 'Intraoperative versus postoperative electrochemotherapy in high grade soft tissue sarcomas: a

preliminary study in a spontaneous feline model', *Cancer Chemother Pharmacol*, **59(3)**: 375–81.
43. Testori A, Tosti G, Martinoli C, et al. (2010), 'Electrochemotherapy for cutaneous and subcutaneous tumor lesions: a novel therapeutic approach', *Dermatol Ther*, **23(6)**: 651–61.
44. Reinhold U (2011), 'Electrochemotherapy for primary skin cancer and skin metastasis related to other malignancies', *Anticancer Drugs*, **22(8)**: 711–8.
45. Curatolo P, Quaglino P, Marenco F, et al. (2012), 'Electrochemotherapy in the treatment of Kaposi sarcoma cutaneous lesions: a two-center prospective phase II trial', *Ann Surg Oncol*, **19(1)**: 192–8.
46. Au J L, Jang S H and Wientjes M G (2002), 'Clinical aspects of drug delivery to tumors', *J Control Release*, **78(1–3)**: 81–95.
47. Jang S H, Wientjes M G and Lu D, et al. (2003), 'Drug delivery and transport to solid tumors', *Pharm Res*, **20(9)**: 1337–50.
48. Spugnini E P, Renaud S M, Buglioni S, et al. (2011), 'Electrochemotherapy with cisplatin enhances local control after surgical ablation of fibrosarcoma in cats: an approach to improve the therapeutic index of highly toxic chemotherapy drugs', *J Transl Med*, **9**: 152.
49. Spugnini E P, Baldi F, Mellone P, et al. (2007), 'Patterns of tumor response in canine and feline cancer patients treated with electrochemotherapy: preclinical data for the standardization of this treatment in pets and humans', *J Transl Med*, **5**: 48.
50. Sersa G, Miklavčič D, Cemazar M, et al. (2008), 'Electrochemotherapy in treatment of tumours', *Eur J Surg Oncol*, **34(2)**: 232–40.
51. Corović S, Al Sakere B, Haddad V, Miklavčič D and Mir L M (2008), 'Importance of contact surface between electrodes and treated tissue in electrochemotherapy', *Technol Cancer Res Treat*, **7(5)**: 393–400.
52. Larkin J O, Collins C G, Aarons S, et al. (2007), 'Electrochemotherapy: aspects of preclinical development and early clinical experience', *Ann Surg*, **245(3)**: 469–79.
53. Miklavčič D, Snoj M, Zupanic A, et al. (2010), 'Towards treatment planning and treatment of deep-seated solid tumors by electrochemotherapy', *Biomed Eng Online*, **9**: 10.
54. Janigro D, Perju C and Fazio V, et al. (2006), 'Alternating current electrical stimulation enhanced chemotherapy: a novel strategy to bypass multidrug resistance in tumor cells', *BMC Cancer*, **6**: 72.
55. Spiegl-Kreinecker S, Buchroithner J, Elbling L, et al. (2002), 'Expression and functional activity of the ABC-transporter proteins P-glycoprotein and multidrug-resistance protein 1 in human brain tumor cells and astrocytes', *J Neurooncol*, **57(1)**: 27–36.
56. Belehradek J Jr, Orlowski S and Ramirez L H, et al. (1994), 'Electropermeabilization of cells in tissues assessed by the qualitative and quantitative electroloading of bleomycin', *Biochim Biophys Acta*, **1190(1)**: 155–63.
57. Beebe S J and Schoenbach K H (2005), 'Nanosecond pulsed electric fields: a new stimulus to activate intracellular signaling', *J Biomed Biotechnol*, **2005(4)**: 297–300.

58. Garon E B, Sawcer D, Vernier P T, et al. (2007), '*In vitro* and *in vivo* evaluation and a case report of intense nanosecond pulsed electric field as a local therapy for human malignancies', *Int J Canc*, **121(3)**: 675–82.
59. Vernier P T, Sun Y, Marcu L, Salem S, Craft C M and Gundersen M A (2003), 'Calcium bursts induced by nanosecond electric pulses', *Biochem Biophys Res Commun*, **310(2)**: 286–95.
60. Vernier P T, Sun Y, Marcu L, et al. (2004), 'Nanoelectropulse-induced phosphatidylserine translocation', *Biophys J*, **86(6)**: 4040–8.
61. Tekle E, Oubrahim H, Dzekunov S M, et al. (2005), 'Selective field effects on intracellular vacuoles and vesicle membranes with nanosecond electric pulses', *Biophys J*, **89(1)**: 274–84.
62. Pakhomov A G, Kolb J, White J, et al. (2007), 'Long-lasting plasma membrane permeabilization in mammalian cells by nanosecond pulsed electric field', *Bioelectromagnetics*, **28**: 655–63.
63. Esser A T, Smith K C, Gowrishankar T R, et al. (2009), 'Towards solid tumor treatment by nanosecond pulsed electric fields', *Technol Cancer Res Treat*, **8(4)**: 289–306.
64. Ibey B L, Mixon D G and Payne J A, et al. (2010), 'Plasma membrane permeabilization by trains of ultrashort electric pulses', *Bioelectrochem*, **79(1)**: 114–21.
65. Nuccitelli R, Chen X, Pakhomov A G, et al. (2009), 'A new pulsed electric field therapy for melanoma disrupts the tumor's blood supply and causes complete remission without recurrence', *Int J Cancer*, **125(2)**: 438–45.
66. Yang W, Wu Y H, Yin D, Koeffler H P, Sawcer D E, et al. (2011), 'Differential sensitivities of malignant and normal skin cells to nanosecond pulsed electric fields', *Technol Cancer Res Treat*, **10(3)**: 281–6.
67. Merla C, Paffi A, Apollonio F, et al. (2011), 'Microdosimetry for nanosecond pulsed electric field applications: a parametric study for a single cell', *IEEE Trans Biomed Eng*, **58(5)**: 1294–302.
68. Nuccitelli R, Pliquett U, Chen X, et al. (2006), 'Nanosecond pulsed electric fields cause melanomas to self-destruct', *Biochem Biophys Res Commun*, **343(2)**: 351–60.
69. Ren W and Beebe S J (2011), 'An apoptosis targeted stimulus with nanosecond pulsed electric fields (nsPEFs) in E4 squamous cell carcinoma', *Apoptosis*, **16(4)**: 382–93.
70. Gowrishankar T R, Esser A T, Smith K C, et al. (2011), 'Intracellular electroporation site distributions: Modeling examples for nsPEF and IRE pulse waveforms', *Conf Proc IEEE Eng Med Biol Soc*, **2011**: 732–5.
71. Al-Sakere B, André F and Bernat C, et al. (2007), 'Tumor ablation with irreversible electroporation', *PLoS One*, **2(11)**: e1135.
72. Joshi R P and Schoenbach K H (2010), 'Bioelectric effects of intense ultrashort pulses', *Crit Rev Biomed Eng*, **38(3)**: 255–304.
73. Stacey M, Fox P, Buescher S and Kolb J (2011), 'Nanosecond pulsed electric field induced cytoskeleton, nuclear membrane and telomere damage adversely impact cell survival', *Bioelectrochem*, **82(2)**: 131–4.
74. Sersa G, Cemazar M, Parkins C S and Chaplin D J (1999), 'Tumour blood flow changes induced by application of electric pulses', *Eur J Cancer*, **35**: 672–7.

75. Jarm T, Cemazar M, Miklavčič D, et al. (1997), 'Antivascular effects of electrochemotherapy: implications in treatment of bleeding metastases', *Expert Rev Anticancer Ther*, **10(5)**: 729–46.
76. Kraus M and Wolf B (1998), 'Physicochemical microenvironment as key regulator for tumor microevolution, invasion and immune response: targets for endocytotechnological approaches in cancer treatment', *Endocytobiol Cell Res*, **12**: 133–56.
77. Jain R K (2001), 'Delivery of molecular and cellular medicine to solid tumors', *Adv Drug Deliv Rev*, **46(1–3)**: 149–68.
78. Höckel M and Vaupel P (2001), 'Tumor hypoxia: definitions and current clinical, biologic, and molecular aspects', *J Natl Cancer Inst*, **93(4)**: 266–76.
79. Molavian H R, Kohandel M, Milosevic M, et al. (2011), 'Deriving mechanisms responsible for the lack of correlation between hypoxia and acidity in solid tumors', *PLoS One*, **6(12)**: e28101.
80. Olszewski-Hamilton U and Hamilton G (2011), 'Dependence of relative expression of NTR1 and EGFR on cell density and extracellular pH in Human pancreatic cancer cell lines', *Cancers*, **3**: 182–97.
81. Adhikari A S, Agarwal N and Iwakuma T (2011), 'Metastatic potential of tumor-initiating cells in solid tumors', *Front Biosci*, **16**: 1927–38.
82. Harris A L (2002), 'Hypoxia – a key regulatory factor in tumour growth', *Nat Rev Cancer*, **2(1)**: 38–47.
83. Cosse J P and Michiels C (2008), 'Tumour hypoxia affects the responsiveness of cancer cells to chemotherapy and promotes cancer progression', *Anticanc Agents Med Chem*, **8(7)**: 790–7.
84. Nordenstrom B E W (1983), *Biologically Closed Electrical Circuits: Clinical, Experimental and Theoretical Evidence for an Additional Circulatory System*. Stockholm: Nordic Medical Publications.
85. Riedl S, Zweytick D and Lohner K (2011a), 'Membrane-active host defense peptides – challenges and perspectives for the development of novel anticancer drugs', *Chem Phys Lipids*, **164(8)**: 766–81.
86. Riedl S, Rinner B, Asslaber M, et al. (2011b), 'In search of a novel target – phosphatidylserine exposed by non-apoptotic tumor cells and metastases of malignancies with poor treatment efficacy', *Biochim Biophys Acta*, **1808(11)**: 2638–45.
87. Acevedo H F, Tong J Y and Hartsock R J (1995), 'Human chorionic gonadotropin-beta subunit gene expression in cultured human fetal and cancer cells of different types and origins', *Cancer*, **76(8)**: 1467–75.
88. Acevedo H F (2002), 'Human chorionic gonadotropin (hCG), the hormone of life and death: a review', *J Exp Ther Oncol*, **2(3)**: 133–45.
89. Acevedo H, Gonzalez N and Moss R (1998), 'Trophoblastic hormones and cancer: a breakthrough in treatment?', paper presented at: *Comprehensive Cancer Care Conference*, Session 205: June 13, 1998. Available at: *http://www.cmbm.org/conferences/ccc98/transcripts/205.html* (accessed May 22, 2003).
90. Van Rinsum J, Smets L A, Van Rooy H and Van Den Eunden D H (1986), 'Specific inhibition of human natural killer cell-mediated cytotoxicity by sialic acid and sialo-oligosaccharides', *Int J Cancer*, **38**: 915–22.

91. Levitan I, Fang Y, Rosenhouse-Dantsker A, et al. (2010), 'Cholesterol and ion channels', *Subcell Biochem*, **51**: 509–49.
92. Joines W T (1984), 'Frequency-dependent absorption of electromagnetic energy in biological tissue', *IEEE Trans Biomed Eng*, **31(1)**: 17–20.
93. Stern R G (1999), 'Carcinogenesis and the plasma membrane', *Med Hypotheses*, **52(5)**: 367–72.
94. Killion J J (1984), 'Electrical properties of normal and transformed mammalian cells', *Biophys J*, **45(3)**: 523–8.
95. Gatenby R A and Gillies R J (2004), 'Why do cancers have high aerobic glycolysis?', *Nat Rev Canc*, **4(11)**: 891–9.
96. Sion A M and Figg W D (2006), 'Lysyl oxidase (LOX) and hypoxia-induced metastases', *Cancer Biol Ther*, **5(8)**: 909–11.
97. Semenza G L (2008), 'Tumor metabolism: cancer cells give and take lactate', *J Clin Invest*, **118(12)**: 3835–7.
98. Calderon-Montano J, Burgos-Moron E, Perez-Guerrero C, et al. (2011), 'Role of the Intracellular pH in the metabolic switch between oxidative phosphorylation and aerobic glycolysis – relevance to cancer', *WebmedCentral Cancer*, **2(3)**: WMC001716.
99. Cairns R A, Khokha R and Hill R P (2003), 'Molecular mechanisms of tumor invasion and metastasis: an integrated view', *Curr Mol Med*, **3(7)**: 659–71.
100. Semenza G L, Artemov D, Bedi A, et al. (2001), 'The metabolism of tumours: 70 years later', *Novartis Found Symp*, **240**: 251–60; discussion 260–4.
101. Liu J (2010), 'The application of metabolic network analysis in understanding and targeting metabolism for drug discovery', [dissertation]. December 3. Available at: *http://etd.library.pitt.edu/ETD/available/etd-12072010-100902/* (accessed Jan 22, 2012).
102. Semenza G L (2003), 'Targeting HIF-1 for cancer therapy', *Nat Rev Cancer*, **3**: 721–32.
103. Garnett M (1998), *First Pulse: A Personal Journey in Cancer Research*. New York: First Pulse Projects.
104. Cardone R A, Casavola V and Reshkin S J (2005), 'The role of disturbed pH dynamics and the Na^+/H^+ exchanger in metastasis', *Nat Rev Canc*, **5(10)**: 786–95.
105. Reshkin S J, Bellizzi A, Caldeira S, et al. (2000), 'Na^+/H^+ exchanger-dependent intracellular alkalinization is an early event in malignant transformation and plays an essential role in the development of subsequent transformation-associated phenotypes', *FASEB J*, **14**: 2185–97.
106. Lopez-Lazaro M (2008), 'The Warburg effect: why and how do cancer cells activate glycolysis in the presence of oxygen?', *Anticancer Agents Med Chem*, **8**: 305–12.
107. Rotin D, Steele-Norwood D, Grinstein S and Tannock I (1989), 'Requirement of the Na^+/H^+ exchanger for tumor growth', *Cancer Res*, **49**: 205–11.
108. Stubbs M, Veech R L and Griffiths J R (1995), 'Tumor metabolism: the lessons of magnetic resonance spectroscopy', *Adv Enzyme Regul*, **35**: 101–15.

109. Nagy I Z, Lustyik G, Nagy V Z, et al. (1981), 'Intracellular Na^+:K+ ratios in human cancer cells as revealed by energy dispersive X-ray microanalysis', *J Cell Biol*, **90(3)**: 769–77.
110. Seyfried T N and Shelton L M (2010), 'Cancer as a metabolic disease', *Nutr Metab (Lond)*, **7**: 7.
111. De Milito A and Fais S (2005), 'Tumor acidity, chemoresistance and proton pump inhibitors', *Future Oncol*, **1(6)**: 779–86.
112. Robey I F, Baggett B K, Kirkpatrick N D, et al. (2009), 'Bicarbonate increases tumor pH and inhibits spontaneous metastases', *Cancer Res*, **69(6)**: 2260–8.
113. Silva A S, Yunes J A, Gillies R J, et al. (2009), 'The potential role of systemic buffers in reducing intratumoral extracellular pH and acid-mediated invasion', *Cancer Res*, **69(6)**: 2677–84.

Electrically-enhanced proliferation control of cancer-stem-cells-like adult human mesenchymal stem cells – a novel modality of treatment

Kavitha Sankaranarayanan, Raja Prabhu Ramachandran and Raji Sundararajan

DOI: 10.1533/9781908818294.127

Abstract: Cancer stem cells (CSCs) have the potential to be chemoresistant and recurrent and/or to become metastatic tumors. It is highly critical to control the proliferation of CSCs using alternative therapies. Towards this end, we studied the proliferation control of cancer-stem-cell like mesenchymal cells using electrical pulses and Bleomycin, a FDA approved (in the US, and the respective medical agencies in the other countries) chemotherapeutic drug. For this purpose, adult human mesenchymal stem cells (hMSCs), derived from bone marrow, were used. Both low intensity, long duration (200 V/cm, 10 ms) and high intensity, short duration (1200 V/cm, 100 μs) pulses were applied and their efficacies studied. The results indicate that electrical pulses-based therapy could be utilized economically and efficiently to control the proliferation of hMSCs and shows promise of an easy transition to use in the clinic.

Key words: stem cells, mesenchymal, cancer stem cells, Bleomycin, viability, colony forming units, electroporation.

6.1 Introduction – stem cells

Stem cells are pluripotent/multipotent cells, which retain the potential to differentiate into different cells of the adult human body [1]. Stem cells are generally classified based on their origin or source, as either adult stem cells (ASCs) or embryonic stem cells (ESCs). While ESCs are obtained from the inner cell mass of the blastocyst from an early embryonic stage, adult stem cells are those originating from adult tissue, where they play a key role in the regeneration and repair of tissues. ESCs are pluripotent in nature, and thus retain the ability to form most types of cells in the adult human body. ASCs are also referred to as multipotent stem cells and they possess the ability to differentiate into certain limited types of cells (Figure 6.1). In other words, ASCs exhibit less plasticity or developmental versatility, because they are further ahead than ESCs in the differentiation pathway. However, ASCs have the advantage of being widely used in the clinical setting, because of their relative ease of isolation from various tissues such as the adipose tissue, bone marrow, etc. The latest set of tissues used for ASC isolation includes menstrual blood and dental pulp, thus increasing the list of source tissues. Small samples of

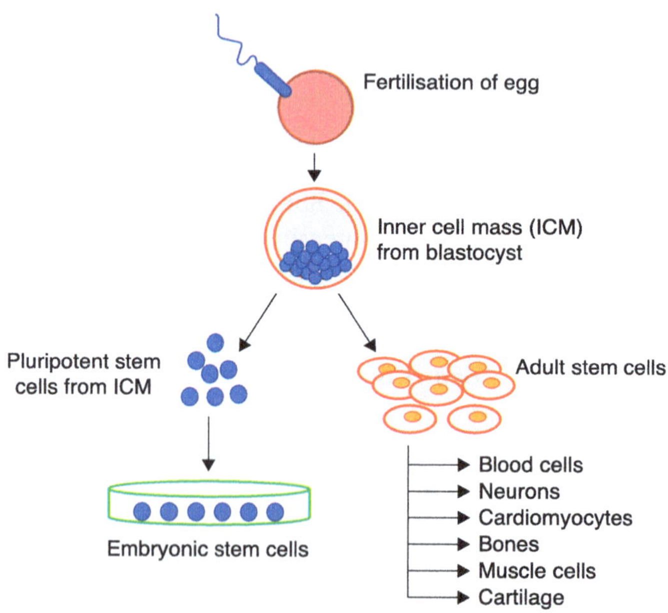

Figure 6.1 Properties of adult stem cells (ASCs)

tissues obtained thus could be used for isolation, culture and expansion of ASCs prior to implantation of the same at the site of interest in the body for therapeutic purposes. It also tends to overcome the ethical issues associated with the use of ESCs.

6.2 Mesenchymal stem cells

Mesenchymal stem cells (MSCs) are a subset of ASCs of mesenchymal origin and can generally be harvested from many different tissues, including bone marrow. MSCs, originating mainly from pericytes, are located in the perivascular district of the bone marrow stroma and are the natural precursors of bone, cartilage and fat, the constituent tissues of bone [1]. They can readily differentiate into cells of mesodermal origin (osteocytes, adipocytes and chondrocytes) and also to transdifferentiate into cells of ecto- and endo-dermal origin. A J Friedenstein first discovered the presence of these cells in the bone marrow and isolated the same based on their plastic adherent, fibroblast-like nature [1, 2]. MSCs have been widely used clinically in the treatment of cardiomyopathy, autoimmune disorders, spinal cord injuries, haematological disorders, cartilage problems, etc. [3].

While the multilineage differentiation of the MSCs have been a major positive trait in the clinical application of these cells, they also have the added advantage of being immuno-privileged, thereby cutting down not only on the risk of immune rejection but also on the cost of immuno-suppressants that are normally used post tissue/cell implantation.

Adult human mesenchymal stem cells (hMSCs) are much favored in stem cell therapy, primarily due to the ease of procurement, being immune privileged, and relaxed ethical constraints associated with the usage of adult stem cells. In addition, adult hMSCs also exert positive paracrine and cytoprotective effects. They can also be useful as a delivery vehicle for macromolecules, including chemotherapeutic agents. Much research is currently being carried out on the usage of stem cells for therapeutic purposes, utilizing their homing capability to target tumors and suppress their growth.

6.3 Cancer and cancer stem cells

Cancer is a leading cause of death worldwide, accounting for 7.6 million deaths (~13% of all deaths) in 2008. More alarming is the projection of cancer deaths in the years to come, with an estimated 13.1 million deaths in 2030 [4]. Cancer kills about one American per minute, with about 1500 deaths per day in the US alone [5]. While novel treatment strategies are on the rise, the statistics suggest adoption of a much more rigorous and novel approach in battling this disease. While early detection and treatment can suffice to save the lives of many, there have been recent highlights on the cause of relapse and metastasis of cancers. With the latest data available, we need to look beyond conventional treatment strategies such as chemotherapy and to supplement or augment the same with newer modes of treatment to completely eradicate the cancer cells/pro-cancerous cells from the system. The property of self renewal without undergoing differentiation or undergoing apoptosis, is one that the stem cells share with cancer cells.

Recent studies have demonstrated the Janus-faced nature of MSCs with evidence supporting both the facts that MSCs home to tumors and suppress their growth [6] and contradict the belief that MSCs may protect tumors by exerting an anti-apoptotic effect, encouraging tumor progression, metastasis and increasing the drug-resistance [7–17].

As aptly described in Wong's review, the 'demonic' side of MSCs is attributed to both an indirect modulation of tumor activity by the MSCs, as well as a direct effect wherein the MSCs themselves undergo a malignant transformation under certain conditions [3]. As reported by many groups, MSCs have been found to inhibit proliferation of glioblastoma 8MGGBA [13], A549 (lung cancer) and Eca-109 (esophageal cancer) [18] cell lines, while exhibiting a protective effect on melanoma A375 cells [8], breast cancer cells [9], acute myeloid leukemia cells 68, U266 and H929 [8].

Chemo-refractiveness of certain types of cancers have also been reported to be augmented by MSCs. Chronic lymphocytic leukemic cells have been protected from dexamethasone, cyclophosphamide and fludarabine [16] induced apoptosis, while U937 cells have been protected from daunoblastina treatment [15] by MSCs. In a separate study, Vianello et al. report on the protection conferred by BMMSCs on chronic myeloid leukemia cells [17]. Another study highlights the expression of stem cell markers such as CD133, CD117, SSEA3 and OCT4 by lung cancer cells resistant to chemotherapy drugs doxorubicin and cysplatin [19]. These cells are capable of self-renewal and were tumorigenic when injected into mice.

MSCs have been shown to also home towards sites of pathology, injury or tumors, and also to release cytokines which seem to favor metastasis, angiogenesis and tumor growth. Cancer cells, on their part have also been shown to release soluble signals such as growth factors and cytokines, which recruit MSCs to tumor stroma [20]. The immunosuppressive nature of MSCs has also been reported to encourage growth of tumors, thereby adding the evidence of MSCs actually contributing to tumor growth and metastasis. Other than the indirect tumor protective properties of MSCs, they have also been shown to be directly involved by their ability to home and interact with the tumor stroma in colon cancer cells [21]. Adhesion molecules expressed by the MSCs are responsible for this property. They include a1b1 and a5b1 integrins, ICAM 1 and VCAM, CD44, etc. [22]. Interaction between mammary cancer cells and MSCs have been shown, proving that there is expression of the chemokine CCL5 by MSCs, which in turn promotes pulmonary metastasis [14]. MSCs also contribute to the formation of stromal elements such as myofibroblasts, extracellular matrices, perivascular structures, etc., once integrated into the tumor stroma [23]. They do this through the expression of pro-angiogenic factors, such as VEGF, angiopoietin, etc., growth factors, such as FGF 2 and 7, and PDGF and cytokines such as IL 6, IL 8/CXCL8 and TNFα [24, 25]. These molecules recruit endothelial progenitors, which promotes vasculogenesis [26].

When in a tumor microenvironment, MSCs tend to undergo a malignant transformation and differentiate into carcinoma associated fibroblasts (CAF) or tumor associated fibroblasts (TAF) [27]. Upon fresh recruitment to the tumor site, they may or may not inhibit the tumor growth, but once primed by the soluble factors in the tumor microenvironment, they could cause further damage by augmenting tumor growth, progress and invasion, form cancer stem cell niches, and differentiate into tumor stroma cells, i.e. CAFs, which can modulate cancer cell behavior.

The MSC-tumor interaction has been deemed to be essential for tumor survival and this property could be exploited by generating treatment strategies to use MSCs as a vehicle to target and kill tumors. Tumoral growth has also been linked to the presence of CSCs. There have been reports that these small groups of cells within the tumor arise due to mutations in stem cells, thereby rendering them tumorogenic. These cells are reported to form a niche within the tumor and bear many characteristics of stem cells. Various researchers have identified a small population of tumorigenic cells with stem-cell-like properties in bladder cancer, melanoma, ovarian cancer, head and neck squamous cell carcinoma, etc. It is reported that while chemotherapy, in general, targets rapidly proliferating cells which form the bulk of the tumor, the CSCs have

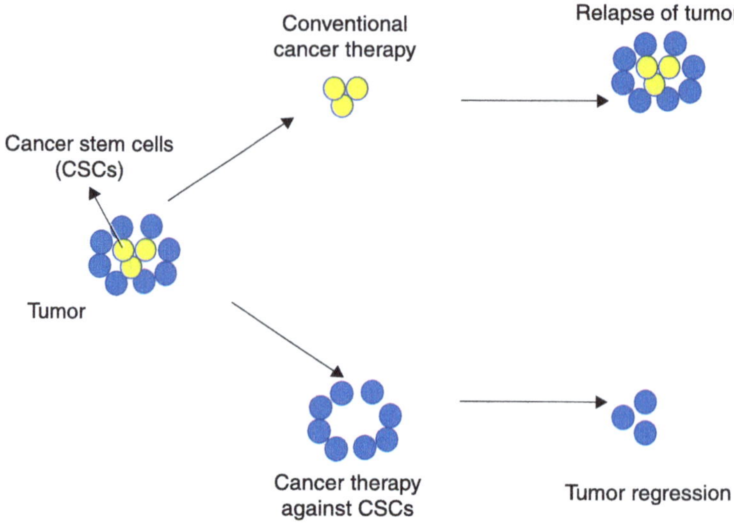

Figure 6.2 Cancer stem cells (CSCs)

mechanisms, such as pumps and other transporters, which can confer resistance to general chemotherapeutic agents. The CSCs then slowly repopulate the tumor, leading to relapse, which is a major issue to be addressed in cancer therapeutics (Figure 6.2).

CSCs are reported to be resistant to a wide variety of chemotherapeutic agents and possess the remarkable ability of recovering from cytotoxic therapy. Glioma stem cells (GSCs) not only play a crucial role in chemoresistance, but they are also vital to the failure of radiation as evidenced from the enriched CSCs in the tumors [28, 29]. Considering the pattern of treatment failure seen with current therapy modalities, alternate treatments that selectively target CSCs, which are functionally distinct and chemo- and radiation-resistant, might provide more success in treating these deadly cancers. There have been various schools of thought regarding this compounding challenge and the recent objective is deriving methods to demolish the population of CSCs within the tumor [30].

CSCs show marked similarity with normal stem cells in terms of cellular and genetic architecture. Adult stem cells, being mulipotent, are capable of renewing themselves indefinitely like CSCs, while retaining the potential to differentiate into different lineages. These cells remain quiescent and become active when the necessity arises, as in the case of tissue damage, disease or a need for regeneration of certain organs. CSCs and MSCs share the common properties of self-renewal, capability of asymmetric division, expression of surface markers such as integrins, CD133, CD 134, CD49e and expression of transcription factors such as OCT and β-atenin, etc. [31]. These data suggest a dual role of MSCs in both tumor inhibition and tumor progression, thereby presenting them as a Janus-faced vehicle, which could, if understood and exploited properly, serve in the development of a novel therapeutic modality for treatment of cancers, particularly the chemo-refractive ones. Genetically engineered MSCs expressing various factors, which could trigger death of cancer cells, could be used to target and kill the same, owing to the homing property of MSCs. While this remains a lucrative topic to be explored, the methods of transfecting these cells with molecules of interest – be it proteins or small drugs – have to be studied in detail.

6.4 Electrochemotherapy

Prevalent methods of tumor therapy include the harsh methods of chemotherapy and radiotherapy which, being nonspecific in nature, tend to harm the surrounding noncancerous tissue. In addition, these methods are associated with very serious side effects [32] and also prove to be too expensive. For example, a new cancer drug costs more than \$30 000 for 8 weeks of chemotherapy for colorectal cancer [33]. Given such a scenario, there has been significant multidisciplinary research targeting the development of a new therapy that would be cost-effective, have minimal side effects, and also more specific in action on the tumorous cells.

Electrochemotherapy (ECT) has proven to be a light at the end of the long tunnel of cancer therapy [34, 35]. ECT involves the usage of electrical pulses of various tolerable intensity and duration on the tumor cells in the presence of chemotherapeutic drugs. One face of ECT involves use of high intensity, short duration pulses to generate pores on the cell membrane through which drugs could pass into the cytoplasm. This technique, called electroporation or electropermeabilisation (EP), aids in opening up the otherwise impermeable cell membrane to chemotherapeutic agents, thereby helping to drastically reduce the concentration of the drug needed. This use of reduced doses of the drugs offer benefits to the patients in many ways, including physical, economical and enhanced quality of care and life. While electroporation using different voltage pulses is commonly used for transfecting bacterial, and even eukaryotic cells with plasmids, and is gaining momentum as an alternative for inoperable, chemo-refractive and radiation-resistive, recurrent tumors [32, 34], the application of electrical pulses to deliver drugs and chemicals into stem cells remain largely unexplored. Various clinical and preclinical trials attest to the efficacy of this technique to treat melanoma and other cancers using chemo drugs and other genes that can target cancer cells and arrest their proliferation and bring about cell death [33–42].

The electroporation technique has been shown to offer 100–1000-fold improved therapeutic benefits compared to using the drug alone, and is gaining acceptance as a viable technique for treatment of tumors of all histological types. The efficacy of EP could be attributed to the redistribution of charges [43] on the membrane surfaces and the variations in the membrane potential of cancer cells compared to normal ones. Electroporation-mediated chemotherapy (ECT) is a viable alternative to conventional cancer treatments, as evidenced by successful Phase I and II clinical trials for various skin cancers, such as lymphomas, squamous cell carcinomas, testicular carcinomas and malignant pleural effusions [40–42]. The treatment of 52 patients using ECT for melanoma and chest wall tumor treatments attest to their success [35]. Clinical trials of delivering vaccines for prostate cancer using electroporation have been conducted at the University of Southampton Medical School [44]. In addition, studies conducted at Stanford Medical School used electroporation to study tumor oxygenation [45]. We also have previously shown that under appropriate conditions, electrical pulses can drive large numbers of molecules into breast cancer cells [46, 47].

We hypothesize that electrical pulses with or without anticancer drugs could effectively be used to control proliferation, as well as enhance intracellular drug delivery in hMSCs. The advantages of ECT would include the generation of a novel strategy, which would be an outpatient-based, efficient, economical, physical technique using short electrical pulses with minimal side effects. Towards this, in this study we utilize electrical pulses to impede the proliferation of mesenchymal stem cells. Also, we look at electroporation as a means of uploading cells with chemotherapeutic agents, which could possibly be extrapolated into a clinical scenario. Responsiveness of CSCs to ECT has not been studied in great detail, primarily due to their small number and consequently difficult techniques of isolation from the primary source. However, their similarity to MSCs makes it more prudent to study the response of relatively easily available MSCs to ECT and extrapolate the same to CSCs.

Since electroporation efficacy depends on the type of cells, the pulse parameters and the molecules to be uploaded, it is necessary to optimize the pulse parameters. For this purpose, various pulse conditions, including low voltage, long duration pulses, are studied for their efficiencies.

6.5 *In-vitro* study of ECT on MSC

In this study, we present a multi-pronged approach to assess the efficacy or response of MSCs to ECT. The cells are treated with electrical pulses of variable voltages and durations to identify the optimal pulse parameters to load the cells with the necessary dosage of drug molecules. Bone marrow MSCs are relatively abundant and easier to isolate than CSCs. Hence this study uses MSCs as the cell of study to standardize the pulse parameters required to arrest proliferation and also to load the cells with chemo drugs.

Another objective of delivering drug molecules into adult hMSCs is to study the possibility of using these cells as carriers of therapeutic molecules to the tumor site, as there have been reports of migration of hMSC towards tumors [48] both *in vivo* and *in vitro*. Thus, adult mesenchymal stem cells possess the ability to home towards tumor cells, making the possibility to practise targeted tumor therapy more realistic [49]. Towards this end, we have developed an *in vitro* model of a combination therapy using adult mesenchymal stem cells.

6.6 Materials and methods

6.6.1 *The cells – hMSCs*

Bone marrow aspirates in heparin-containing syringes were obtained from adult patients (of both genders – 42-year-old male, 54-year-old male, 55-year-old female, 67-year-old female and 70-year-old male) undergoing cardiac surgery at the Frontier Lifeline Hospital, Chennai, after obtaining informed consent and Institutional Ethics Committee clearance.

The aspirates were collected from the sternum or the breast bone using a long aspiration needle. Bone marrow mononuclear cells (MNCs) were isolated using the Ficoll-Hypaque (Himedia) (d = 1.077 g/cm^3) gradient technique and plated on high-glucose DMEM (Dulbecco's modified Eagle's medium) containing 4.00 mM L-glutamine, 4500 mg/L glucose, 110 mg/L sodium pyruvate (Hyclone) and 100 U/mL penicillin,

100 mg/mL streptomycin and 0.25 mg/mL amphotericin (antibiotic–antimycotic solution – Himedia). The sample was centrifuged at 1800 rpm for 20 minutes after a 1:1 dilution with DMEM. The upper layer was aspirated leaving behind the mononuclear cell layer at the interphase. The Buffy coat was transferred to another tube and the volume was made up to 10 mL with PBS. The contents were centrifuged at 1800 rpm for 10 minutes. After 2–3 washes with PBS, the supernatant was discarded and the pellet was then used for expansion. The cells were seeded at approximately 1×10^6cells/mL density in 90 mm cell culture dishes containing low glucose DMEM, 10% FBS, L-Glutamine and antibiotics. The cell cultures were maintained at 37 °C in a humidified 5% CO_2 incubator. MSCs form a subset of the isolated MNCs and are adherent in nature, which is used as a means of isolation of the same for further studies. Non-adherent cells were removed after 3 days and the medium replaced. Adherent hMSCs grew as CFU-Fs (colony-forming unit-fibroblast) and were regularly passaged upon becoming 80–90% confluent. The cells were passaged by adding 0.25% trypsin-EDTA (Himedia). The cells are centrifuged again at 2000 rpm for 5 minutes and resuspended to required volume for experimentation. The morphology of adult hMSCs isolated from a 42-year-old male patient is shown in Figure 6.3.

6.6.2 The chemo drug - Bleomycin

Bleomycin, a glycopeptidic antibiotic with potent anti-tumor activity, isolated from the bacterium *Streptomyces verticillu*, was first discovered in 1966 and obtained FDA approval in the US as early as 1973. Bleomycin induces breakage of the DNA strands and inhibits Thymidine nucleotide incorporation during DNA synthesis [50]. Bleomycin inflicts direct damage to the DNA by generating chromosomal gaps, deletions and DNA fragmentation. The activity of Bleomycin is largely dependent on the presence of O_2 and metal ions (primarily Cu and Fe), with which it chelates forming a pseudoenzyme [51]. The pseudoenzyme then reacts with O_2 to form highly reactive free radicals, which then bind to G-C rich regions of DNA and cleaves it. The cytotoxicity of Bleomycin is specific for the G2 phase of the cell cycle. The chelations also trigger peroxidation of cellular lipids and induce oxidation of other molecules within the cell [52].

The molecular mass of Bleomycin is 1415.55 g and its potency is measured in units of antimicrobial activity, classified as an 'antitumor

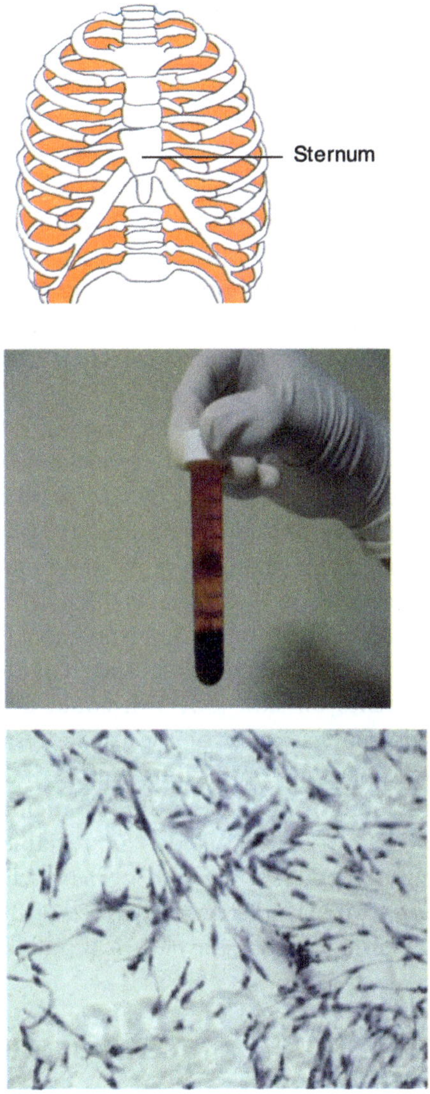

(a)

(b)

(c)

Figure 6.3 Isolation of adult hMSCs from human sternum: (a) sternum; (b) Ficoll Tube indicating the various layers' formation (the top layer is plasma, middle is Buffy coat and the bottom is red blood cells); and (c) morphology of adult mesenchymal stem cells (42-year-old male) antibiotic'. Bleomycin can be used in combination with surgery or radiotherapy, or in the palliative treatment of a number of cancers. It is used in the treatment of squamous cell cancers, melanoma, sarcoma, testicular cancer and both Hodgkin's and non-Hodgkin's lymphoma [33–35, 53].

A single molecule of Bleomycin ($C_{55}H_{84}N_{17}O_{21}S_3$) [54] can cause around 8–10 DNA breaks, which explains its high cytotoxicity when present inside the cell. Cell death due to Bleomycin happens in one of two ways. If only a few thousand Bleomycin molecules are present in the cell, the cell is arrested during cell division, becomes enlarged and two or more large nuclei and micronuclei are observed. The cell then dies in a slow process, over one or more days. If instead, the cell contains several millions of Bleomycin molecules, it is killed within a few minutes. Thus, Bleomycin is an extremely toxic agent once inside the cell, but this very high intrinsic cytotoxicity is restricted by the inability of Bleomycin to freely diffuse through the plasma membrane. It is this caveat of using Bleomycin that is overcome by using it in combination with electrical pulses, thereby increasing its cytotoxicity [33]. In addition, the side effects of Bleomycin, including skin rash, hives, changes in skin sensation, hyper-pigmentation (darkening of skin and nails), hair loss, mouth ulcers and hard patches on skin [55, 56], could be brought down by determination and administration of the exact dosage of the drug as needed. The dosage depends on various factors including tumor size, stage, height, weight and overall health of the patient.

The benefit of electroporation is the appreciable reduction in the dosage of the drug needed, as illustrated in many clinical trials. In this study, hMSCs were treated with very low doses of Bleomycin (0.1, 1.0, 5.0 and 10.0 μM) before applying the electrical pulses.

6.6.3 The technique – electroporation

A BTX ECM 830 (Genetronics Inc., San Diego, CA), square wave electroporator with 0.4 cm (electrode gap) cuvette was used. The electroporation buffer used was DMEM. The sample to be electroporated is taken in the cuvette and made up to a volume of 200 μL using media; 100 μL of the cell suspension with 100 μL of media was used as a negative control for electroporation in the absence of Bleomycin. For electroporation in the presence of the drug, 100 μL of cell suspension was made up to 200 μL with DMEM, after addition of appropriate volume of the drug.

Both low intensity, long duration pulses (200 V/cm, 40 ms) and high intensity, short duration pulses (1200 V/cm, 100 μs [8–10]) were used for the ECT procedure. In addition, to study the effect of intermediate values for these parameters, 450 V/cm, 25 ms [57] and 500 V/cm, 10 and 20 ms pulse were also used. Each trial using the above-mentioned parameters uses 8 pulses at 1 second intervals (1 Hz frequency), unless otherwise mentioned.

6.6.4 The assays – viability assay and the colony forming unit assay

Cell counting and viability assay

The morphology and viability of cells before and after electroporation is estimated using dye exclusion assay. While an intact cell membrane does not allow penetration of Trypan Blue dye into the cytoplasm, any condition which alters the cell membrane permeability, including cell death (which breaks down cell membrane integrity), causes the dye to enter the cell. Thus, cells with the dye are deemed unviable, while those excluding the dye are viable. After electroporation, cells are counted to determine the viability. Both live and dead cells are counted using the Neubauer chamber and viability estimated using the above described Trypan Blue exclusion assay.

For this, 500 μL of the solution from the cell culture dishes was centrifuged at 2000 rpm for 5 minutes. The pellet containing the hMSCs was collected and 10 μL of the solution containing the cells was diluted in 50 μL of PBS and mixed homogeneously; 10 μL of this suspension was mixed with an equal volume of Trypan Blue and 10 μL of the prepared sample was transferred under the glass cover of a Neubauer hemocytometer (Nexcelom Bioscience LLC, MA) followed by cell imaging and counting. The Trypan Blue, being an exclusion dye, stains only dead cells, while the live cells appear as bright spots. Cell viability is computed as the ratio of number of live cells to the total number of cells (Live and Dead). The cells were imaged and counted using an Olympus CKX41, phase contrast, inverted florescent microscope. Images were captured using Motix software.

Colony forming unit assay

The colony forming unit (CFU) is a measure of viable colonogenic cell numbers in CFU/mL. These are an indication of the number of cells that remain viable enough to proliferate and form small colonies. Isolated hMSCs were plated in a 6-well cell culture plate along with 2–3 mL of DMEM medium. Then 100 μL of the sample, along with 700 μL of DMEM medium are taken in a 4-mm gap cuvette and electroporated at 1200 V/cm, using eight 15 μs pulses at 1 second intervals. After electroporation, the cells were seeded onto tissue culture dishes in DMEM with antibiotics and maintained at 37 °C in a 5% CO_2 atmosphere. They were undisturbed for 14 days until colonies of cells appeared. The dishes were washed with PBS and cells were stained with 0.1% Crystal Violet in Methanol and incubated at room temperature for 10 minutes. The staining solution was discarded, and the dishes were observed under a microscope. A collection of cells of around 100 mm^2 (~50 cells) was counted as 1 colony.

6.7 Results and analyses

Based on our previous experience, the data available in the literature, and a number of trials, various pulse parameters (electric field intensity, E = Voltage applied V/electrode gap (0.4 cm), pulse width) were studied. It was reported by Hui et al., that ET = k, a constant, where E is the electric field intensity and T is the time duration of the pulses [58]. Hence, when the electric field intensity is higher, the pulse duration should be lower and vice versa. In our study, a pulse of 1200 V/cm and duration ≥5 ms pulses resulted in arcing and frothing of the media with 100% cell mortality. However, lower pulse intensity coupled with shorter time duration would be insufficient to open up transient pores on the plasma membrane. An optimization of the pulse parameters (voltage and duration) for hMSCs had been performed as the pulse parameters vary with types of cells. In the case of 500 V/cm and 200 V/cm, the microsecond pulses did not generate significantly different results from the non-electroporated samples (data not shown), hence 10 and 20 ms durations were used. This is in line with the clinical study of Li's group [59].

Table 6.1 shows the various pulse parameters studied. Here, 8 pulses were applied at 1 Hz and the electroporation was carried out at room temperature.

6.7.1 Colony forming units

hMSCs have the tendency to form colonies upon plating in suitable culture conditions. The number of such colonies is a direct indicator of the

Table 6.1 Pulse parameters studied

Electric field intensity, V/cm	Pulse width
1200	100 μs
500	5 ms
500	10 ms
500	20 ms
450	25 ms
200	5 ms
200	10 ms
200	20 ms

clonogenic ability of the cell, which is a measure of its proliferative capacity (Figure 6.4). When freshly isolated hMSCs were used for the CFU assay after electroporation (1200 V/cm, 100 μs), a significant reduction in the number of CFUs was observed (Figure 6.5). The same number of cells was used in both the control as well as electroporated samples; hence it could be safely inferred that electroporation exerts a negative effect on the colonogenic ability of hMSCs. This indicates that there is a reduction in the number of healthy MSCs capable of proliferation after electroporation.

6.7.2 Viability study

As described earlier, the live cells exclude Trypan Blue and appear white, whereas the dead cells pick up the dye and appear blue in color (Figures 6.6(a–d)). The fraction of cells in the total cell count that have not taken up the dye is a direct indication of the viability of the hMSCs. It was observed that there were a higher number of dead cells after electroporation. Overall, more dead cells were seen when subjected to the EP procedure, thereby indicating that the electrical pulses may be responsible for breaking the integrity of the plasma membrane and thus killing the cell. Figure 6.7 shows the morphological changes.

Dose curve

All or most of the chemo drugs have a lifetime limit that must not be exceeded. For example, no more than 400 units of Bleomycin should be

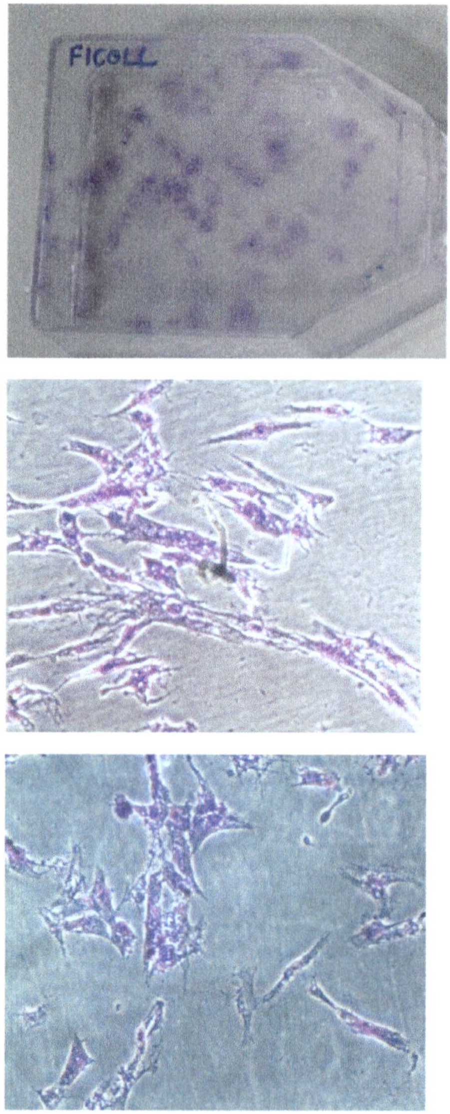

(a)

(b)

(c)

Figure 6.4 (a) CFU assay indicating the colonies; (b) microscopic view of colonies in a control sample of a patient; and (c) microscopic view of colonies in a pulsed sample of the same patient

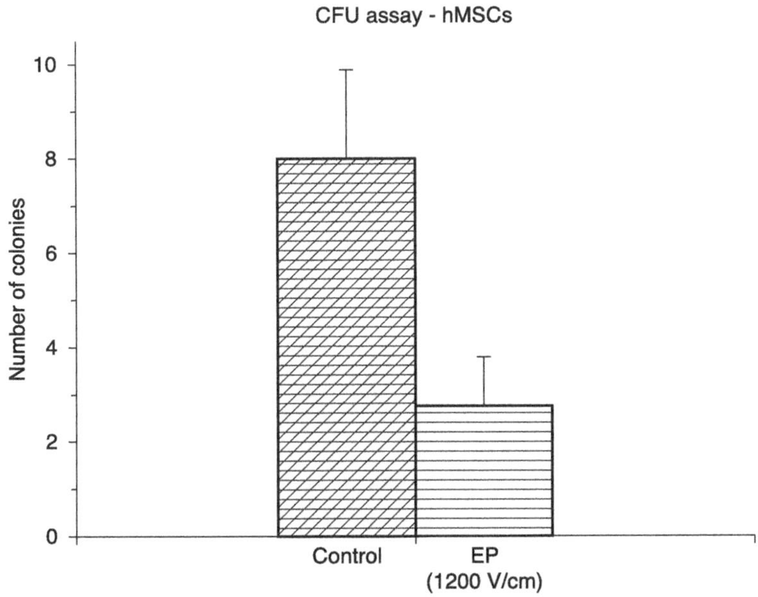

Figure 6.5 **Reduction in the number of CFUs of stem cells of patients due to electroporation**

administered to a patient in her/his entire lifetime. Hence, a dose curve study was performed to identify the efficacy of various doses of the drug, such as at 0.1, 1.0 and 10.0 μM concentrations and their viabilities for various pulse parameters. Figure 6.8(a) shows the variation in the cell viability for these doses under 500 V/cm, 5 ms pulse conditions.

It is seen that as the concentration of Bleomycin increases, the number of live cells decrease, even when a constant pulse protocol is used. It is realized that doses as low as 0.1 μM can also be effective. As the Bleomycin concentration was changed from 0.1 to 10 μM, there was a 15–20% reduction in viability. This could be explained by the fact that at 500 V/cm pulse intensity and duration, the maximum number of drug molecules that could enter the cell have done so, causing this effect. It also shows the nonlinearity in the variation of viability with respect to increase in the dose of the drug.

The dose study was repeated for other pulse parameters and Figure 6.8(b) shows the cell viabilities for these conditions. The 10 μM concentration of the drug produces maximum mortality in all the 3 pulse parameters and a very high mortality of around 55% is achieved when

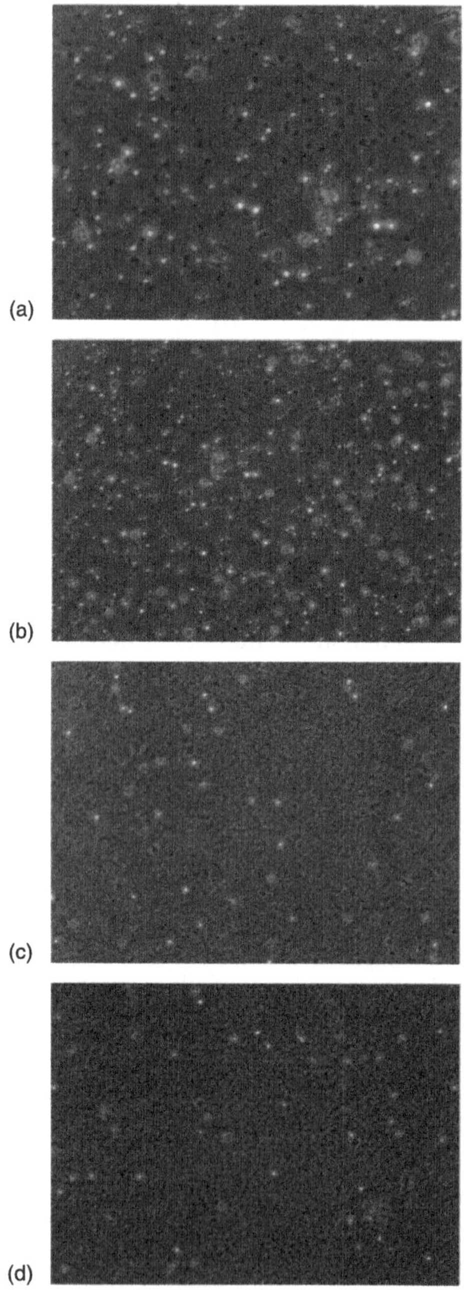

Figure 6.6 Trypan Blue exclusion assay showing the live cells (white) as bright spots and dead cells as dark spots

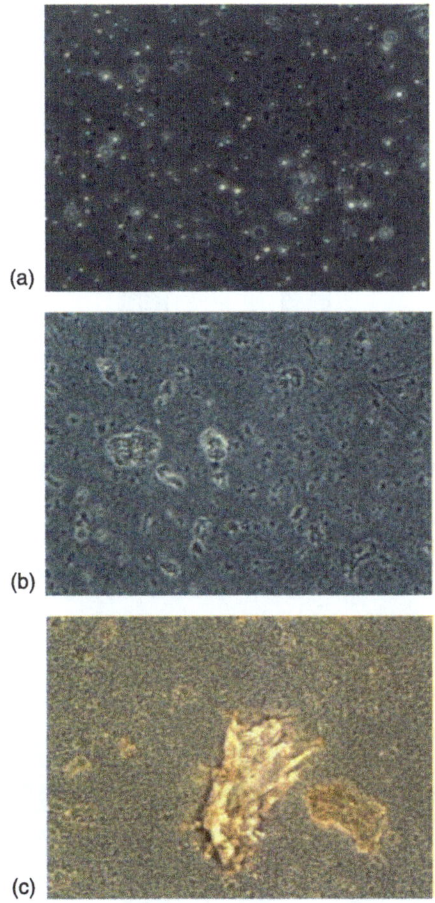

Figure 6.7 Morphology changes of electroporated cells pulsed at 500 V/cm, 40 ms compared to control

pulsed under the conditions of 450 V/cm, 25 ms, 8 pulses. It is seen that as the applied voltage intensity increases, there is a corresponding decrease in the number of live cells for a given concentration of Bleomycin. The results obtained clearly indicate that the cytotoxicity of the drug is increased greatly in the presence of the pulse. Table 6.2 gives a summary of the various conditions studied and their corresponding mortalities.

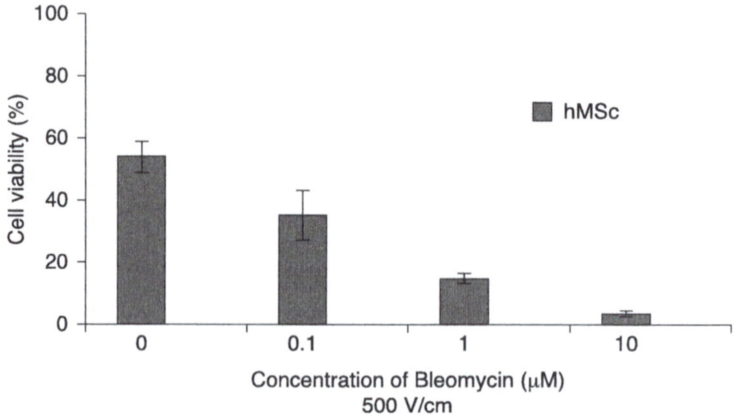

Figure 6.8 (a) Dose curve of Bleomycin at 500V/cm, 5ms pulse set

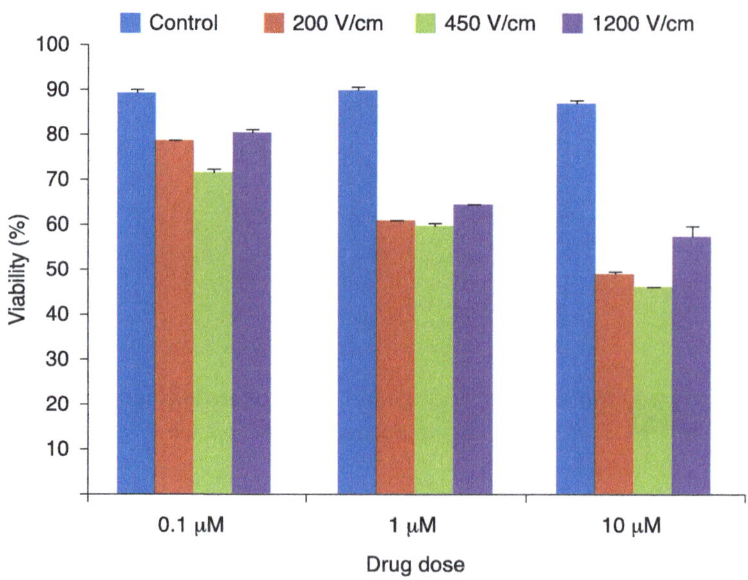

Figure 6.8 (b) Dose curves for the three pulse parameters tested along with control

Effect of voltage, number of pulses and pulse duration on cell viability

Different voltage pulses, 200 V/cm, 5 ms; 450 V/cm, 25 ms; 500 V/cm, 5 ms; and 1200 V/cm 100 μs were used to electroporate the cells, and the

Table 6.2 Mortality variations for various pulse conditions

Electric field intensity V/cm	Pulse width T	Bleomycin dose	Mortality
Control (none)	–	–	6.3%
Bleomycin only	–	1.0 μM	13.0%
EP only 200 V/cm	40 ms	–	16.7%
EP only 450 V/cm	25 ms	–	18.2%
EP only 1200 V/cm	100 μs	–	15.4%
200	40 ms	0.1 μM	22.3%
450	25 ms	0.1 μM	27.3%
1200	100 μs	0.1 μM	20.0%
200	40 ms	1.0 μM	38.5%
450	25 ms	1.0 μM	39.5%
1200	100 μs	1.0 μM	34.0%
200	40 ms	10.0 μM	50.0%
450	25 ms	10.0 μM	54.5%
1200	100 μs	10.0 μM	40.0%

viabilities were estimated 24 and 48 hours after the electroporation procedure. As the voltage increases from 200–1200 V/cm (keeping the duration constant at 5 ms), there is a significant reduction in the number of viable cells (Figure 6.9(a)), especially at 1200 V/cm, compared to the control. The effect of voltage (with constant pulse duration) on cell viability with time reveals that while 8 pulses of 200 V/cm at 5 ms duration (1 s interval) showed more viable cells 24 hours after pulse delivery, they declined by 48 hours post pulse delivery. The number of viable cells after 500 V/cm and 1200 V/cm electroporation also showed reduction after 24 and 48 hours. This indicates that pulses as low as 200 V/cm and 5 ms duration could be sufficient to trigger cell death in hMSCs. The effect was seen to increase after 48 hours of incubation of the treated cells, indicating that the process may not only have killed the cells immediately but also triggered apoptotic pathways which gradually killed the cells much later, after the actual delivery of the pulse.

As expected, maximum cell death was observed upon the delivery of the 1200 V/cm pulse for the same time duration, compared to 200 and

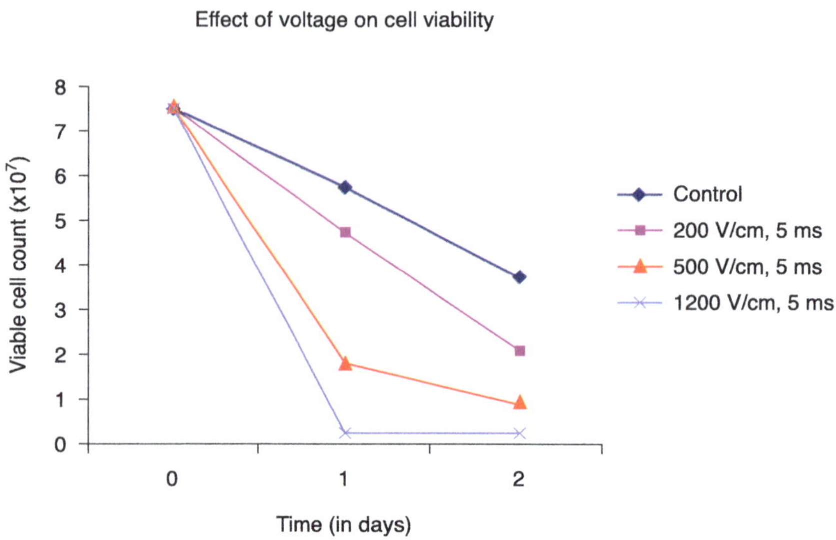

Figure 6.9 **(a) Effect of voltage on cell viability with constant pulse duration**

500 V/cm pulses. However, administering this voltage for the long duration of 5 ms resulted in arcing at times; hence the 100 μs pulse was used to minimize errors due to inhomogeneous results if arcing occurred. This slight time linearity given to the 1200 V/cm is seen in the results obtained (Figure 6.9(b)), wherein the least viability was observed with 500 V/cm, 5 ms pulses, due to longer duration.

Figures 6.9(c–e) shows the mortality with time for other pulse conditions, with and without the drug. Again, there is reduced proliferation with electroporation and the anticancer chemo drug, Bleomycin.

Effect of pulse duration on cell viability

A pulse of 500 V/cm, when administered to the hMSCs for different time durations, exhibited more viable cells when the pulse duration was the lowest (100 μs) compared to the same pulse given for 5 ms and 40 ms (Figure 6.10). We could thus infer that the longer the pulse duration administered, the more the number of cells that would be killed by the pulse. A comparison with ME180, human cervical cancer cells, indicates that the hMSCs are more susceptible to the pulse than the cervical cancer cells.

Figures 6.11(a–c) illustrate the effect of various pulse conditions. Figure 6.12 shows a summary of cell viability for a number of pulse

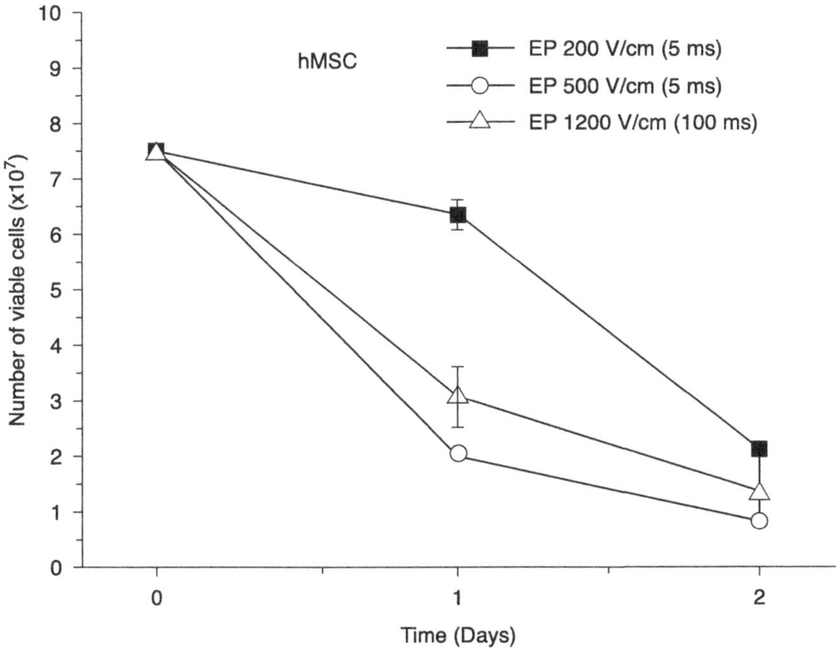

Figure 6.9 (b) Effect of voltage on cell viability with 1200 V/cm, 100 μs compared to 5 ms pulses at 200 and 500 V/cm

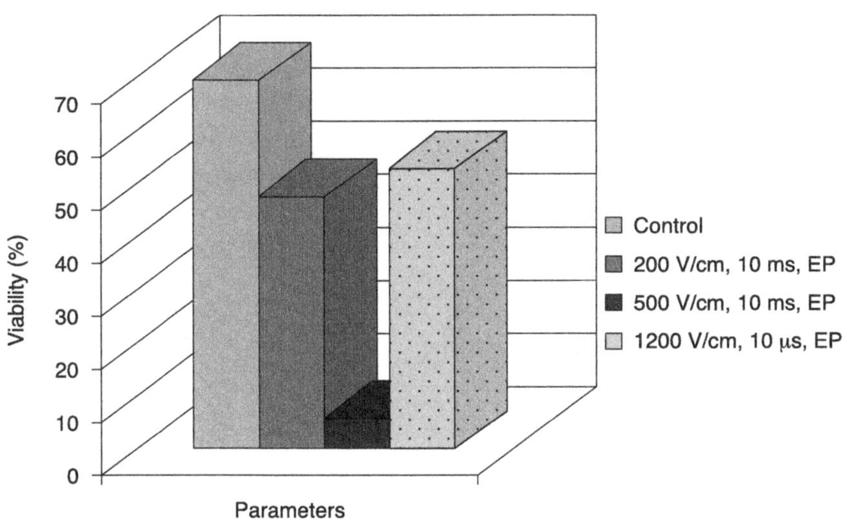

Figure 6.9 (c) Effect of voltage on cell viability with low voltage 10 ms and high voltage 100 μs pulse durations

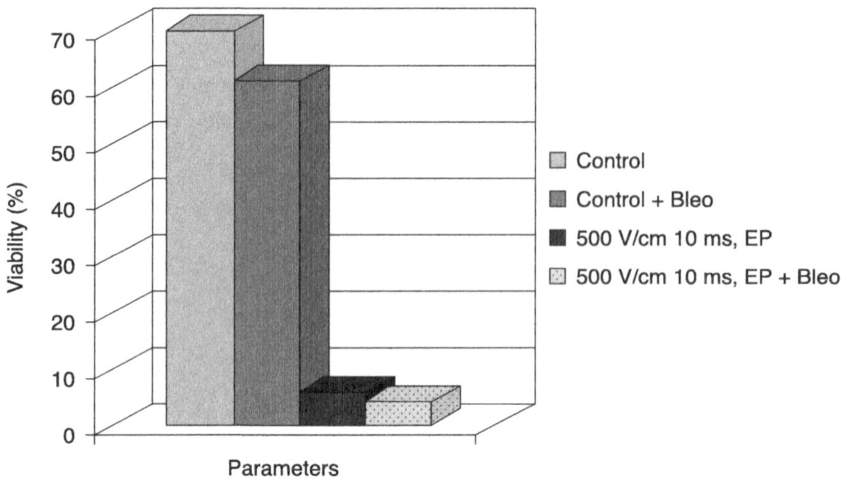

Figure 6.9 (d) Effect of voltage on cell viability with low voltage 10 ms and high voltage 100 μs pulse durations

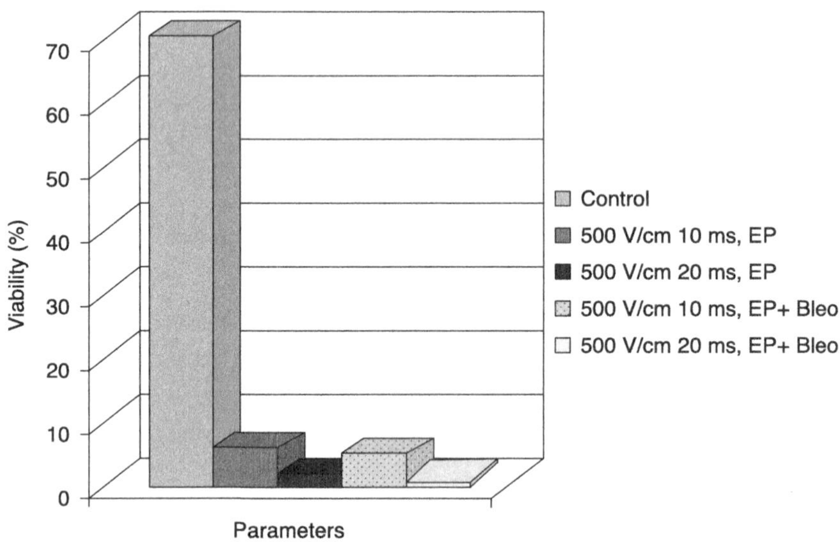

Figure 6.9 (e) Effect of voltage on cell viability with 10 and 20 ms pulses at 500 V/cm magnitude

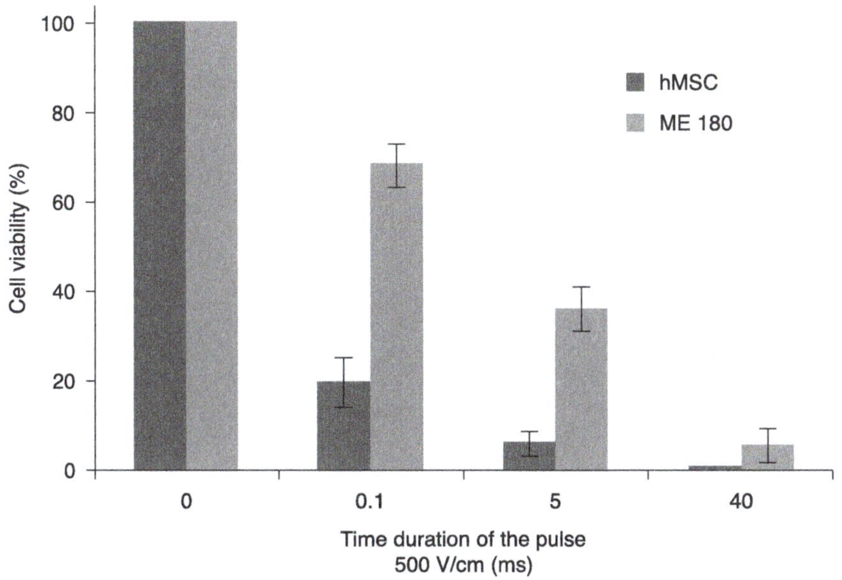

Figure 6.10 Comparison of effect of pulse durations on hMSC and ME 180 cervical cancer cells

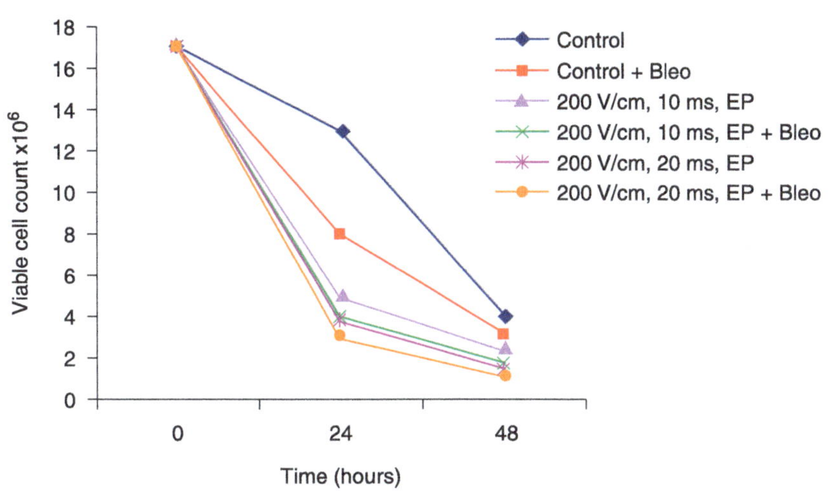

Figure 6.11 (a) Effect of variation of duration for 200 V/cm pulses with and without Bleomycin

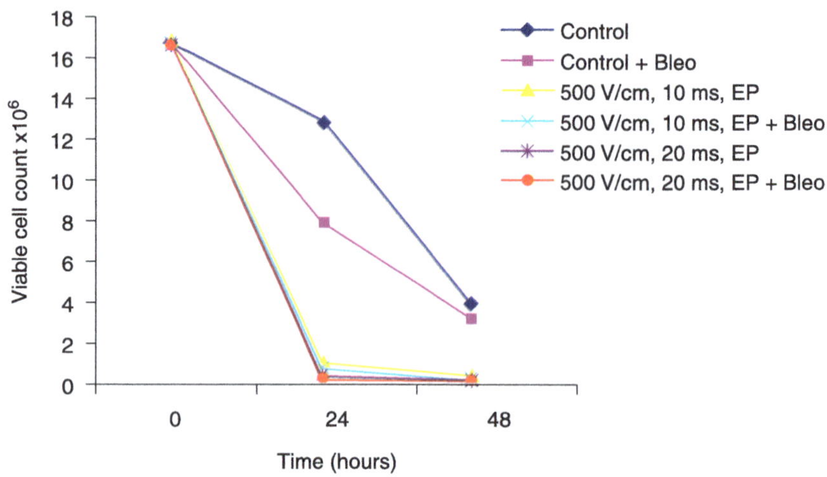

Figure 6.11 (b) Effect of variation of duration for 500 V/cm pulses with and without Bleomycin

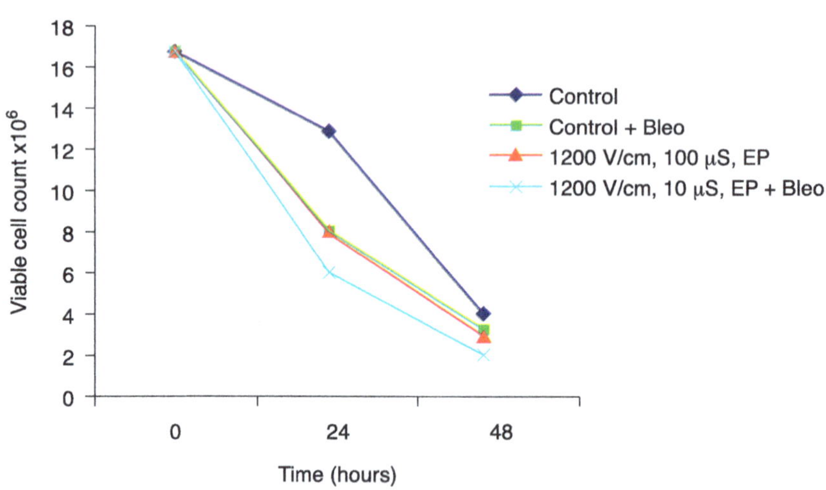

Figure 6.11 (c) Effect of 1200 V/cm pulses with and without Bleomycin

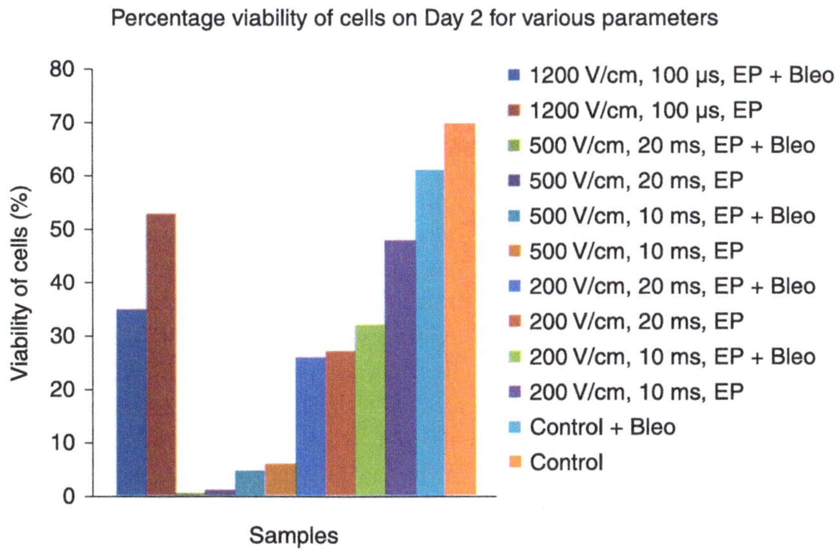

Figure 6.12 **The strong effect of Bleomycin, the cancer drug, with and without electroporation on reducing the viability of the mesenchymal stem cells on Day 2 (the legend represents samples from right to left, right most is the control with the highest live cells)**

parameters, with and without electroporation and with and without the drug at 48 hours. The pulse parameter, 500 V/cm, 20 ms pulses with the drug killed almost all of the cells in 48 hours.

6.8 Discussion and conclusions

The use of electrical pulses for active delivery of drug/gene molecules to the interior of cells, especially adult stem cells, offers exciting prospects. The pulse parameters explored here rely on the cell's natural ability to heal itself after electroporation. This is because membrane disruption and resealing is common *in vivo*, as cells normally heal smaller wounds naturally in day-to-day life. Towards this end, our study investigates the application of pulse power technology (short duration (milli- and micro-second) electrical voltage pulses) for achieving proliferation control of hMSCs, in an attempt to understand the scale-up of these to application on CSCs. In addition, application of these pulses also aids in the uploading of drug molecules into adult mesenchymal stem cells, which could serve as a Trojan Horse ferrying the deadly drugs to the cancer cells by means of their homing properties. Drug delivery is one of the most challenging issues of the medical community, and the need for alternate treatments, especially non-chemical, non-viral, physical ones, cannot be overstated.

The results of this study indicate that electrical pulses could effectively and economically be used to enhance drug transport across the normally impermeable plasma cell membranes and, thus to reduce proliferation in hMSCs. The results obtained in this study also correlate qualitatively well with previous results that cell viability decreases in a negative linear fashion with increase in energy delivered (data not shown) [60,61].

With respect to concentration of the drug, it is seen that as the concentration increases from 0.1–10 μM, the viability drops drastically in all the samples. The variation in the viability is more predominant in higher concentrations than in lower concentrations, as can be seen from Table 6.2.

Our findings demonstrate that pulses as low as 200 V/cm and ms durations can be used to trigger cell death in hMSCs and also deliver extracellular molecules such as Bleomycin into them, thereby increasing the rate of cell death. In addition, we can utilize the unique property of hMSCs homing towards tumor cells, aided by signaling molecules secreted by the tumors to recruit MSCs to enable tumor growth and progression by regenerating connective tissue. There have been reports of MSC migration to commercially available breast cancer cell lines, but not on primary culture from fresh specimens [49]. The MSCs could thus be used as a vehicle for anti-tumor therapeutics (genes or other molecules) in a Trojan Horse approach. hMSCs also resemble CSCs, and this *in vitro* study on hMSCs could be used to study the effect of Bleomycin, as well as electrical pulses on curbing proliferation in CSCs.

6.9 Future directions

This study describes the effect of short duration electrical pulses on critical parameters of stem cells including proliferation, viability, ability to form colonies, etc. This could also be effectively translated into a strategy of using ECT to kill CSCs within tissues. While CSCs may be resistant to drugs by means of the presence of transporters and pumps, the response of the cell to killer electrical pulses may not be variant from that of the other stem cells. This study, when carried out further, could help in understanding this better and in devising techniques to completely remove tumors with minimal chances for relapses.

Usage of MSCs to specifically target cancer cells and deliver drugs/ biomolecules could be used in designing novel strategies in cancer therapeutics. They could also be bio-engineered to express genes, which could trigger cell death to the cancerous cells once in contact with them. This study describes briefly the electrical parameters required to effectively load molecules into human mesenchymal stem cells. This could be of great significance in understanding and designing appropriate protocols for use in clinical practice.

The findings reported here are the preliminary results of a vast study designed to use adult hMSCs as vehicles in targeted tumor therapy. We optimistically hope that the new knowledge and research outlined in this paper will help contribute to new therapies for cancer, heart diseases, diabetes, and a whole variety of other diseases.

6.10 Acknowledgments

The authors are very grateful to Raghavan Kumar, Sinthu Muralitharen and S Mansoor Usman of BS Abdur Rahman University (Crescent Engineering College), Chennai, India for conducting the experiments and to Frontier Lifeline scientists for their help with the technical details and acquiring cell lines during the course of experiments.

6.11 References

1. Afanasyev B V, Elstner E E and Zander A R (2009), 'A J Friedenstein, founder of the mesenchymal stem cell concept', *Cell Thera Transpl*, **1**(3): 35–8.
2. Friedenstein A J, Piatetzky S, II and Petrakova K V (1966), 'Osteogenesis in transplants of bone marrow cells', *J Embryol Exp Morphol*, **16**(3): 381–90.
3. Wong R S Y (2011), 'Mesenchymal stem cells: angels or demons?', *J Biomed Biotech*, Article ID 459510, 8 p.
4. Globocan 2008, IARC (2010). Available at: *http:// globocan. iarc.fr/ factsheets/populations/ factsheet.asp? uno=900#KEY*
5. Cancer and You (2009). Available at: *http://www.jimmy.org/research/ cancer-facts.html? tmpl+complement &print=1 &page =Apr 22 2009*
6. Ohlsson L B, Varas L, Kjellman C, Edvardsen K, Lindvall M (2003), 'Mesenchymal progenitor cell-mediated inhibition of tumor growth *in vivo* and *in vitro* in gelatin matrix', *Exp Mol Pathol*, **75**(3): 248–55.

7. Ramasamy R, Lam E W, Soeiro I, Tisato V, Bonnet D and Dazzi F (2007), 'Mesenchymal stem cells inhibit proliferation and apoptosis of tumor cells: impact on *in vivo* tumor growth', *Leukemia*, **21(2)**: 304–10.
8. Wang X, Zhang Z and Yao C (2010), 'Survivin is upregulated in myeloma cell lines cocultured with mesenchymal stem cells', *Leuk Res*, **34(10)**: 1325–9.
9. Patel S A, Meyer J R, Greco S J, Corcoran K E, Bryan M and Rameshwar P (2010), 'Mesenchymal stem cells protect breast cancer cells through regulatory T cells: role of mesenchymal stem cell-derived TGF-beta', *J Immunol*, **184(10)**: 5885–94.
10. Konopleva M, Konoplev S, Hu W, Zaritskey A Y, Afanasiev B V and Andreeff M (2002), 'Stromal cells prevent apoptosis of AML cells by up-regulation of anti-apoptotic proteins', *Leukemia*, **16(9)**: 1713–24.
11. Wei Z, Chen N, Guo H, et al. (2009), 'Bone marrow mesenchymal stem cells from leukemia patients inhibit growth and apoptosis in serum-deprived K562 cells', *J Exp Clin Cancer Res*, **28**: 141.
12. Tian L L, Yue W, Zhu F, Li S and Li W (2011), 'Human mesenchymal stem cells play a dual role on tumor cell growth *in vitro* and *in vivo*', *J Cell Physiol*, **226(7)**: 1860–7.
13. Kucerova L, Matuskova M, Hlubinova K, Altanerova V and Altaner C (2010), 'Tumor cell behaviour modulation by mesenchymal stromal cells', *Mol Cancer*, **9**: 129.
14. Karnoub A E, Dash A B, Vo A P, et al. (2007), 'Mesenchymal stem cells within tumour stroma promote breast cancer metastasis', *Nature*, ***449(7162)***: 557–63.
15. Lin Y M, Zhang G Z, Leng Z X, et al. (2006), 'Study on the bone marrow mesenchymal stem cells induced drug resistance in the U937 cells and its mechanism', *Chin Med J (Engl)*, **119(11)**: 905–10.
16. Kurtova A V, Balakrishnan K, Chen R, et al. (2009), 'Diverse marrow stromal cells protect CLL cells from spontaneous and drug-induced apoptosis: development of a reliable and reproducible system to assess stromal cell adhesion-mediated drug resistance', *Blood*, **114(20)**: 4441–50.
17. Vianello F, Villanova F, Tisato V, et al. (2010), 'Bone marrow mesenchymal stromal cells non-selectively protect chronic myeloid leukemia cells from imatinib-induced apoptosis via the CXCR4/CXCL12 axis', *Haematologica*, **95(7)**: 1081–9.
18. Levina V, Marrangoni A M, DeMarco R, Gorelik E and Lokshin A E (2008), 'Drug-selected human lung cancer stem cells: cytokine network, tumorigenic and metastatic properties', *PLoS One*, **3(8)**: e3077.
19. Hall B, Dembinski J, Sasser A K, Studeny M, Andreeff M and Marini F (2007), 'Mesenchymal stem cells in cancer: tumor-associated fibroblasts and cell-based delivery vehicles', *Int J Hematol*, **86(1)**: 8–16.
20. Shinagawa K, Kitadai Y, Tanaka M, et al. (2010), 'Mesenchymal stem cells enhance growth and metastasis of colon cancer', *Int J Cancer*, **127(10)**: 2323–33.
21. Liu C H and Hwang S M (2005), 'Cytokine interactions in mesenchymal stem cells from cord blood', *Cytokine*, **32(6)**: 270–9.

22. Karnoub A E, Dash A B, Vo A P, et al. (2007), 'Mesenchymal stem cells within tumour stroma promote breast cancer metastasis', *Nature*, **449(7162)**: 557–63.
23. Fritz V and Jorgensen C (2008), 'Mesenchymal stem cells: an emerging tool for cancer targeting and therapy', *Curr Stem Cell Res Ther*, **3(1)**: 32–42.
24. Zacharek A, Chen J, Cui X, et al. (2007), 'Angiopoietin1/Tie2 and VEGF/Flk1 induced by MSC treatment amplifies angiogenesis and vascular stabilization after stroke', *J Cereb Blood Flow Metab*, **27(10)**: 1684–91.
25. Bruno S, Bussolati B, Grange C, et al. (2006), 'CD133^{+} renal progenitor cells contribute to tumor angiogenesis', *Am J Pathol*, **169(6)**: 2223–35.
26. Croitoru-Lamoury J, Lamoury F M, Zaunders J J, Veas L A and Brew B J (2007), 'Human mesenchymal stem cells constitutively express chemokines and chemokine receptors that can be upregulated by cytokines, IFN-beta, and Copaxone', *J Interferon Cytokine Res*, **27(1)**: 53–64.
27. Mishra P J, Humeniuk R, Medina D J, et al. (2008), 'Carcinoma-associated fibroblast-like differentiation of human mesenchymal stem cells', *Cancer Res*, **68(11)**: 4331–9.
28. Dey M, Ulasov I V, Tyler M A, Sonabend A M and Lesniak M S (2011), 'Cancer stem cells: the final frontier for glioma virotherapy', *Stem Cell Rev*, **7(1)**: 119–29.
29. Eramo A, Ricci-Vitiani L, Zeuner A, et al. (2006), 'Chemotherapy resistance of glioblastoma stem cells', *Cell Death Differ*, **13(7)**: 1238–41.
30. Gupta P B, Onder T T, Jiang G, et al. (2009), 'Identification of selective inhibitors of cancer stem cells by high-throughput screening', *Cell*, **138(4)**: 645–59.
31. Jorgensen C (2009), 'Link between cancer stem cells and adult mesenchymal: implications for cancer therapy', *Regen Med*, **4(2)**: 149–52.
32. *www.stemline.com/images/cscLarge.gif* (2011)
33. Gothelf A, Mir L M and Gehl J (2003), 'Electrochemotherapy: results of cancer treatment using enhanced delivery of bleomycin by electroporation', *Cancer Treat Rev*, **29(5)**: 371–87.
34. Larkin J O, Collins C G, Aarons S, et al. (2007), 'Electrochemotherapy: aspects of preclinical development and early clinical experience', *Ann Surg*, **245(3)**: 469–79.
35. Campana L G, Mocellin S, Basso M, et al. (2009), 'Bleomycin-based electrochemotherapy: clinical outcome from a single institution's experience with 52 patients', *Ann Surg Oncol*, **16(1)**: 191–9.
36. Heller L C and Heller R (2006), '*In vivo* electroporation for gene therapy', *Human Gene Therapy*, **17**: 890–7.
37. Gehl J and Geertsen P F (2000), 'Efficient palliation of haemorrhaging malignant melanoma skin metastases by electrochemotherapy', *Melanoma Res*, **10(6)**: 585–9.
38. Heller R, Jaroszeski M J, Reintgen D S, et al. (1998), 'Treatment of cutaneous and subcutaneous tumors with electrochemotherapy using intralesional bleomycin', *Cancer*, **83(1)**: 148–57.
39. Dev S B, Rabussay D P, Widera G and Hoffmann G A (2000), 'Medical applications of electroporation', *IEEE Trans. Plasma Science*, **28(1)**: 206–23.

40. Mir L M, et al. (1991), 'Electrochemotherapy, a novel antitumor treatment: first clinical trial', *CR Acad Sci Paris*, **313**: 613–18.
41. Heller R H (2002), 'Overview of electroporation', *Tech Canc Res Treat*, **1**(5): 317–18.
42. Heller R, Gilbert R and Jaroszeski M J (2000), 'Clinical trials for solid tumors using electrochemotherapy', *Methods Mol Med*, **37**: 137–56.
43. Weaver J C and Schoenbach K H (2002), 'Bio-dieelctrics – Editorial', *IEEE Trans DEI*, **10**(5): 1223.
44. Mansell P (2007), 'Inovio's electroporation tech delivers the goods in trial', Available at: *www.in- pharmatechnologist.com/content/view/print/118017*, 24 May 2007.
45. Maxim P G, Carson J J, Ning S, et al. (2004), 'Enhanced effectiveness of radiochemotherapy with tirapazamine by local application of electric pulses to tumors', *Radiat Res*, **162(2)**: 185–93.
46. Camarillo I, Nichols M, Zheng M, Sonnenburg D and Sundararajan R (2008), 'Electro-endocrine combination therapy for aggressive breast tumors', *J Electrostat*, **66**: 99–106.
47. Xiao F, et al. (2009), 'Electrically-enhanced chemodrug delivery to human breast cancer cells', *Electrostatic Society of America Annual Meeting*, Boston, June.
48. Gao H, Priebe W, Glod J and Banerjee D (2009), 'Activation of signal transducers and activators of transcription 3 and focal adhesion kinase by stromal cell-derived factor 1 is required for migration of human mesenchymal stem cells in response to tumor cell-conditioned medium', *Stem Cells*, **27**(**4**): 857–65.
49. *http://clincancerres.aacrjournals.org/cgi/content/full/13/17/5020* (2012).
50. Takimoto C H and Calvo E (2008), 'Principles of oncologic pharmacotherapy', in: Pazdur R, Wagman L D, Camphausen K A and Hoskins W J (eds), *Cancer Management: A Multidisciplinary Approach*, 11th edition.
51. Dorr R T (1992), 'Bleomycin pharmacology: mechanism of action and resistance, and clinical pharmacokinetics', *Semin Oncol*, **19(2 Suppl 5)**: 3–8.
52. Larkin J O, et al. (2007), 'Electrochemotherapy: aspects of preclinical development and early clinical experience', *Ann Surg*, **245**(**3**): 469–79.
53. Sersa G, Cemazar M and Rudolf Z (2003), 'Electrochemotherapy: advantages and drawbacks in treatment of cancer patients', *Cancer Therapy*, **1**: 133–42.
54. Brunner K W and Sonntag R W (1975), 'Bleomycin in testicular teratomas', *Z Krebsforsch Klin Onkol Cancer Res Clin Oncol*, **84**(**3**): 291–7.
55. McKinnell R G, Parchment R E, Perantoni A O and Pierce B G (2003), *The Biological Basis of Cancer*, New York: Cambridge University Press.
56. Love S M and Linsey K (2005), *Dr Susan Love's Breast Book*. Cambridge, MA: Perseus Publishing.
57. Cutrera J, Torrero M, Shiomitsu K, Mauldin N and Li S (2008), 'Intratumoral bleomycin and IL-12 electrochemogenetherapy for treating head and neck tumors in dogs', *Methods Mol Biol*, **423**: 319–25.

58. Hui S W and Li L H (2000), '*In vitro* and *ex vivo* gene delivery to cells by electroporation', in: *Electrochemotherapy, Electrogenetherapy, and Transdermal Drug Delivery*, Jaroszeski M J, Heller R and Gilbert R (eds), New Jersey: Humana Press.
59. Li S, Zhang X and Xia X (2002), 'Regression of tumor growth and induction of long-term antitumor memory by interlukin 12 electrogenetherapy', *J Natl Cancer Inst*, **94**, 762–8.
60. Chen G Y, Conner A J, Wang J, Fautrier A G and Field R J (1998), 'Energy dissipation as a key factor for electroporation of protoplasts', *Mol Biotechnol*, **10(3)**: 209–16.
61. Canatella P J, Karr J F, Petros J A and Prausnitz M R (2001), 'Quantitative study of electroporation-mediated molecular uptake and cell viability', *Biophys J*, **80(2)**: 755–64.

7

An *in vitro* study of electroporation of leukemia and cervical cancer cells

Raja Prabu Ramachandran, S. Madhivanan, Raji Sundararajan, Cindy Wan-Ying Lin and Kavitha Sankaranarayanan

DOI: 10.1533/9781908818294.161

Abstract: Cancer therapies mainly target the uncontrolled proliferation of the cancerous cells. A translational therapy for inoperable, chemo- and radio-resistant tumors is electroporation-based chemotherapy, known as electrochemotherapy, a physical means of using electrical pulses in conjunction with minimal doses of the chemo drug to achieve proliferation control. It is applicable to all histologies of tumors and its efficacy depends on a variety of factors, including the electric field intensity, the pulse duration and the number of pulses. In this chapter, we describe the optimization of pulse parameters for effective electroporation of leukemic and cervical cancer cell lines. Human promyelocytic leukemia (HL60), human acute myeloid leukemia (KG1) and human cervical (ME180) cancer cell lines were subjected to electroporation in the presence of various doses of FDA approved chemo drug, Bleomycin. The effect of curcumin and electroporation is also tested for HL60 cells. The results indicate that by using electroporation, chemo drug molecules could be uploaded into these cells to control proliferation. This promises to be a very useful tool for treating patients suffering from chemo-refractive tumors.

Key words: Leukemia, cervical cancer, HL60 cell line, KG1 cell line, ME180 cell line, Bleomycin, curcumin, dose curve, voltage pulses, viability.

7.1 Introduction

About every four minutes, two people are diagnosed with a blood cancer and about every ten minutes, two die due to a blood cancer, in the US [1]. Leukemia is one of the blood cancers that affects the bone marrow, which is the source of blood cells and occurs at all ages. It is characterized by a rapid increase in the number of immature blood cells in the blood stream, thereby affecting regular functions. In addition, the disease is subdivided based on the type of blood cells affected – lymphoblastic (cancer of lymphocyte forming cells) and myeloid leukemia (cancer of cells that become red blood corpuscles (RBCs)), some types of white blood corpuscles (WBCs) and platelets. Leukemia is one of the leading causes of mortality around the world, as evidenced by the global statistics conducted by the Leukemia and Lymphoma Society [2]. Table 7.1 shows the incidence of the disease and the death rate for various ages [3]. Figure 7.1 shows the 5-year survival rate for the periods, from 1960–63 versus 1975–77 versus 2001–07 [1]. Leukemia accounts for about 30% of cancer cases in children aged 0–14 years. There are more than 232 000 people in the US living with leukemia and approximately 44 000 new cases were diagnosed in 2011 [4]. The standard of care currently used is radiotherapy and chemotherapy. With the current poor cure rate, the need for alternate therapies cannot be overstated.

While leukemia affects both genders and at all ages, cervical cancer is a gynecologic cancer (any cancer that starts in a woman's reproductive

Table 7.1 Leukemia statistics – incidence and death rate at various ages [3]

Age	% Diagnosed	% Died
<20	11.4	3.3
20–34	5.2	3.5
35–44	5.9	3.9
45–54	9.7	6.7
55–64	13.9	11.8
65–74	20.5	23.4
75–84	23.2	31.1
85 and above	10.1	16.4

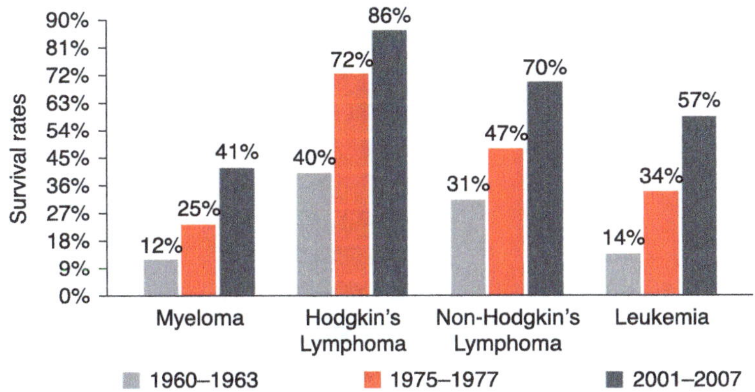

Figure 7.1 Five-year relative survival rate for the periods, from 1960–63 vs. 1975–77 vs. 2001–07 [1]

organs). It is the cancer that begins in the cervix, the lower part of the uterus [5]. It is almost always caused by human papilloma virus (HPV) infection [6], with virtually all cervical cancer cases (99%) being linked to infection with the HPV. The E6 and E7 genes present in the genome of HPV suppress the activity of two tumor suppressor genes, Rb and p53, thereby triggering uncontrolled proliferation of the same [7].

All women are at risk and it occurs most often in women over age 30. In the US each year, it affects about 12 000 women. Worldwide, it is the third most commonly diagnosed cancer and the fourth leading cause of cancer death in women, accounting for 9% (529 800) of the total new cancer cases and 8% (275 100) of the total cancer deaths among females in 2008 (Figure 7.2 [8]). More than 85% of these cases (453 300/529 800) and deaths (242 000/275 100) occur in developing countries. India, the second most populous country in the world, accounts for 27% (77 100) of the total cervical cancer deaths. Worldwide, the highest incidence rates are in Eastern, Western and Southern Africa, as well as South-Central Asia and South America. Rates are lowest in Western Asia, Australia/New Zealand and North America. The high incidence rate of cervical cancer in developing countries could be attributed to the lack of screening, its awareness and the socio-economical status, lacking resources and infrastructure, and the culture. The health care infrastructure in these countries does not support Papanicolaou testing or other types of screening tests. In addition, the vaccines do not provide protection for 30% of chronic infections by HPV types, other than HPV 16, 18, 6 and

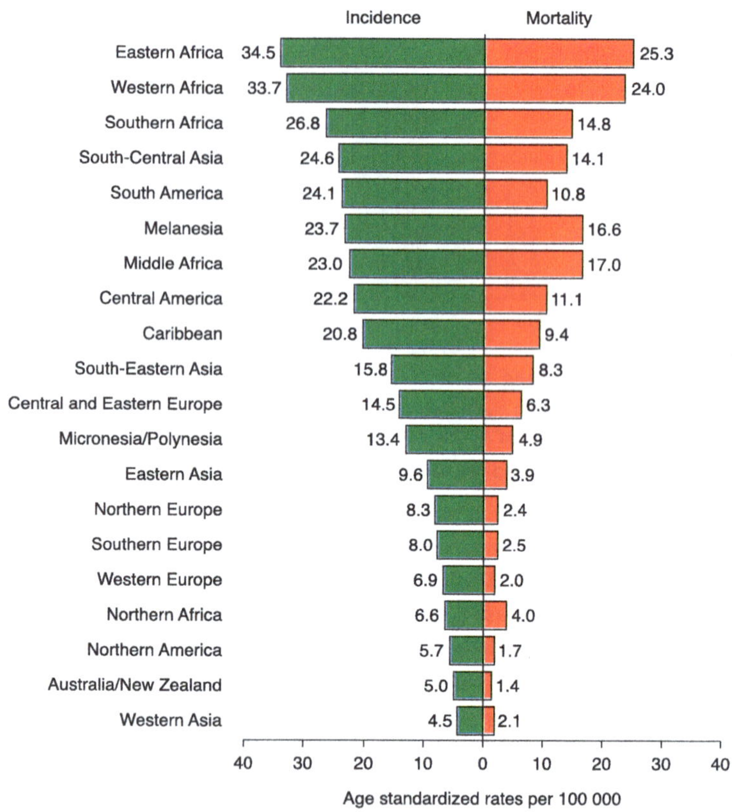

Figure 7.2 Age-standardized cervical cancer incidence and mortality in various countries [8]

11 that cause cervical cancer [8–10]. Normally, they are treated with surgery, chemotherapy, radiotherapy and hormonal therapy. However, these do not work well for all patients, as seen from the high death rate. Hence, additional/alternate, efficient and economical therapies are critical to save these women.

Cancerous cells such as HL60 develop resistance to a variety of cytotoxic drugs, making treatment of such types of cancers difficult. The drug, which by itself is relatively less permeable to the cell membrane, is not retained in the cancerous cell as a result of an enhanced drug efflux mechanism acquired by these cancerous cells [11–13]. This has been demonstrated to be due to the over-expression of a specific type of membrane phosphoglycoprotein and a drug resistance associated ATP binding protein on the cell surface. It has been demonstrated that as HL60 cells divide and 'age', there is an increase in the cell cholesterol content, which in turn modulates a decrease in membrane lipid dynamics [14].

Most leukemia and lymphoma cells grown as suspension culture are resistant to most gene transfer methods, thereby making electroporation the method of choice. The cell membrane, comprising a lipid bilayer, is interspersed with protein molecules – some of the transmembrane spanning the entire lipid bilayer (i.e. ion channels, H^+ pumps, etc.) are only partially embedded in the cell membrane and are exposed only at either end. In addition to the lipids and proteins, carbohydrates are also present in the cell membrane, as either glycoproteins or glycolipids. The lipid and protein compositions of the cell membrane vary from cell to cell and this bestows unique membrane properties to different types of cells, based on their function. Although, in previous research, the efficiency of electroporation for the hematopoietic cells was reported to be roughly just 1% [15], with a variety of low voltage, long duration and high voltage and short duration pulses, it is possible to identify the optimal pulse parameters. This is because of the mechanical property of the membrane as a major determinant in the efficiency of application of electrical pulses to a cell for electroporation-based methods for gene/molecule/drug transfer. It is largely determined by the cytoskeletal architecture of a cell, which is a variant in cancerous cells [16]. Electrodeformation, or the mechanical response of the cell to electrical fields, vary from cell to cell [17] and would account for the necessity for different parameters while working with various types of cells. It has been reported that there is a change in the membrane properties of the myeloid cells during maturation as well as differentiation [18]. Whether the changes in membrane properties affect the parameters required for electrochemotherapy of the different type of cells is what was probed into by this study.

Professor McConnel's group in Stanford had demonstrated that weak non-uniform external electric fields have disruptive effects on biological membranes [19]. This effect is primarily due to the effect of the fields on the lipid layer of the cell membranes, thereby affecting the membrane proteins impregnated on the same, which in turn results in abnormal cellular function. The size of the pores thus formed remains a function of the duration of the electric field pulse. Nanosecond pulses have been shown to result in a temporary externalization of the phosphatidyl serine residues [20], a release of intracellular calcium [21,22] and induction of apoptosis [23,24].

In the presence of a strong electric field, there is a significant increase in permeability and conductance of the lipid membrane. If the applied electrical field is of not too high an amplitude and time duration, then the cell membrane is capable of returning to its previous state, as in reversible electroporation. At higher amplitudes or time durations, the membrane loses its ability to reseal and self-destructs, as in the process of irreversible electroporation. Initially it was not understood as to where these pores were formed in the lipid bilayer – the lipid matrix, integral membrane protein region or the contact region between the two. However, studies on human erythrocytes and lymphocytes along with an artificial bilayer revealed that the electrical breakdown of the membrane arises in its lipid matrices [25].

A study of tumor treating fields (TTFs) of low intensity, intermediate frequency, alternating current on a variety of human and rodent tumor cell lines exhibit varying rates of reduction of cell proliferation, both *in vitro* and *in vivo*. This could be due to the difference in the property of the cell membrane, though the basic lipid matrix largely remains the same [26]. The study demonstrated a specific intensity and frequency dependence of the TTF inhibition on different types of cancerous cells. It also revealed that the electric fields disturbed the normal polymerization-depolymerization processes of the microtubules of the mitotic spindle during mitosis and hence arrests the proliferation of actively dividing cells such as cancer cells.

Large cells have been shown to be more susceptible to electrical breakdown, which might hint at a higher risk of breakdown with an increase in the membrane surface area. Hence the parameters for successful electroporation may not only be dependent on the type of cellular lipids and the other membrane components, but also its size. In this connection we have probed into the parameters for pulsing HL60, KG1 and ME180, which are of different cellular origins for increased uptake of chemo drugs [27–33]. This method of treatment could be a very economical palliative treatment strategy, owing to the absence of any rigorous time-consuming procedures, as well as a significant reduction in the dose of the drug required and could be out-patient-based, ideally suited for developing countries. The process is simpler for tumors on the skin and for other surface lesions, while endoscopic ECT procedures need to be used for deep tumors. It is safe and the entire process can be completed in a few minutes per nodule. The side effects of the drug are also drastically reduced, since the concentration of the chemo drug used is also minimal, with up to 1000 times the efficacy for the same dose [29].

7.2 Materials and methods

7.2.1 *Cell lines*

HL60 (human promyelocytic leukemia cell line)

HL60 is a human caucasian promyelocytic leukemia cell line (Figure 7.3(a)), derived from a 36-year-old woman with acute PML at the National Cancer Institute. HL60 cells are predominantly a neutrophilic promyelocyte, used for laboratory research on how certain kinds of blood cells are formed. PML, or a cancer of the blood forming tissue viz. the bone marrow, is caused due to a non-inherited (or somatic) mutation in the PML gene. The disease is characterized by an accumulation of immature white blood cells (promyelocytes) in the bone marrow. The PML gene codes for a tumor suppressor, which prevents uncontrolled proliferation of cells and induces apoptosis in such cells, providing a check point for healthy cell division. Mutations in this gene results in an uncontrolled proliferation of the white blood cells leading to PML [34, 35]. HL60 proliferates continuously in suspension culture in a RPMI medium supplemented with fetal bovine serum, L-glutamine, HEPES and antibiotics [36]. A hemocytometer was used to obtain a final concentration of 1×10^6 cells/ml.

KG1 (human acute myeloid leukemia cell line)

The KG1 cell line was derived from a bone marrow aspirate of a 59-year-old man with acute myelogenous leukemia, which developed from

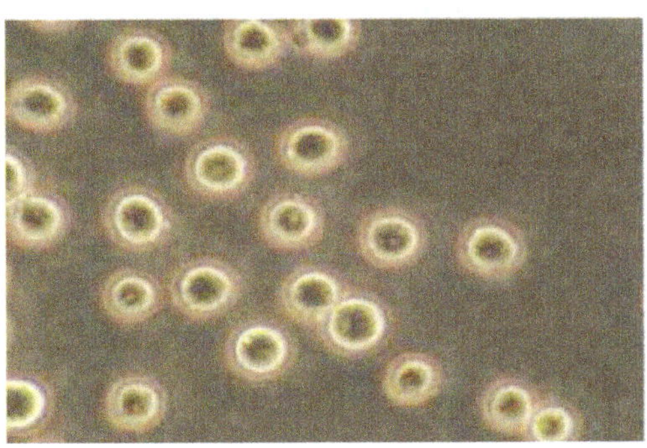

Figure 7.3 (a) Microscopic view of HL 60 cell line

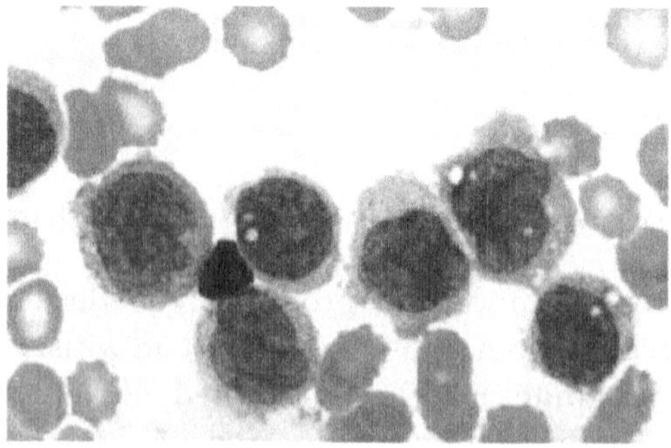

Figure 7.3 (b) Microscopic view of KG1 cell line

erythroleukemia (Figure 7.3(b)) [37]. In medical terms, 'acute' refers to a sudden onset, rapidly progressing disease that could turn fatal if not treated, as is a typical of KG1. Myeloid leukemias refer to leukemias in non-lymphocyte white blood cells, red blood cells or platelet-making cells [38]. Acute myeloid leukemia refers to a malignant disorder where there is a maturational arrest of bone marrow cells at early stages of differentiation [39]. Chromosomal abnormalities and other genetic translocations are involved in the onset of this disease. The production of normal blood cells, including RBCs, neutrophils and thrombocytes, are affected (reduced) resulting in anemia, neutropenia and thrombocytopenia (blood conditions involving lower than normal numbers of these cells). The presence of thrombocytopenia or a reduction in the count of platelets explains the abnormal bleeding which is one of the symptoms of the disease. The cells lose their ability to undergo apoptosis and this culminates in them accumulating in the spleen, bone marrow, liver and in the blood. These cells are responsive to the colony stimulating factor (CSF). The cells are rounded and myeloblast in nature and grow in suspension culture. They are capable of spontaneous differentiation into granulocytes and macrophages and represent an early stage of hematopoietic differentiation.

The cells were cultured in RPMI 1640 media with 10% fetal bovine serum (ATCC) and incubated in a 5% CO_2 atmosphere at 37°C. A hemocytometer was used to obtain a final concentration of 1×10^6 cells/mL.

ME180 (human cervical cancer cell line)

ME180, the human cervix-derived epithelial cell line (Figure 7.3(c)), is used as the model for studying cervical cancer characteristics. Isolated

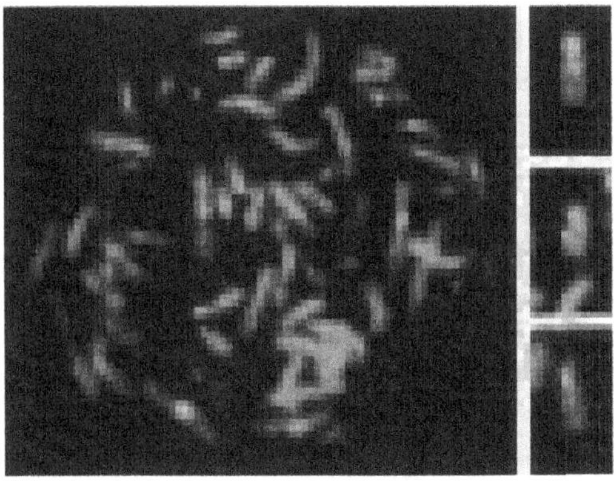

Figure 7.3 (c) Microscopic view of ME180 human cervical cancer cell line

from an omental metastasis of a spreading cervical carcinoma of a 66-year-old Caucasian female, this line was derived from a highly invasive squamous cell carcinoma with irregular cell clusters and no significant keratinization. The cells contain HPV DNA, with greater homology to HPV-39 than HPV-18 [40]. The cells were cultured RPMI 1640 media with 10% fetal bovine serum (ATCC) and incubated in a 5% CO_2 atmosphere at 37 °C. A hemocytometer was used to obtain a final concentration of 1×10^6 cells/mL.

7.2.2 Anticancer drugs

Bleomycin

Bleomycin (C55H84N17O21S3) is FDA approved in the US and by the respective agencies in other countries as an anti-cancer chemotherapy drug. It is classified as an 'antitumor antibiotic'. Bleomycin can be used in combination with surgery or radiotherapy, or in the palliative treatment of a number of cancers. It is used in the treatment of squamous cell cancers, melanoma, sarcoma, testicular cancer, Hodgkin's and non-Hodgkin's lymphoma [28]. Figure 7.4 shows the chemical structure of Bleomycin.

Bleomycin inflicts direct damage to the DNA [41]. It is a powerful drug and its administration causes a number of side effects such as skin rash, hives, changes in skin sensation, hyper-pigmentation (darkening of skin and nails), hair loss, mouth ulcers, hard patches on skin [42], etc. Thus

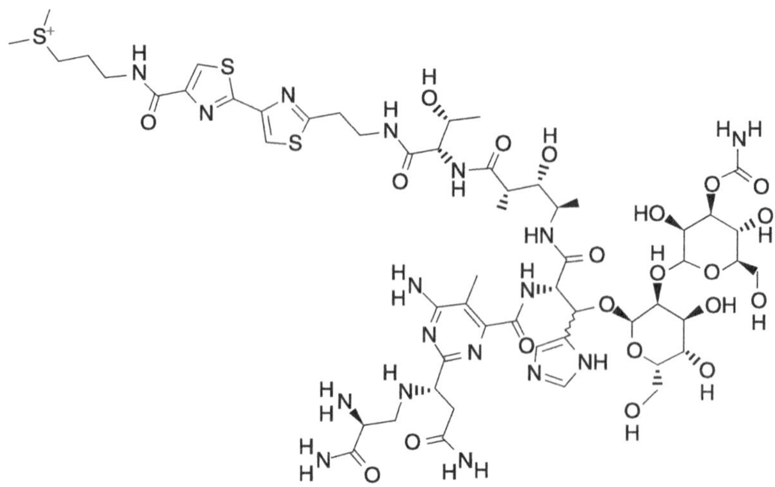

Figure 7.4 Chemical structure of Bleomycin

the dose should always be optimal and determined based on the patient's response. In our study, we aim to examine the effect of reducing the concentration of the drug so as to reduce the side effects and alleviate the pain suffered by cancer victims.

Curcumin

Curcumin, the yellow pigment of turmeric, is a hydrophobic phytochemical exerting a number of therapeutic effects, including the inhibition of cancer cell proliferation. It also has both anti-inflammatory and antioxidant effects. It inhibits production of the free radicals H_2O_2 and NO by macrophages [43, 44] and astrocytes [45] *in vitro*. The pharmacological safety and efficacy of curcumin makes it a potential compound for treatment and prevention of a wide variety of human diseases.

In spite of its efficacy and safety, curcumin has not yet been approved as a therapeutic agent, and the relative bioavailability of curcumin has been highlighted as a major problem [46]. For this purpose, we intend to upload it using electrical pulses and study its efficacy under these conditions in the HL60 cells.

Curcumin (diferuloylmethane) with a purity of more than 98% was purchased from Sigma-Aldrich Corporation (St Louis, MO), and was dissolved in dimethyl sulfoxide (DMSO) as a 10 mM stock solution and stored at −20 °C.

7.2.3 *Electroporation*

Using Bleomycin and HL60, KG1, and ME180 cell lines

Both high intensity, short duration (1200 V/cm, 100 μs) [27–33] and low intensity, long duration pulses (200 V/cm, 40 ms) [47] were used. In addition, medium intensity, 450 V/cm, 25 ms pulses were also used [48]. These pulses were chosen based on previous research and our experience. The choice of pulse parameters is critical as, if the pulse intensity is too high and/or too long, it kills the cells, and if it is too low and/or too short, pores will not open up to allow the drug to enter into the cells [49–52]. A BTX ECM 830 (Genetronics, Inc., San Diego, CA), square wave electroporator with 0.4 cm (electrode gap) cuvettes was used to conduct the study (Figure 7.5).

The electroporation buffer used was RPMI 1640 media with 10% fetal bovine serum (ATCC). The sample to be electroporated was taken in the cuvette and generally fixed at a total volume of 200 μL. For those experiments requiring electroporation without any drug, 100 μL of the cell suspension is used along with 100 μL of media and electroporated. For electroporation with the drug, 100 μL of cell suspension, 50 μL of appropriate drug stock and 50 μL of media was used. The drug is made up to required concentration by serial dilution technique.

The pulse parameters used are given in Table 7.2 and the doses used were 0.1, 1 and 10 μM of Bleomycin, very small compared to the conventional amount used in the clinics. In each case, 8 pulses at 1 second intervals (1 Hz) between pulses were used.

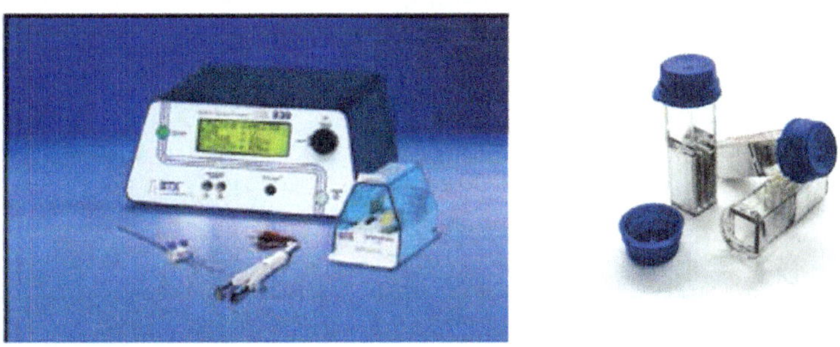

Figure 7.5 Electroporation apparatus and cuvettes

Table 7.2 Pulse parameters studied

Electric Field intensity, V/cm	Pulse width
1200	100 μs
450	25 ms
200	40 ms

Using curcumin and HL60

In the case of curcumin, 400 μL cells were used with 5 μL of curcumin at a concentration of 100 μM. Table 7.3 shows the various pulse parameters studied.

7.2.4 Cell counting and viability

After electroporation, cells are counted to determine the viability. Trypan Blue assay was used to count the cells [53].

For curcumin tests, iNcyto, disposable hemocytometer was used. For other tests, an Olympus CKX41, Phase contrast, inverted florescent microscope was used to image and count the cells. Images were captured using Motix software.

Table 7.3 Curcumin pulse parameters for HL60 cells

Sample No.	Description	Electric field intensity, V/cm	Pulse Duration	Number of pulses
1	Control	–	–	–
2	Drug only	–	–	–
3	Drug + Pulse	1200 V/cm	100 μs	8
4	Drug + Pulse	1000 V/cm	100 μs	8
5	Drug + Pulse	200 V/cm	5 ms	4
6	Drug + Pulse	500 V/cm	20 ms	2

7.3 Results and analysis

7.3.1 Dose curve

In order to prescribe the optimal amount of drug, so as to achieve maximum cell death without causing harmful side effects on the patient, a proper drug dose is required. There is only a thin line of seperation between optimal dose and drug overdose. In the case of drugs such as Bleomycin, this becomes even more vital, as the patient is only allowed to take a lifetime dose of 400 units.

In this study, the viability was studied for the doses of 0.1, 1.0 and 10.0 μM, for all 3 cell lines and for all the 3 pulse parameters. The cell viability decreases as the drug concentration increases, which is shown in Figure 7.6. While very low doses such as 0.1 μM can also be effective, the 1 μM dose was much more effective as we moved from one pulse parameter to the other in all the 3 cell lines. When the dose was changed from 1–10 and 1–0.1 μM, there was a nonlinear swing of 15–20% viability on both sides. Figure 7.6 shows the drug dosage curve for the HL60 cell line. Similar data were obtained for other cell lines too.

7.3.2 Viability study

Figure 7.7 shows the microscopic pictures illustrating the viability of control (no treatment), 450 V/cm, 25 ms and 1200 V/cm, 100 μs samples

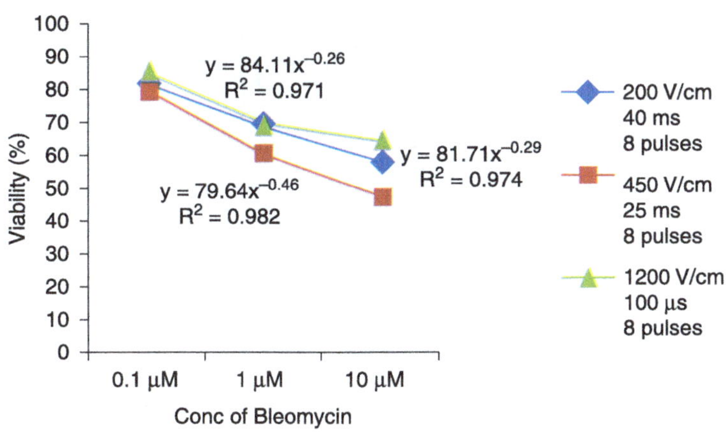

Figure 7.6 Bleomycin dose curve for HL60 cell line

of ME180 cervical cancer cells. There is reduction in the number of live cells due to electroporation compared to control cells. Similar results were also obtained for other cells (data not shown). Table 7.4 shows the mortality of the 3 types of cells studied. Figure 7.8 shows a comparison of the viability of the 3 cell lines for various pulse parameters at 1 μM dose.

The results indicate that among the various electroporation parameters, the cell mortality is found to be the highest at 450 V/cm, 25 ms, 8 pulses condition for all the 3 concentrations of Bleomycin for all the 3 cell

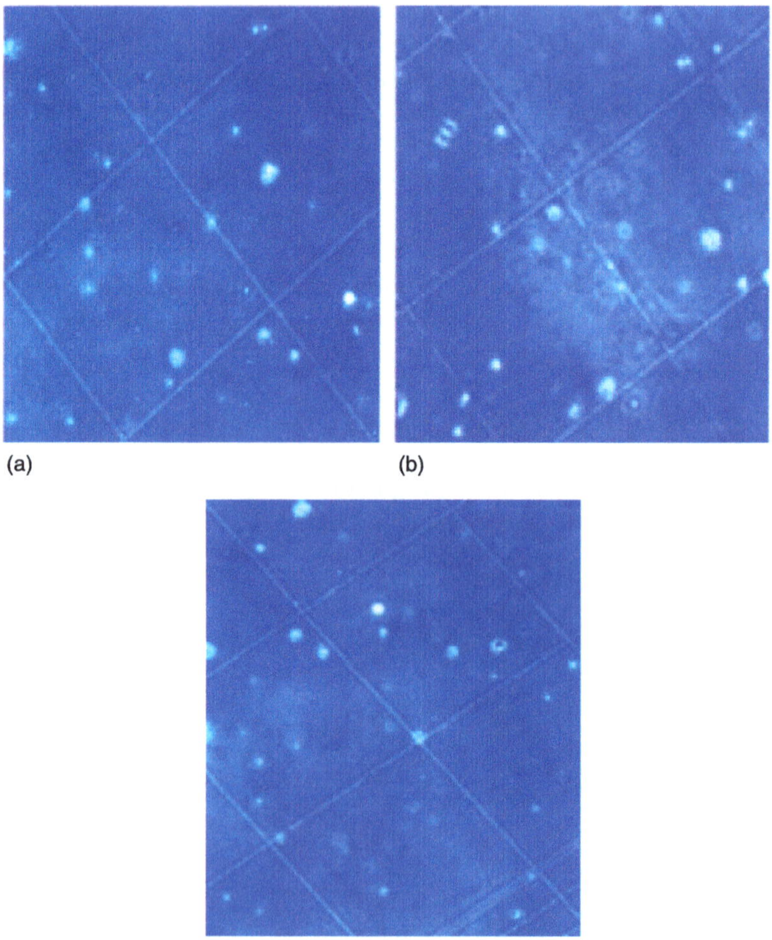

(a) (b) (c)

Figure 7.7 Viability of ME180 cervical cancer cells – control, 450 V/cm, and 1200 V/cm samples respectively from left to right

Table 7.4 Viability study of the three cell lines

Electric Field Intensity V/cm	Pulse Width	Bleomycin Dose	HL60	KG1	ME180
			Mortality, %		
Control	–	–	11.0	12.0	7.7
Bleomycin only	–	1.0 μM	14.8	14.7	13.4
EP only 200	40 ms	–	18.0	14.8	14.2
EP only 450	25 ms	–	18.8	17.3	20.0
EP only 1200	100 μs	–	14.2	13.7	10.0
200	40 ms	0.1 μM	26.4	19.2	18.8
450	25 ms	0.1 μM	30.0	21.5	20.6
1200	100 μs	0.1 μM	20.6	15.0	18.0
200	40 ms	1.0 μM	34.0	31.4	30.0
450	25 ms	1.0 μM	38.2	40.0	35.0
1200	100 μs	1.0 μM	23.0	31.8	26.9
200	40 ms	10.0 μM	42.9	42.1	42.2
450	25 ms	10.0 μM	56.5	53.4	53.0
1200	100 μs	10.0 μM	39.2	35.8	35.5

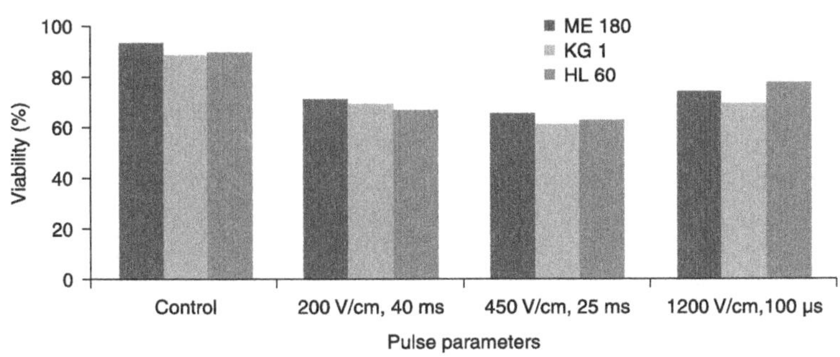

Figure 7.8 Viabilities of the three cell lines at various pulse parameters for 1 μM concentration of Bleomycin for ME180 (left), KG1 (middle) and HL60 (right) cells at various conditions

lines. Comparatively 200 V/cm, 40 ms shows a lower cell mortality and 1200 V/cm, 100 μs mortality values were even smaller. At 10 μM concentration, the viability decreased to around 56.5% for 450 V/cm pulses in HL60 cell line, which is almost a 50% decrease compared to control, whose viability is 89.0%.

Table 7.4 shows the pulse parameters studied for curcumin at 100 μM concentration using HL 60 cells and Figure 7.9 shows the corresponding viabilities at various periods, including immediately after pulsing (0 h), after 24 hours and 48 hours. The results show that with electrical pulses, the efficacy of curcumin is felt immediately compared to the control (with no treatment) and with drug only (no pulses). In general, all of the pulse parameters are effective after 24 and 48 hours, including the drug-only sample. Considering that the bioavilability of the turmeric is rapid and it is metabolized quickly, the use of electrical pulses is a viable option for enhancing the efficacy of targeted locoregional delivery.

Sample 1-control (no treatment), 2-drug only, 3–1200 V/cm, 100 μs, 8 pulses, 4–1000 V/cm, 100 μs, 8 pulses, 5–200 V/cm, 5 ms, 4 pulses and 6–500 V/cm, 20 ms, 2 pulses. All at 1 second interval

7.3.3 Equivalent energy and charge

In the absence of the measurement of the resistance or the current values, the energy applied during each pulse application could be computed as V^2TN J/Ω, where V is the applied voltage in V (E V/cm*0.4 cm), T is the

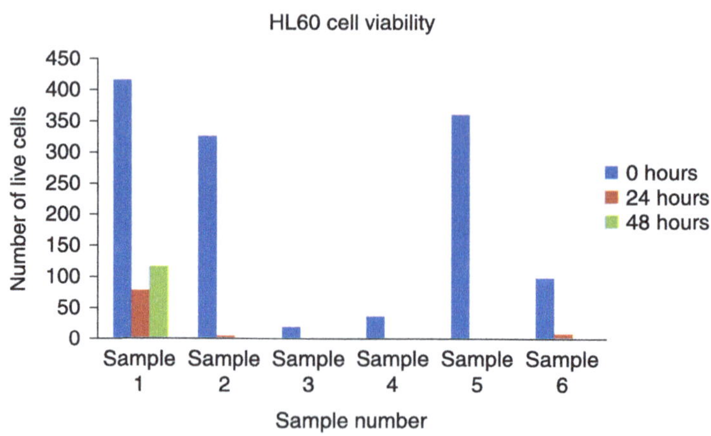

Figure 7.9 Viabilities of the HL60 cell line at various pulse parameters for 100 μM concentration of curcumin. Sample 1-control (no treatment), 2-drug only, 3–1200 V/cm, 100 μs, 8 pulses, 4–1000 V/cm, 100 μs, 8 pulses, 5–200 V/cm, 5 ms, 4 pulses and 6–500 V/cm, 20 ms, 2 pulses. All at 1 second intervals

pulse width in s, and N is the number of pulses. The equivalent of charge is calculated as ETN C/Ω. Energy and charge are indicators of the current applied and hence the duration for which the voltage is applied. The higher the energy, the lower the viability and vice versa. Thus a negative linear relationship is obtained for each of the pulse parameters studied in all the three cell lines. Use of energy as a parameter also has an attractive serendipity, in that the medical community would understand the terminology and the technique easily, since energy in joules is used for defibrillation and other techniques. Figures 7.10(a) and (b) show the energy and charge applied and the corresponding viability for the various samples for the HL60 and KG1 cell lines. A comparison of previously published results indicates good qualitative correlation in trends (Figures 7.11(a) and (b)) [54, 55].

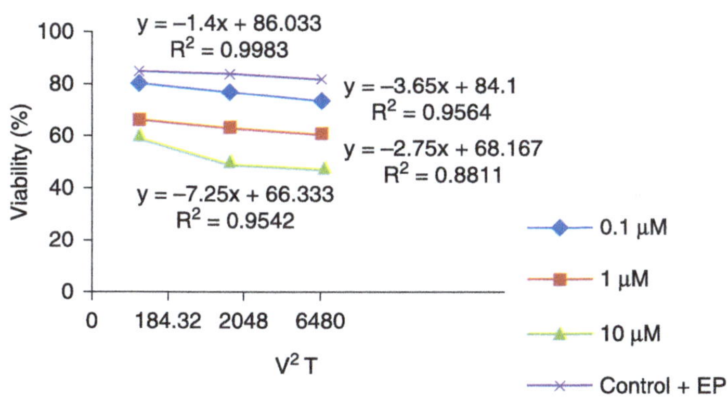

Figure 7.10 (a) Energy (J/Ω) vs. viability plot for HL60 cell line

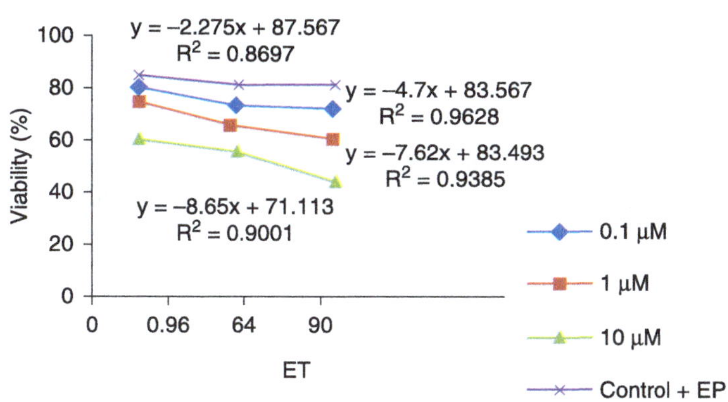

Figure 7.10 (b) Charge equivalent (Q/Ω) vs. viability plot for KG1 cell line

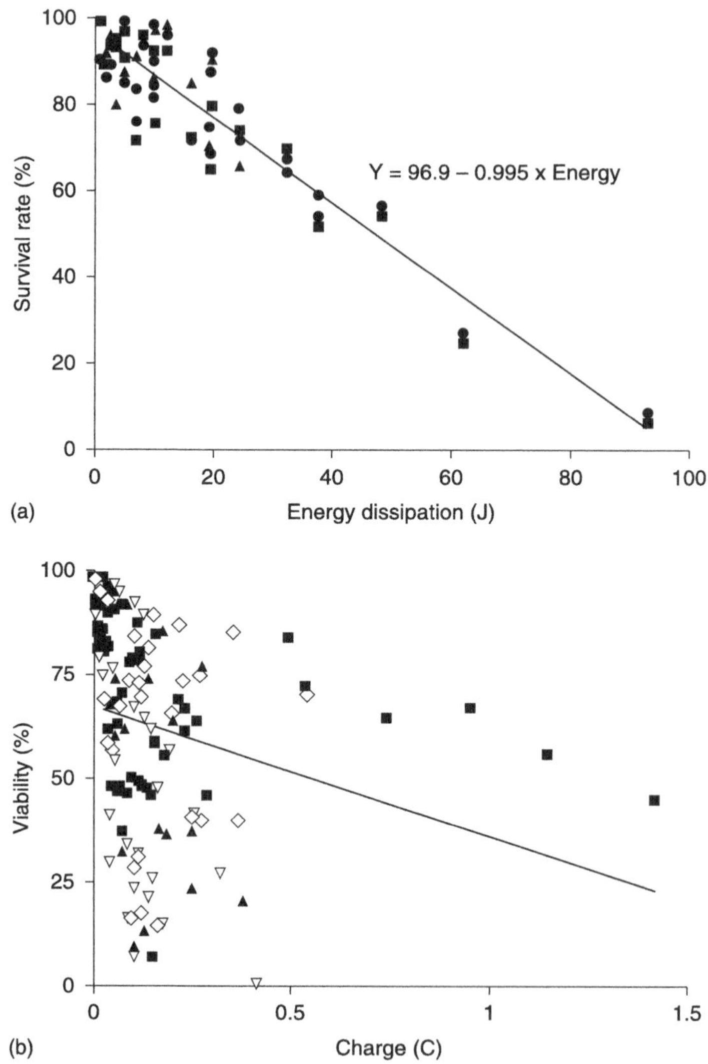

Figure 7.11 Negative linear relationship between viability vs. electroporation energy applied (a) [54] and charge delivered (b) [55]

7.4 Conclusions

Electrochemotherapy is an extremely effective, physical technique that enables cytotoxic drugs to have direct access to the cytosol [56]. This budding mode of therapy uses electrical pulses and chemo drugs to trigger cell death in cancerous cells. The efficacy of this technique depends on a variety of factors, including electric field intensity, pulse duration and the number of pulses. In this study, HL60, KG1 and ME180 cancerous cell lines were subjected to electroporation in the presence of 1 μM Bleomycin at 1200 V/cm, 100 μs; 450 V/cm, 25 ms; and 200 V/cm, 40 ms.

HL60 cells were also studied at 100 μM curcumin concentration for their efficacy in preventing proliferation of the HL60 leukemia cells. The results indicate the promise of using electro-curcumin therapy for the proliferation control of HL60 leukemia cell lines. Electrical pulse intensities, as low as 200 V/cm, but long duration (40 ms) and 450 V/cm, 25 ms offer lower viability, indicating the efficacy of electrochemotherapy using Bleomycin chemodrug.

The dose curve helps to optimize the dose levels for a given viability (and hence cell death) of the tumors to be treated. The energy applied and the charge accumulated with each pulse is also calculated. The energy calculations indicate that the higher the energy value, the lower is the viability or with increase in energy delivered, there is increase in cell death. It is possible to use a single, simple, key parameter, such as energy to describe the efficacy of optimization of electrical parameters.

It is interesting to observe that when using Bleomycin, the viabilities of the three cell lines at various pulses parameters are almost similar. This could indicate that the properties and composition of the cell membrane of these three types of cells may be similar. It may be noteworthy that KG1 and HL60 are cancers of myeloid origin and hence arise from the same precursor. This means that the overall membrane composition of these blood-forming cells would be similar, which can account for the similarity seen in the electroporation viability values. With similar viability also observed in the cervical cancer cell line, we may conclude that the membrane properties of these three cancer cells, in general, may be similar with particular respect to amenability to these parameters of electroporation at these doses.

Considering that 20 million people were living with cancer in 2010 [10] and the trend seems to rise every year, and that the currently administered doses of chemo drugs are costly with severe side effects, not just from chemo drugs but also from other drugs given for coping with the chemo and surgery, there is a critical need for effective, safe, physical and affordable alternative treatments. Since this method allows specific entry of poorly permanent chemotherapeutic drugs into cells at specific tissue locations without affecting neighboring regions, electrochemotherapy has the potential to be a viable tool for the treatment of patients suffering from chemo-refractive tumors.

7.5 Acknowledgments

The authors are very grateful to M Sriram Kumar, V Madan Kumar, S Vignesh and P Sadasivam of BS Abdur Rahman University (Crescent Engineering College), Chennai, India for conducting the experiments and to Frontier Lifeline scientists for their help with the technical details and acquiring cell lines during the course of experiments.

7.6 References

[1] Facts 2012 (2012), Leukemia & Lymphoma Society – fighting blood cancers.

[2] *www.lls.org*. 2012.

[3] Leukemia Statistics (2012). Available at: *http://leukemia.emedtv.com/leukemia/leukemia-statistics.html*

[4] Leukemia stages and survival rates (2012). Available at: *http://www.thesurvivorsclub.org/health/surviving-cancer/leukemia-stages-and-survival-rates*, June.

[5] Basic information about gynecologic cancers. Available at: *www.Cdc.gov/cancer/gynecologic/basic_info/index.htm*

[6] Schiffman M, Castle P J, Jeronimo J J, Rodriguez, A and Wacholder C S (2007), 'Human papillomavirus and cervical cancer', *Lancet*, **370(9590)**: 890–907.

[7] Massimi P and Banks L (1997), 'Repression of p53 transcriptional activity by the HPV E7 proteins', *Virology*, **227(1)**: 255–9.

[8] Globocan (2008), Cervical cancer incidence and mortality worldwide in 2008, Summary.

[9] Ferlay J, Shin H R, Bray F, Forman D, Mathers C, and Parkin D M (2010), 'Estimates of worldwide burden of cancer in 2008: Globocan 2008', *Int J Cancer*, **127(12)**: 2893–917.

[10] *http://www.who.int/whosis/whostat/2010/en/index.html*

[11] Inaba M, Kobayashi H, Sakurai V and Johnson R K (1979), 'Active efflux of Daunorubicin and Adriamycin in sensitive and resistant sub-lines of P388 leukemia', *Cancer Res*, **39**: 2200–3.

[12] Skovsgaard T M (1978), 'Mechanisms of resistance to Daunorubicin in Ehrlich ascites tumor cells', *Cancer Res*, **38**: 1785–91.

[13] McGrath T and Center M S (1988), 'Mechanisms of multidrug resistance in HL60 cells: evidence that a surface membrane protein distinct from p-glycoprotein contributes to reduced cellular accumulation of drug', *Cancer Res*, **48**: 3959–63.

[14] Levi M, et al. (1997), 'K562 and HL60 cells membrane ageing during cell growth is associated with changes in cholesterol concentration', *Mech Age Devel*, **97(2)**: 109–19.
[15] Pachernick J, et al. (2002), 'Conditions for gene transfection into the HL60 human leukemia cell line by electroporation', *Folia Biologica (Praha)*, **48**: 154–6.
[16] Cross S E, Jin Y S, Rao J and Gimzewski J K (2007), 'Nanomechanical analysis of cells from cancer patients', *Nat Nanotechnol*, **2**: 780–3.
[17] Chen J, et al. (2011), 'Electrodeformation for single cell mechanical characterization', *J Micromech Microeng*, **21**: 054012 (11 pager).
[18] Meng X, Riordan N H, Riordan H D, Mikirova J, Jackson J, et al. (2004), 'Cell membrane fatty acid composition differs between normal and malignant cell lines', *PRHSJ*, **2(June)**: 103–6.
[19] *news.stanford.edu/pr/94/940201Arc4464.html*
[20] Vernier P T, Sun Y H, Marcu L, Craft C M and Gundersen M A (2004), 'Nanosecond pulsed electric fields perturb membrane phospholipids in T lymphoblasts', *FEBS Lett*, **572**: 103–8.
[21] Buescher E S and Schoenbach K H (2003), 'Effects of submicrosecond, high intensity pulsed electric fields on living cells – intracellular electromanipulation', *IEEE Transactions on Dielectrics and Electrical Insulation*, **10**: 788–94.
[22] White J A, Blackmore P F, Schoenbach K H and Beebe S J (2004), 'Stimulation of capacitative calcium entry in HL-60 cells by nanosecond pulsed electric fields', *J Biol Chem*, **279**: 22964–72.
[23] Beebe S J, Fox P M, Rec L J, Somers K, Stark R H and Schoenbach K H (2002), 'Nanosecond pulsed electric field (nsPEF) effects on cells and tissues: apoptosis induction and tumor growth inhibition', *IEEE Transactions on Plasma Science*, **30**: 286–92.
[24] Beebe S J, Fox P M, Rec L J, Willis E L and Schoenbach L H (2003), 'Nanosecond, high-intensity pulsed electric fields induce apoptosis in human cells', *FASEB J*, **17**: 1493–5.
[25] Chernomordik L V, Sukharev S I, Popov S V, Pastushenko V F, Sokirko A V, et al. (1987), 'The electrical breakdown of cell and lipid membranes: the similarity of phenomenologies', *Biochimica et Biophysica Acta*, **902**: 360–73.
[26] Kirson E D, Gurvich Z, Schneiderman R, Dekel E, Itzhaki A, et al. (2004), 'Disruption of cancer cell replication by alternating electric fields', *Cancer Res*, **64**: 3288–95.
[27] Sersa G, Cemazar M and Rudolf Z (2003), 'Electrochemotherapy: advantages and drawbacks in treatment of cancer patients', *Cancer Thera*, **1**: 133–42.
[28] Gothelf A, Mir L M and Gehl J (2003), 'Electrochemotherapy: results of cancer treatment using enhanced delivery of Bleomycin by electroporation', *Cancer Treat Rev*, **29(5)**: 371–87.
[29] Heller R, Gilbert R and Jaroszeski M J (2000), 'Clinical trials for solid tumors using electrochemotherapy', in: *Electrochemotherapy, Electrogenetherapy, and Transdermal Delivery*, Jaroszeski, M, Heller R and Gilbert R (eds), New Jersey: Humana Press, 137–56.

[30] Dev S B, Rabussay D P, Widera G and Hofmann G A (2000), 'Medical applications of electroporation', *IEEE Transactions on Plasma Science*, **28(1)**: 206–23.

[31] Campana L G, Valpione S, Mocellin S, Sundararajan R, Granziera E, et al. (2012), 'Superficial disease control in melanoma patients treated with electrochemotherapy in a palliative setting', *Brit J Surg*, **99**: 821–30.

[32] Campana L G, Mocellin S, Basso M, Puccetti O, De Salvo G L, et al. (2009), 'Bleomycin-based electrochemotherapy: clinical outcome from a single Institution's experience with 52 patients', *Ann Surg Oncol*, **16**: 191–9.

[33] Larkin J O, et al. (2007), 'Electrochemotherapy: aspects of practical development and early clinical experience', *Ann Surg*, **245(3)**: 465–79.

[34] *http://omim.org/entry/102578*

[35] Salomoni P and Pandolfi P P (2002), 'The role of PML in tumor suppression', *Cell*, **108(2)**: 165–70.

[36] *http://bioinformatics.istge.it/hypercldb/cl3022.html*

[37] *http://bioinformatics.istge.it/hypercldb/cl1694.html*

[38] *http://www.cancer.org/Cancer/Leukemia-AcuteMyeloidAML/DetailedGuide/leukemia-acute-myeloid-myelogenous-what-is-aml*

[39] *http://emedicine.medscape.com/article/197802-overview#a0101*

[40] *http://www.atcc.org/ATCCAdvancedCatalogSearch/ProductDetails/tabid/452/Default.aspx?ATCCNum=HTB- 33&Template=cellBiology*

[41] Branner K W and Sonntag R W (1975), 'Bleomycin in testicular teratomas', *J Cancer Res Clin Oncol*, **84(3)**: 291–7.

[42] Cancer Drug Manual (the Manual) (1994), British Columbia Cancer Agency (BCCA) & Compendium of Pharmaceuticals and Specialties, 2006. Blenoxane®. Canadian Pharmacists Association.

[43] Joe B and Lokesh B R (1994), 'Role of capsaicin, curcumin and dietary n-3 fatty acids in lowering the generation of reactive oxygen species in rat peritoneal macrophages', *Biochim Biophys Acta*, **1224**: 255–63.

[44] van Meeteren M E, Hendriks J J, Dijkstra C D, van Tol E A (2004), 'Dietary compounds prevent oxidative damage and nitric oxide production by cells involved in demyelinating disease', *Biochem Pharmacol*, **67**: 967–75.

[45] Soliman K F and Mazzio E A (1998), '*In vitro* attenuation of nitric oxide production in C6 astrocyte cell culture by various dietary compounds', *Proc Soc Exp Biol Med*, **218**: 390–7.

[46] Anand P, Kunnumakkara A B, Newman R A and Aggarwal B (2007), 'Bioavailability of curcumin: problems and promises', *Mol Pharm*, **4(6)**: 807–18.

[47] Xiao F, et al. (2009), 'Electrically-enhanced chemodrug delivery to human breast cancer cells', *Electrostatic Society of America Annual Meeting*, Boston, June.

[48] Cutrera J, Torrero M, Shiomitsu K, Mauldin N and Li S (2008), 'Intratumoral Bleomycin and IL-12 electrochemotherapy for treating head and neck cancer tumors in dogs', *Methods in Mol Biol*, **423**: 319–25.

[49] Neumann E, Sowers A E and Jordan C A (eds) (1989), *Electroporation and Electrofusion in Cell Biology*. New York: Plenum.

[50] Zimmermann U and Neil G A (eds) (1996), *Electromanipulation of Cells*. Boca Raton, FL: CRC Press.

[51] Lynch P T and Davey M R (eds) (1996), *Electrical Manipulation of Cells.* New York: Chapman & Hall.
[52] Okino M and Mohri H (1987), 'Effects of a high-voltage electrical impulse and an anticancer drug on *in vivo* growing tumors', *Jpn J Cancer Res*, 78: 1319–21.
[53] *http://www.protocolonline.org./prot/Cell_Biology/Cell_Culture/Generalprocedures/CellCountingandViabilityAssay/index.html*
[54] Chen G Y, Conner A J, Wang J, Fautrier A G and Field R J (1998), 'Energy dissipation as a key factor for electroporation of protoblasts', *Mol Biotech*, 10: 209–16.
[55] Cantella P J, Karr J F, Petros J A and Prausnitz M R (2001), 'Quantitative study of electroporation-mediated molecular uptake and cell viability', *Biophysical J*, 80: 755–64.
[56] Rols M-P, Golzio M and Delteil C (2000), '*In vitro* delivery of drugs and other molecules to cells', in: *Electrochemotherapy, Electrogenetherapy, and Transdermal Drug Delivery*, Jaroszeski M J, Heller R, and Gilbert R (eds), New Jersey: Humana Press.

Low voltage nanosecond electroporation for breast cancer treatment: an *in vitro* study

Jason Harper, Funian Xiao, Ganesh Venkatraman, Ignacio G. Camarillo and Raji Sundararajan

DOI: 10.1533/9781908818294.185

Abstract: Increasing breast cancer mortality worldwide indicates the critical need for alternate therapies. New technologies are needed for improving treatment outcomes and reducing breast cancer related deaths. To treat and enhance the efficacy of drug delivery, electroporation (EP) was used to improve the uptake of hormone (endocrine) drugs. Typically, 100 microsecond (μs) pulses are used for electro-chemotherapy. Use of nanosecond (ns) pulses, at very high intensities, such as 10–300 kV/cm is gaining momentum. Considering that it is always safe to use low voltage for patient treatments, we hypothesized that the use of low intensity, ns pulses for longer durations may result in efficient transfer of external drug moieties into cytosol. Towards that end, we have designed and built a MOSFET-based, ns pulse generator and studied the efficacy of various low electric field intensities on MCF-7 cells for efficient uptake of the FDA approved, anti-hormone drug Tamoxifen. Results indicate that low voltage, ns pulses are effective in killing breast cancer cells, demonstrating that low voltage, ns electro-endocrine-therapy has a promising potential for the treatment of breast cancer.

Key words: nanosecond pulses, electroporation, breast cancer, Tamoxifen, proliferation control.

8.1 Introduction

About 40000 women die each year of breast cancer in the US [1] and the trend seems to rise every year all over the world [2]. Currently, chemotherapy is the mainstay for treatment of breast cancers, but it causes severe side effects. Besides, treatment regimens are also expensive. More importantly, drug-induced resistance is an important concern to be addressed. Thus, there is a critical need for effective, safe, physical, affordable alternative treatments. Electrical voltage pulses could be used for manipulation of cancer cells to augment treatments. This technique is exceptional where conventional techniques are not applicable.

The attractiveness of this technique is that it is applicable to all types of cancer histologies. Since charge distribution contributes to the cytotoxicity, it is possible to reap enormous benefits from applications of electrical pulses for targeted macromolecular delivery, where they redistribute the charges and the synergy of the drug and the pulses result in an improved outcome compared to the drug alone. This electrical phenomenon results in dosage reduction and the associated side effects and cost. This physical technique, known as electroporation or electropermeabilization (EP), utilizes precisely controlled electric fields of short duration and high intensities to open up transient pores (aqueous pathways) through semi-permeable membranes and tissues, allowing targeted delivery of therapeutic materials including drugs, antibodies and genes (DNA) [3–5]. EP has been shown to offer a 100–1000-fold improved therapeutic benefit compared to using a drug alone, and is gaining acceptance as a viable technique to enhance the efficacy of drug delivery for cancer treatment, gene transfer and similar applications in biology, biotechnology and medicine [6–8].

Electroporation-mediated chemotherapy – Electrochemotherapy (ECT), is a viable alternative to conventional cancer treatments, as evidenced by successful Phase I and II clinical trials for various cancers, such as lymphomas, squamous cell carcinomas, testicular carcinomas and malignant pleural effusions [3, 7, 8]. In breast carcinoma infiltrating the chest wall, when surgery, radiotherapy and/or chemotherapy failed, ECT was found to be effective for the treatment [9, 10].

Clinical trials for EP typically used 100 μs pulses and electric strengths of 1200 or 1300 V/cm [3–10]. Most studies have used 6 or 8 pulses with an interval of 1 second (1 Hz) [3, 4, 7]. Recently, ultra-short pulses of nanosecond (ns) durations, such as 10–300 ns and 10–300 kV/cm, have also been attempted for electroporation [11–14]. For example, three 60 ns, 60 kV/cm pulses at 1 second intervals [14] were applied in a nanopulse electroporation study using jurkat cells. In addition, 300 ns,

26 kV/cm and 10 ns, 150 kV/cm pulses were also studied. Vernier et al. studied jurkat cells with 10 ns and 20–40 kV/cm pulses [15,16].

In another study, fifty 7 ns, 25 kV/cm pulses were applied. Ten 30 ns, 25 kV/cm pulses were used in another study [16]. Nanosecond electroporation was also performed by Beebe et al. [17]. In this case, cells were exposed to a single, 10 ns pulse at 150 kV/cm. Use of 20 kV/cm, 300 ns pulses without drug (electrical ablation, similar to hyperthermia) is reported by Nuccitelli et al. for treating melanomas [11]. Cutaneous papilloma and squamous cell carcinoma were also treated using nanosecond pulses [18]. However, due to their high cost and for safety reasons, in addition to the complexity of the design at high voltages [19], it is desirable to use low voltage, ns pulses for the delivery of therapeutic molecules. A 20 ns, 1000 V, nanopulser was built and the efficacy of these pulses at various low voltages were studied for the first time on human breast cancer cells. The objective was to explore the application of ns pulses for proliferation control of breast cancer cells using the most commonly used hormone drug, Tamoxifen [20].

8.2 Materials and methods

8.2.1 Design and development of nanopulser

A 0–1000 V, compact (4' × 4' × 2'), MOSFET-based nanosecond pulser (ns pulser), which could operate using a 9.6 VDC rechargeable battery, with variable pulse durations of 20 ns to 20 μs and with variable pulse repetition rate (PRF) of 1 kHz–6 kHz, was developed in our lab. It has three blocks:

1. high voltage source block, which converts the battery's 9.6 VDC to a variable high voltage output and stores this charge for use when the pulsing begins;
2. the switching block, which is a MOSFET based low-side switch, which applies short nanosecond pulses to the load; and
3. the pulse generation block, which generates a short nanosecond pulse and supplies enough current to adequately turn the MOSFET on fast enough to apply a nanosecond pulse to the load.

Voltages were applied to produce 200 V/cm for 10 minutes, 500 V/cm for 5 minutes and 1200 V/cm for 5 minutes; Figure 8.1 shows the circuit schematic. Figure 8.2 shows a view of the nanopulser designed and built and Figure 8.3 shows a 20 ns waveform obtained at 1000 V.

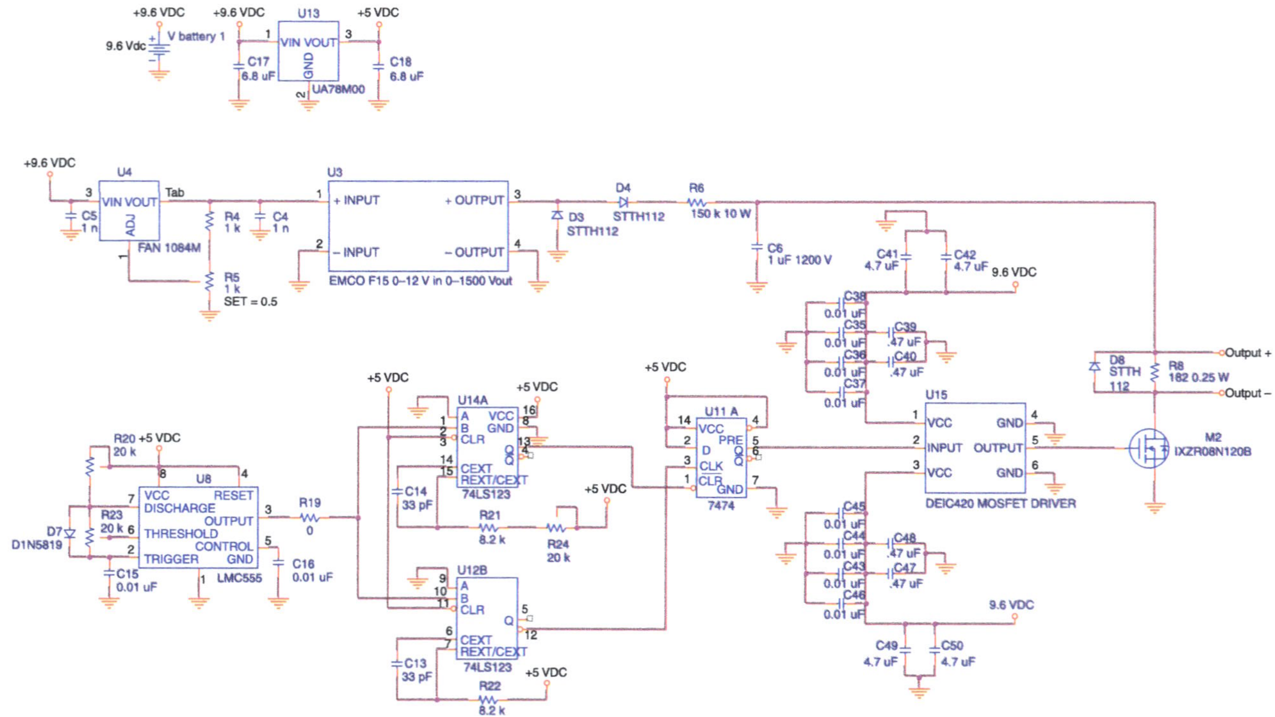

Figure 8.1 Overall miniature high voltage MOSFET-based nanosecond pulse generator circuit schematic

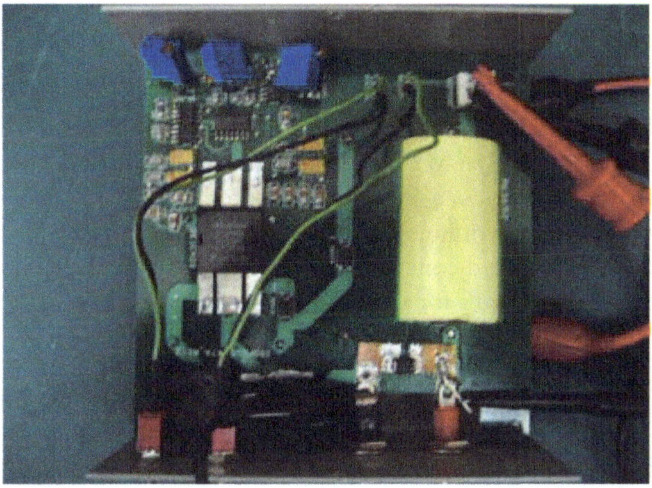

Figure 8.2 Miniature (4" × 4" × 2") nanosecond pulse generator (nanopulser) designed and developed in our lab

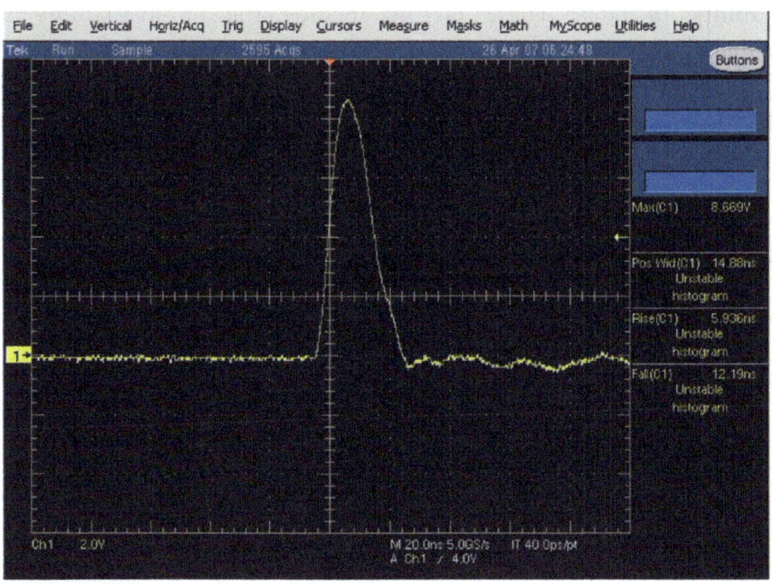

Figure 8.3 1000 V, 20 ns load pulse

8.2.2 MCF-7 breast cancer cells

To test the newly developed ns pulser, we performed *in vitro* studies on MCF-7 breast cancer cells. For electroporation, MCF-7 cells were grown to confluence in RPMI1640 media, supplemented with 10% FBS and 1% Pencillin/Streptomycin, in a 5% CO_2 incubator, at 37 °C. Prior to EP, cells were detached from confluent culture plates, resuspended in a cuvette and treated with desired concentrations of the drug. The cell density was maintained at 1×10^6/mL for the EP process.

8.2.3 Tamoxifen – endocrine (hormone) drug

Tamoxifen is one of the most effective endocrine agents for treatment of breast cancer [19]. Its structure is shown in Figure 8.4 [21]. It interacts between biologically active estrogen with the estrogen receptor in estrogen receptor positive (ER^+) cells. The interaction is due to the majority of pharmacological effects occurring both *in vivo* and *in vitro*. In addition, Tamoxifen has several pharmacological factors to influence its ability *in vivo*. These factors include the bio-availability and pharmacokinetics of Tamoxifen, the metabolism of Tamoxifen, and the ability of these compounds to interact with estrogen receptors and form inappropriate complexes, which are no longer able to regulate the estrogen responsive genes involved in the growth pathway.

8.2.4 Electroporation

The cells were resuspended in growth medium along with Tamoxifen in a cuvette. In the present study, we investigated the efficacies of nanosecond

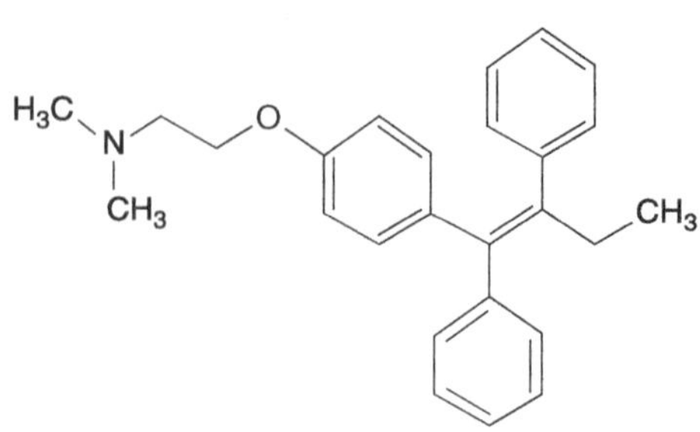

Figure 8.4 Tamoxifen anticancer drug structure

pulses. Voltages were applied as 20 ns pulses at intensities of 200 V/cm for 10 minutes; and 500 V/cm and 1200 V/cm for 5 minutes, each.

8.2.5 Measurement of cell viability

After electroporation, cells suspended in media were stained with a 1:1 concentration of Trypan Blue. The dead cells were stained blue, so they could be distinguished from the clear, live cells and were counted using the Cellometer (Nexcelom Bioscience LLC., MA).

8.3 Results and analysis

Figure 8.5 shows the dose curve using 200 V/cm, 20 ns pulses for 10 minutes on the MCF-7 breast carcinoma cells. It follows the inverse, fractional power law pattern, typical of biodielectrics, such as MCF-7 cells, indicating the reduced viability with increase in drug dose.

Figure 8.6 shows the viability for various doses for pulse parameters (n = 3). There is a drastic reduction in the viability at 1 and 5 μM concentrations of Tamoxifen compared to the control and drug alone. The viability of the control is around 89% in all the 3 cases and it is 37.5%, 26.6% and 33.1% at 1 μM for the 200 V/cm, 1200 V/cm and 500 V/cm pulses respectively. The reductions were 36%, 57.5% and

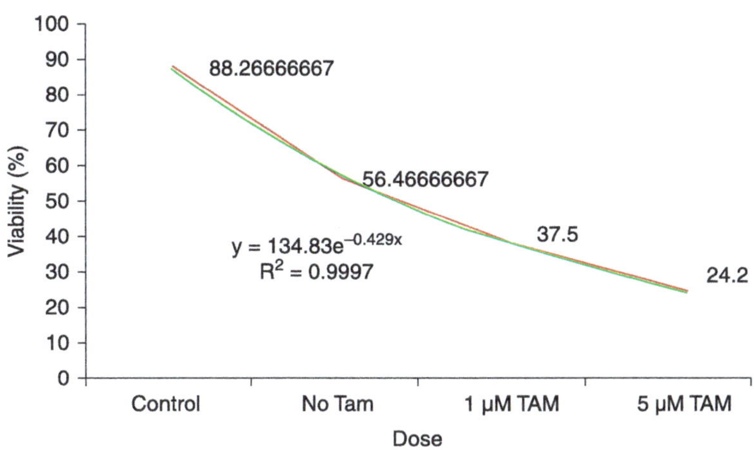

Figure 8.5 Dose curve using 200 V/cm, 20 ns pulses applied for 10 minutes on MCF-7 human breast cancer cells

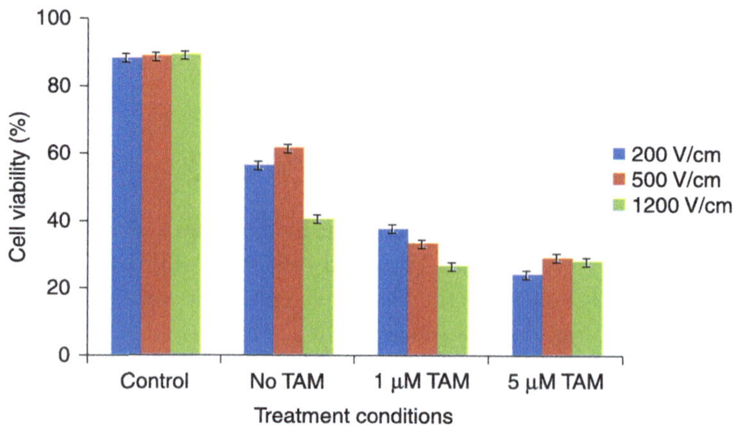

Figure 8.6 Dose curves for the three different voltage parameters studied

72.5% respectively for no drug, and 1 and 5 μM drug doses compared to the control when 200 V/cm pulses were applied for 10 minutes. These indicate the excellent efficacy of these pulses at that dose. At 500 V/cm and 1200 V/cm, there is in general reduction in the viability, when applied for 5 minutes each, compared to 200 V/cm applied for 10 minutes.

The 200 V/cm, 20 ns pulses for 10 minutes have the highest viability compared to the other pulses for 5 minutes each. This could be due to the lower intensity of electrical field used. A similar look at 5 μM indicates that the percentage viabilities are almost comparable at all the voltages. The viabilities were 24.2%, 27.93% and 29.1% at the dose of 5 μM for the voltages. This could be due to the dominance of the higher dose at 5 μM. The viabilities at both 1 μM and 5 μM appear to be close to each other, except for 200 V/cm, indicating that a low dose of 1 μM will suffice to effectively kill cells selectively at these voltages. The pulse parameters, the duration for a given voltage or the voltage for a given duration of pulses could be tailored to obtain the desired viability for a given cell type. The energy levels of these pulses are extremely low and hence no heat or any other adverse effects are expected using these ns pulses.

8.4 Discussion and conclusions

Once a disease of the industrialized Western world, today breast cancer occurs worldwide [2]. It is the second most common cancer, with more than one million women suffering from it. Despite improvements in treatment, the recurrence rate is high [22]. As it is regarded as the number one killer of women aged between 15 and 54, additional therapies are critically needed [1]. In the present study, we have given a proof of concept for using low voltage, ns pulses for effective intracellular delivery of Tamoxifen into human breast cancer cells. The effect of electrical pulse induced drug delivery was very potent as compared to the drug alone. This can be inferred from the cell viability data measurements shown. The improved efficacy illustrates that low voltage, high frequency, nanosecond pulses can transiently and reversibly disrupt cell membranes causing calcium release, apoptosis and other events normally associated with cell signaling [13]. When applied in physiological media, they can produce significant voltages across intracellular membranes, such as the nucleus, mitochondria, storage vacuoles, golgi compartments and the endoplasmic reticulum, without irreversibly porating the plasma membrane.

The primary effect of nanopulses is the perturbation of internal organelles with little or no effect to the plasma membrane [17, 19]. Thus, ns pulses can physically manipulate intracellular structures in a precise manner with a wide range of amplitudes, durations and patterns, extending the reach of the electric field past the cytoplasmic membrane to the internal membranes of the cell. The extent of permeation and the resultant physiological and morphological changes are different from those seen with conventional electroporation [15, 18]. This is due to the frequency effect in bio-dielectrics, such as cells and cell systems [23–26].

This is in line with the 1985 prediction of Schwan [26] that electric fields of appropriate magnitudes and high frequencies could be used for coupling electrical energy into the interior of cells, which was realized using advanced nanotechnology for design and development of pulsers [27]. This is important as chemicals (drugs) alone could not cure all the tumors, as we have been seeing in real life for many cancer patients. This is because cells have intrinsic electrical properties and a combination of electrical and chemical dosages will elicit the synergy of these two powerful techniques, giving the patients a greater favorable outcome. In addition, considering that the development of cancer is due to certain chemicals, viruses and bacteria [28], and the modification or redistribution of charges [29] of the cell surface resulting in alterations in the cell membrane potential and organelle membrane electrical potentials, the functions and compositions of membranes, intercellular mineral content (of sodium, calcium, potassium, etc.), energy production and genetic expression [28], and electrical pulse-mediated therapy seems to be appropriate and needed.

This also means that therapeutic methods such as nanosecond electroporation, which manipulate the contents of intracellular organelles, ions and minerals, can result in alterations in normal functions, and reduce their proliferations, as evidenced by our results due to the charging of the membranes by the applied nanosecond external electric field. This enables the enhanced efficacies of therapies that exploit these different electrical properties of tumor cells, such as electrochemotherapy and electrogenetherapy [5, 9]. Toxic compounds that disrupt the electrical potential of cell membranes will deactivate the transport chain and disturb oxygen-dependent energy production. Cells will then revert to fermentation, which is a less efficient primeval form of energy production. When these cytotoxic drugs are accompanied by electrical pulses, they can enhance the transport and disable the proliferation of the cells. Our results shown here demonstrate that nanosecond pulses of even low magnitudes can be used for injecting membrane impermeable molecules into mammalian cells.

We chose Tamoxifen to study because endocrine (hormonal) therapies are generally milder to take with fewer side effects than chemotherapy [20]. Tamoxifen is the world's largest selling breast cancer drug used for treatment of primary and metastatic diseases in pre- and post-menopausal women. In 1998, it was also approved by the FDA for prevention, to reduce the incidence of breast cancer in high-risk women. For the past 20 plus years, Tamoxifen has been the drug of choice in hormonal therapy for breast cancer in women. It is a 20 mg/day oral pill that works as a Selective Estrogen Receptor Modulator (SERM) blocking estrogen, the hormone that can cause breast cancer. In clinical studies or experiments with human cells in culture, it has been demonstrated that Tamoxifen exhibits anti-estrogenic (inhibition) and in some cases estrogenic (stimulation) activity, depending on tissue type. In breast cancer cells, Tamoxifen suppresses estrogen stimulated gene expression that leads to cytostatic (cessation of growth) or apoptotic (cell death) effects, with higher Tamoxifen concentrations triggering apoptosis [30].

For Tamoxifen to be effective, tumors should have ER^+ cells. The majority of breast cancer patients are diagnosed with ER^+ tumors, and many of these women benefit from Tamoxifen. However, the response rate of Tamoxifen is only 21–41% [32]. Thus more than 60% of patients do not respond to Tamoxifen. Patients receiving Tamoxifen experience significant side effects that include hot flushes, increased risk of thromboembolic events and increased risk of endometrial cancer [21]. These side effects result from estrogenic affects of orally delivered Tamoxifen that travels via the blood stream and acts on estrogen receptor containing tissues other than the breast, such as the uterus [20]. Hence, an alternate route is highly desired.

Our study assessed and demonstrated, for the first time, the prospects of using low electrical voltage, nanosecond pulses of the orders of 200 V/cm, 500 V/cm and 1200 V/cm, for 10 and 5 minutes to enhance the effect of minute doses of chemotherapeutics. Our results indicate that the MCF-7 cell line is receptive to low voltage, ns electroporation. Considering that the previous studies utilized 30–300 kV/cm, the significance of our study could be appreciated.

A comparison of ns pulse results after milliseconds and micro-second pulse results of our previous study [33, 34] indicate that there is good correlation in impeding the proliferation of these cells. In [33], Camarillo, et al. studied the effects of combining Tamoxifen with EP cells at micro- and milli-seconds, MCF-7 were exposed to Tamoxifen at 1 and 5 μM and at different conditions, such as 100 V/cm, 1 ms; 250 V/cm, 200 μs, 400 V/cm, 2 5 ms; 2 pulses, 500 V/cm, 260 μs; and 1200 V/cm, 100 μs. A significant reduction in cell growth occurred with each Tamoxifen concentration at voltages corresponding to 400, 100 and 1200 V/cm treatment. In [34], Xiao studied the effect of ms pulses of various durations, such as 10 ms, 20 ms, 30 ms and 40 ms, to optimize the ms pulse duration with a 5 μM Tamoxifen dose. Figure 8.7 illustrates the results indicating the successful proliferation control at a very low electric field intensity of 200 V/cm.

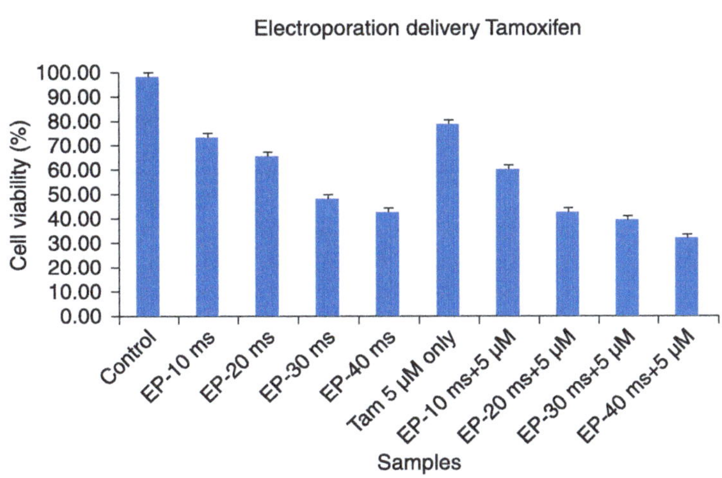

Figure 8.7 Electroporation effect using 200 V/cm, ms pulse durations at 5 μM dosage of Tamoxifen [34]

Our results demonstrate that it is possible to impede the proliferation of breast cancer cells by applying low voltage, nanosecond pulses. It is also demonstrated that the combination therapy, electroporating cells using low voltage, ns pulses along with the FDA-approved cancer drugs, such as Tamoxifen administration, has great potential as a therapy for treating cancer tumors. Also, using local Tamoxifen via EP, in addition to suppressing cell growth, may actively stimulate different mechanisms that lead to cell death. Thus, our studies suggest clinical use of low voltage ns EP combined with Tamoxifen could lead to a more effective anti-tumor response, with low doses and less side effects.

8.5 Acknowledgment

The authors are extremely grateful to Mr. Drew Campbell for his contributions towards the building of the nanopulser.

8.6 References

1. *http://womenshealth.about.com/cs/breastcancer/a/breastcancerfacts.htm*, December 2010.
2. Kingsbury K (2007), 'The changing face of breast cancer', Special Report, *Time Magazine*, October 15.
3. Heller R, Jaroszeski M, Grass L, Messina J, Rapaport D, et al. (1996), 'Phase I/II trial for the treatment of cutaneous and subcutaneous tumors using electrochemotherapy', *Cancer*, 77: 964–71.
4. Gehl J and Geertse P F (2000), 'Efficient palliation of haemorrhaging malignant melanoma skin metastases by electrochemotherapy', *Melanoma Res*, **10**: 1–5.
5. Mir L M, et al. (1999), 'High efficiency gene transfer into skeletal muscle mediated by electric pulses', *Proc Natl Acad Sci USA* **96**: 4262–7.
6. Dev S B, Rabussay D P, Widera G and Hofmann G A (2000), 'Medical applications of electroporation', *IEEE Trans Plasma Sci*, **28**: 206–23.
7. Heller R, Gilbert R and Jaroszeski M J (2000), 'Clinical trials for solid tumors using electrochemotherapy', in: *Electrochemotherapy, Electrogenetherapy, and Transdermal Delivery*, Jaroszeski M, Heller R and Gilbert R (eds), New Jersey: Humana Press, 137–56.
8. Gothelf A, Mir L and Gehl J (2003), 'Electrochemotherapy: results of cancer treatment using enhanced delivery of bleomycin by electroporation', *Cancer Treat Rev*, **29**: 371–87.
9. Campana L, *et al.* (2009), 'Bleomycin-based electrochemotherapy: clinical outcome from a single Institution's experience with 52 patients', *Ann Surg* Oncol, **16**: 191–9.

10. Larkin J O, et al. (2007), 'Electrochemotherapy: aspects of practical development and early clinical experience', *Ann Surg*, **245**(**3**): 465–79.
11. Nuccitelli R, et al. (2010), 'Optimized nanosecond pulsed electric field theory can cause murine malignant melanomas to self-destruct with a single treatment', *Intl J of Cancer*, **127**(**7**): 1727–36.
12. Nuccitelli R. et al. (2006), 'Pulsed electric fields cause melanomas to self-destruct', *Biochem Biophys Res Commun*, **343**: 351–60.
13. Schoenbach K H, Beebe S J and Buescher E S (2001), 'Intracellular effect of ultrashort electrical pulses', *Bioelectromagnetics*, **22**: 440–8.
14. Beebee S J, Fox P M, Rec L J, Willis E L and Schoenbach K H (2003), 'Nanosecond, high-intensity pulsed electric fields induce apoptosis in human cells', *FASEB J*, **17**: 1493–5.
15. Vernier P T, Li A, Marcu L, Craft C M and Gundersen M A (2003), 'Ultrashort pulsed electric fields induce membrane phospholipids translocation and caspase activation: differential sensitivities of Jurkat T lymphoblasts and rat glioma C6 cells', *IEEE Trans Dielectrics and Electrical Insulation*, **10**: 795–809.
16. Vernier P T, Sun Y, Marcu L, Salemi S, Craft C and Gundersen M A (2003), 'Calcium bursts induced by nanosecond electric pulses', *Biochem Biophys Res Comm*, **310**: 286–95.
17. Beebe S J, Blackmore P F, White J, Joshi R P and Schoenbach K H (2004), 'Nanosecond pulsed electric fields modulate cell function through intracellular signal transduction mechanisms', *Physil Meas*, **25**: 1077–93.
18. Yin D et al. (2012), 'Cutaneous papilloma and squamous cell carcinoma therapy utilizing nanosecond pulsed electric fields (nsPEF)', *PLOS ONE*, 7(8): on-line, 1–11.
19. Schoenbach K H, Joshi R P, Kolb J F, Nianyong C, Stacey M, et al. (2004), 'Ultrashort electrical pulses open a new gateway into biological cells', *Proceedings of the IEEE*, **92**: 1122–37.
20. Love S M and Lindsey K (2005), *Dr Susan Love's Breast Book*. Cambridge MA, Perseus Publishing.
21. *www.drugbank.ca/drugs/DB00675*, December 2010.
22. Gonzalez-Angulo A M, Morales-Vasquez F and Hortobagyi G N (2007), 'Overview of resistance of systemic therapy in patients with breast cancer', *Adv Exp Med Biol*, **608**: 1–22.
23. Schwan H P (1989), 'Dielectrophoresis and rotation of cells', in: *Electroporation and Electrofusion in Cell Biology*, Neumann E, Sowers A E and Jordan C A (eds). New York: Plenum Press, 3–21.
24. Ellappan P and Sundararajan R (2005), 'A simulation study of the electrical model of biological cells', *J Electrostat*, **63**: 297–307.
25. Schwan H P (1989), 'Dielectric properties of tissues and biological materials: a critical review', *Crit Rev Biomed Eng*, **17**: 25–104.
26. Schwan H P (1985), 'Dielectric properties of cells and tissues', in: *Interactions between Electromagnetic Fields and Cells*, Chiabrera A, Nicolini C and Schwan H P (eds). New York: Pergamon Press 75–97.
27. Sundararajan R (2008), 'Nanosecond electroporation: another look', *Mol Biotechnol*, **41**: 69–82.
28. Haltiwanger S (2010), 'The electrical properties of cancer cells'. Available at: *http://www.royalrife.com/haltiwanger1.pdf*, July 2010.

29. Weaver J C (2000), 'Electroporation of cells and tissues', *IEEE Trans Plasma Sci*, **28(1)**: 24–33.
30. Zheng A, Kallio A and Harkonen P (2007), 'Tamoxifen-induced rapid death of MCF-7 breast cancer cells is mediate via extracellularly signal-regulated kinase signaling and can be abrogated by estrogen', *Endocrinology*, **148**: 2764–77.
31. Berger K and Bostwick A (1998), *A Woman's decision – Breast care, Treatment and Reconstruction*, 3rd edition. New York: St Martin's Griffin.
32. Gonzalez-Augulo A M, Morales-Vasquez F and Hortobagyi G N (2007), 'Overview of resistance to systemic therapy in patients with breast cancer', in: *Breast Cancer Chemosensitivity*, Yu D and Hung M-C (eds). New York: Springer Science + Business Media, LLC, Landes Bioscience, 1–22.
33. Camarillo I G, Nichol, M, Zheng M, Sonnenburg D and Sundararajan R (2008), 'Electro-endocrine therapy for aggressive breast tumors', *J Electrostat*, **66**: 99–106.
34. Xiao, F (2009), 'Enhanced anticancer drug delivery using electrical pulses', MS Thesis. ECET Dept., Purdue University, December.

Low and high voltage electroporation of *in vitro* human ovarian adenocarcinoma cells

Raji Sundararajan, Arutselvan Natarajan, Kavitha Sankaranarayanan, Lisa M. Reece and Brent Vernon

DOI: 10.1533/9781908818294.199

Abstract: The pitiful, overall cure rate of only approximately 30% of ovarian cancer for several years now indicates the complexity of the organ and the inadequacy of current standard of care. Often the tumors are also chemo-resistant. Thus, there is a critical need for alternate therapies. Electroporation, a physical technique of applying high intensity electrical voltage pulses of appropriate magnitude for short durations (microseconds) that transiently creates pores or permeabilizes cell membranes, enabling macromolecules to pass through which otherwise are impermeable, is a viable alternative. This technique, which has a wide scope in *in vitro* biotechnology to deliver molecules inside any living cell, is also currently being exploited for therapeutic purposes. Here we report our results on the application of various low and high voltage pulses of different durations and their effect on the viability of SKOV-3, human ovarian cancer cells. This treatment modality, using electrical pulses only paves the way for adapting this technique for alternate clinical treatment of ovarian cancer patients.

Key words: ovarian cancer, adenocarcinoma, SKOV–3 cells, fluorescence microscopy, voltage pulses, live/dead assay, light microscopy, electroporation.

9.1 Introduction

Ovarian cancer is the deadliest of all types of gynecologic malignancies worldwide [1–4]. This is the only cancer that has a similar death rate in both the developed and the developing countries. For example, the death rate is 64% (64 500 deaths for 100 300 new cases) in the developed countries and is 60% (75 700 deaths for 125 200 new cases) in the developing countries [1]. In the US, typically about 20 000 new cases occur each year and about 14 000 die. According to the National Cancer Institute (NCI), in 2009, 21 550 new cases were reported with 14 600 deaths (68%) in the US. This could be due to the complex biology of the organ and its cancer [5]. For example, the ovary is the most complex among the organs of the body and its functions are achieved by numerous cell types. So, any of these cells can undergo malignant transformations, although 90% of all ovarian cancers are due to the surface epithelia cells, caused by the modified peritoneal mesothelium [5]. This could be caused by incessant or repeated ovulation contributing to the initiation of the disease. The frequent manifestation of the drug resistance phenotype is yet to be fully understood to overcome the high death rates. Table 9.1 shows various classifications of ovarian tumors [5].

The prevalent methods of treating ovarian cancer are surgery, chemotherapy and radiation. In most cases the treatment proves futile due to late diagnosis, metastases of the tumor and development of

Table 9.1 Various classifications of ovarian tumors [5]

Tumor	Type of cell origin
Serous	Surface epithelium
Mucinous	Surface epithelium
Endometroid	Surface epithelium
Clear cell	Surface epithelium
Unclassified	Surface epithelium
Gonadal stromal	Granulose cells
	Sertoli cells
	Leydig cells
	Stromal cells
Germ cell	Germ cells

resistance to drugs [5, 6]. The most adopted therapeutic measure for treating ovarian cancers, apart from surgical removal of the tumor, includes chemotherapy using drugs such as cisplatin, paclitaxel and bleomycin. These compounds have been shown to be non-specific in nature, exerting their anticancerous effect by affecting the mechanism of cell division. This in turn also has an effect on the normal cells of the body other than the tumorous cells. In addition, these drugs have less transmembrane permeability, necessitating the usage of very high doses. It is reported that the drug resistance could be due to the following three reasons, in addition to other known and unknown ones:

1. due to blocking of drug entry into the cells or reduced uptake of the drug, and increased export of the drug;
2. inactivation or sequestering of the drug; and
3. if the drug reaches the DNA, the damage inflicted is tolerated or repaired with fidelity and in a timely manner [5].

In addition, due to the drug-resistance and non-responsiveness to a particular drug, the regimen of choice calls for a combination or cocktail of these potent drugs, thereby also increasing their drug resistance, and/or collective side effects. Examples of the various serious side effects of these drugs include nausea, hair loss, vomiting and heart problems, among others [7,8]. All these maladies could be alleviated with the development of physical/electrical techniques to improve the transmembrane permeability of these drugs or by new techniques to drive the molecules inside the cell.

Electroporation or electropermeabilization (EP) is the process of applying high intensity, short duration electrical pulses of appropriate magnitude and duration to generate transient pores in the lipid bilayer of the cell [9–14]. These voltage pulses, when applied to the cell, result in an accumulation of charges across the membrane and the resultant high field strength induces transient permeability [15, 16]. This enables the uptake of molecules by the cell, which would otherwise have not crossed the bilayer. Electroporation-enhanced plasmid vector cells, therapeutic genes and chemotherapeutics delivery have become a promising technology for *in vitro* and *in vivo* applications. For example, one of the recent studies showed that the gene delivery for ovarian cancer therapy utilized electroporation to deliver single chain (scFv)-based chimeric immune receptor (CIR) RNA to human peripheral blood leukocytes (PBLs), which could be used for rapid generation and high number of therapeutic antigen-specific T cells for adoptive immunotherapy [17]. The current gene transfer technology mainly relied on retrovirus and lentivirus vectors, which require a long time to obtain a sufficient number of transduced cells and be stably incorporated into the genome. To overcome these limitations, electroporation can be one of the most attractive non-invasive approaches to deliver the drug concentration at a tumor site to increase antitumor responses *in vivo*. In one study, electroporation was used to enhance the delivery of cisplatin to tumors, which increased the radiosensitizing effect of cisplatin [18].

This treatment strategy, when used with chemo drugs, is called electrochemotherapy (ECT), which utilizes the use of high intensity electrical pulses in synergy with chemo drugs or, in other words, reversible electroporation to permeabilize the membrane to the otherwise membrane-impermeant drugs. This has the potential of decreasing the effective concentration of the drug to be used, thereby bringing down the side effects. In addition, the short, quick bursts of electrical pulses in the order of 1200 V/cm (100 μs duration) trigger apoptotic signals in the cells, which results in programmed cell death or apoptosis. Thus ECT attacks these tumor cells in a two pronged manner:

1. by initiating apoptosis; and
2. by transiently permeabilizing the cell to take up molecules of the drug used for chemotherapy.

ECT proves to be an effective therapeutic intervention for targeting and electrocuting tumorous cells without affecting the surrounding normal tissue, thereby reducing the adverse side effects resulting from drug overload. Successful clinical trials for head and neck carcinoma and other tumors [9] attest to the efficacy of this technique. Occasionally, higher field strengths applied also result in localized burning of the tissue, muscle contractions and twitching due to the electrical stimuli (however, they cease once the pulses applications are completed) [17]. The voltage at which effective uptake of the drug molecules, or ablation of the cells (when no drug is used, known as irreversible electroporation) takes place, is largely dependent on the exact constitution of the cell membrane, which varies from cell to cell based on their bilayer composition. In other words, the EP parameters, such as voltage (electric field strength) and duration (pulse length), depend on the type of cell among other variables. Thus, it is mandatory to precede ECT with an *in-vitro* standardization of the pulses, which could be used to target a specific tumor.

In this study, we aimed to optimize the electrical parameters for SKOV–3, a human ovarian cancer cell line, without any drug. By a series of experiments, the optimal conditions of pulse parameters to be used have been characterized using currently used magnitudes as well as lesser magnitudes (and longer durations). Furthermore, we studied if the resulting effects were comparable to those of short duration, high voltage pulses. It is always safer to use low magnitude pulses than higher magnitude pulses. This also serves to eliminate the use of high voltage pulse generators, which are not only costly but also can cause muscle contractions and localized electrode burns at times. For this purpose, the effects of reduced field strengths were studied using ET = K, a constant [19], where E is the electric field strength (V/cm) and T is the pulse duration (μs) on SKOV-3 cells. This *in vitro* study could be used as a preliminary report for clinicians aiming to apply ECT to treat ovarian cancers, because when using the intact animal, only a limited level of complexity could be achieved in the study of the surface epithelium and hence a controlled cellular environment is a good model [5].

9.2 Materials and methods

9.2.1 Cell line

SKOV-3 (ATCC #-HTB–77, Manassas, VA) adenocarcinoma (human ovarian cancer) cells were cultured in modified McCoy's 5A Media (ATCC) and supplemented with 10% fetal bovine serum (FBS, Gibco, Life Technologies, Carlsbad, CA) and maintained at 37 °C in a 5% CO_2 atmosphere. The medium was replaced periodically.

9.2.2 Electroporation

Cells were washed in phosphate buffered saline (PBS, 1X, Gibco, Life Technologies, Carlsbad, CA), trypsinized and counted using a hemocytometer, then 800 μL aliquots (~20 000–1 000 000 cells) were placed in 0.4 cm (electrode gap) cuvettes for subsequent electroporation. The cell counts used were dependent on the type of assay. A fixed number of pulses of specific voltages and durations were applied at 1 second intervals in serum-free media. The cells were left undisturbed in the cuvettes for 30 minutes after pulsing, and then transferred to separate wells in a 96-well assay plate, T-12.5 cell culture flask, or 7 mL centrifuge tube, depending on the assay to be completed.

The cells were electroporated with a BTX ECM830 square (also known as rectangular) wave electroporator (Genetronics, San Diego, CA). A number of voltages, pulse durations and number of pulses were employed for various analyses. Using ET = K, for 500 V/cm, from 1300 V/cm, and 100 μs, the duration obtained is 260 μs. The interval between pulses for all the cases was 1 second (1 Hz frequency).

9.2.3 Live/dead assay

For this assay, the cells were placed in 7 mL centrifuge tubes, where FBS and 7 mL of modified McCoy's 5A Media were added. The cells were then incubated at 37 °C in a 5% CO_2 atmosphere for 2 hours. The tubes were rocked to reduce clumping and to promote a good oxygen and nutrient environment. The centrifuge tube was filled to capacity in order to minimize shear stress on the cells. Following incubation, aliquots of cells were removed and stained using the exclusion dye, 0.4% Trypan Blue (Sigma-Aldrich, St Louis, MD) and counted using a Fisher Scientific (Pittsburgh, PA) hemocytometer and a 5-grid counting system. Counts for live, dead and total number of cells were made for each electroporation treatment. Two replicates for each treatment were counted and the average taken.

9.2.4 Fluorescent microscopy

The cells were placed in 96-well plates and maintained as described above for 2 days. The cells were then washed with PBS and stained with Propidium Iodide (PI, red fluorescence) at a concentration of 1 mg/mL in PBS (Life Technologies) and fluorescein diacetate (FDA, green fluorescence) at a concentration of 4.2 mg/mL in dimethyl sulfoxide (DMSO, Sigma-Aldrich). Following this, the cells were imaged at 200× using an inverted microscope (Leica DM IRBE, fitted with a Optronics 750D color digital camera, Germany) with fluorescence, phase contrast and modulation contrast capabilities. The cells were visualized using a 480–510 nm excitation filter and a 535–585 nm emission filter for FDA and 605–660 nm emission filter for PI.

9.2.5 Light microscopy

The cells for this assay were placed in a T-12.5 cell culture flask. FBS and 2 mL of modified McCoy's 5A Media were added and the cells were then incubated at 37 °C in a 5% CO_2 atmosphere on a rocker for 2 days. The cells were then photographed at 10× at four points in the flask (Nikon TMS-F inverted microscope (Melville, NY) fitted with Cohu (Poway, CA) high performance CCD camera). Cells were counted and a density per square millimeter was determined. Two flask replicates for each electroporation treatment were made and the average was taken.

9.2.6 MTT assay

The treated cells (electroporated and appropriate controls) for this assay were plated on a 96-well plate and incubated at 37 °C in a 5% CO_2 atmosphere for 2 days. Three replicates were used for the experiment. After 2 days, additional cells (2000–60 000 cells) of the same type were placed in empty wells in the assay plate and used to quantify the number of cells alive in the experimental wells. Twenty μL of CellTiter 96® AQueous One Solution Reagent (Promega, Madison, WI) was added to each of the wells and then incubated at 37 °C in a 5% CO_2 atmosphere for another 90 minutes.

After 90 minutes, the absorbance at 490 nm was read and recorded using a 96-well plate reader. A standard graph was plotted using the absorbance values of the wells containing known quantities of cells. Using the equation from the regression line of plotted data and the recorded absorbance of the observations, the number of cells in the observation wells was determined.

9.2.7 Flow cytometry

To quantify the viability of electroporated SKOV-3 cells, flow cytometry was performed using 7-amino actinomycin D (7-AAD, Vioprobe, PharMingen International, San Diego, CA), which binds to DNA, like propidium iodide [17,19,20]. Peak emission of 7-AAD is approximately 685 nm. The cells were treated per manufacturer's guidelines to conduct the flow cytometry study. They were pulsed at 500 V/cm, 260 μs and 1300 V/cm, 100 μs. Six pulses were applied in each case. Flow cytometry was utilized to gate on the target cells and measure 7-AAD negative versus 7-AAD positive cells at 488 nm excitation. A viability study was conducted using 300 000 cells per assay. Percentage viability was calculated using System II software (Tucker-Davis Technologies, Alachua, FL). Samples were analyzed using an EPICS XL (Beckman-Coulter, Inc., Miami, FL).

9.2.8 Statistical analysis

Standard ANOVA testing was performed using Minitab version 13 (Minitab Inc., PA) using the voltages, pulse durations and number of pulses for all samples.

9.3 Results and analysis

9.3.1 Effect of voltage (electric field strength) on cell viability (live/dead assay)

The ovarian cancer cells were exposed to various voltages, such as 200 V, 400 V, 520 V and 600 V corresponding to the electric field strengths of 500 V/cm, 1000 V/cm, 1300 V/cm and 1500 V/cm, respectively. The duration of the pulses (100 μs), number of pulses (4 pulses) and the interval (1 s) were kept constant, as they can affect the electroporation efficacy. The percentage of viable cells in the pulsed samples was calculated by counting the number of live cells and comparing them with the unpulsed control samples. Figure 9.1 illustrates the percentage viability observed after treatment with various electrical pulses. The percentage

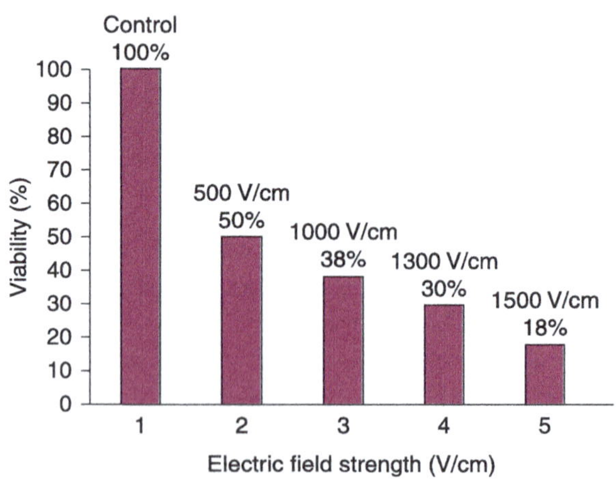

Figure 9.1 Viability of SKOV-3 cells at various voltages (average of 2 sets of experiment). The pulse duration, number of pulses and interval remain the same at 100 μs, 4 pulses at 1 Hz

viability decreased with the increase in voltage, with about 82% cell mortality noted at 1500 V/cm and just 50% mortality at 500 V/cm pulses. The finding that higher electric field strengths result in higher mortality in cells correlates very well with reported results in the literature by previous researchers [12,17,18]. Given the fact that each cell has a different composition of the cell membrane (variations of hydrophilic/polar heads and nonpolar tails and amphipathic proteins), the electric field strength needed to permeablilize the same varies and hence needs to be optimized for each type of treatment.

9.3.2 Fluorescence microscopy

Figures 9.2(a-d) show the fluorescence microscopy data for the SKOV-3 cells for the electric fields of 500 V/cm, 1000 V/cm and 1300 V/cm, along with control (unpulsed sample). The pulses were 100 μs long and 4 pulses were applied in each case at 1 Hz frequency. The red dots correspond to

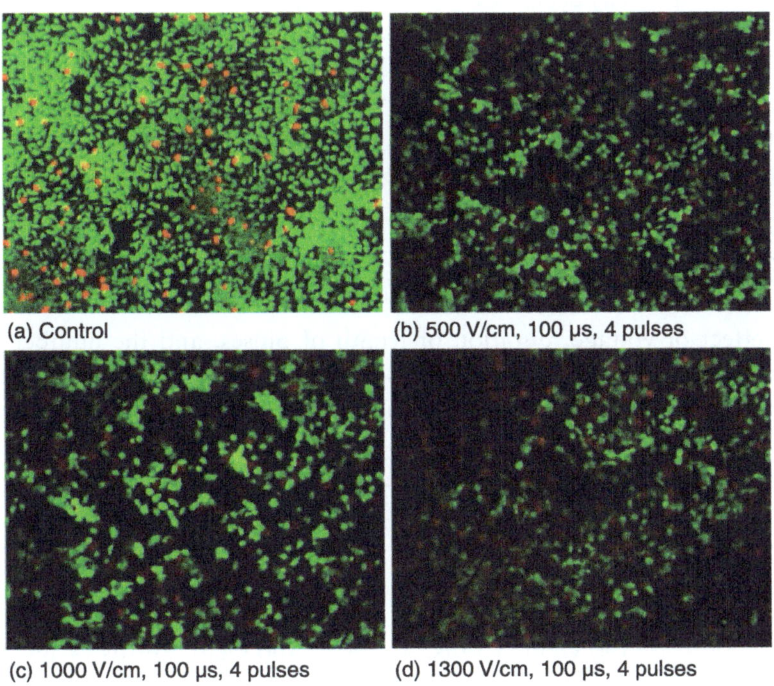

Figure 9.2 Fluorescence micrographs illustrating the voltage (electric field strength) effect on SKOV-3 cells

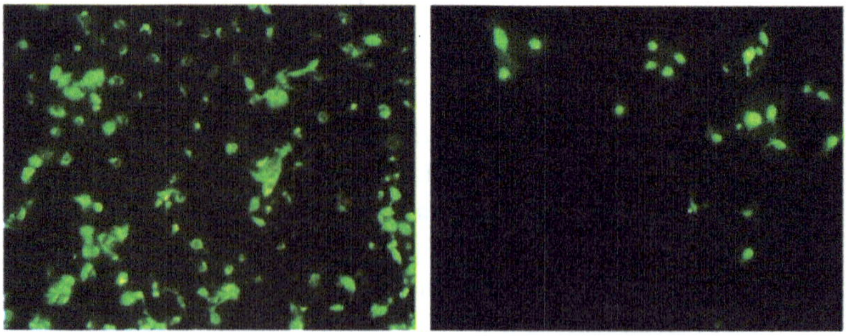

(a) Control (b) 3250 V/cm, 100 μs, 8 pulses

Figure 9.3 Fluorescence micrographs illustrating the effect of very high electric field strength (irreversible electroporation effect)

the dead cells and the green ones represent the live cells. The control sample has more live cells than the other electroporated samples at each intensity of field strength, and with an increase in the field strength, the number of live cells decreases.

Figure 9.3 shows the effect when a very high field strength of 3250 V/cm, (100 μs, 8 pulses, 1 Hz) was applied. It killed almost all the cells and the percentage viability is very low (<10%). This indicates that it is possible to eradicate a tumor or the malignant cells using just pulses only, without drugs [21].

9.3.3 Light microscopy

The effect of voltage, duration or length of pulses, and the number of pulses on the cell viability was also studied using light microscopy. Figures 9.4(a and b) show a comparison of these data at various conditions. The cells consistently showed higher viability with lower voltages, shorter durations and less number of pulses. There is close correlation between 500 V/cm, 260 μs and 8 pulses at 1 s intervals and 1300 V/cm, 100 μs and 8 pulses at 1 s intervals ($p = 0.5894$ (Table 9.2)).

In the figure, the conditions from left to right are: Control (0 V/cm, 0 μs, 0 pulses), 500 V/cm, 100 μs, 4 pulses, 500 V/cm, 260 μs, 4 pulses, 1300 V/cm, 100 μs, 4 pulses, 1300 V/cm, 260 μs, 4 pulses, 500 V/cm, 100 μs, 8 pulses, 500 V/cm, 260 μs, 8 pulses, 1300 V/cm, 100 μs, 8 pulses, and 1300 V/cm, 260 μs, 8 pulses.

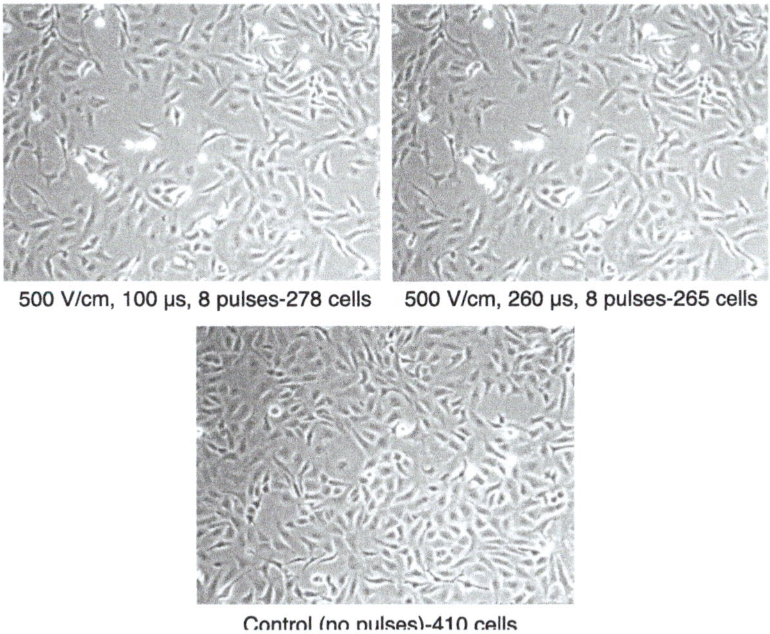

Figure 9.4 (a) Light microscopy pictures illustrating the cell density variation at various electroporation conditions

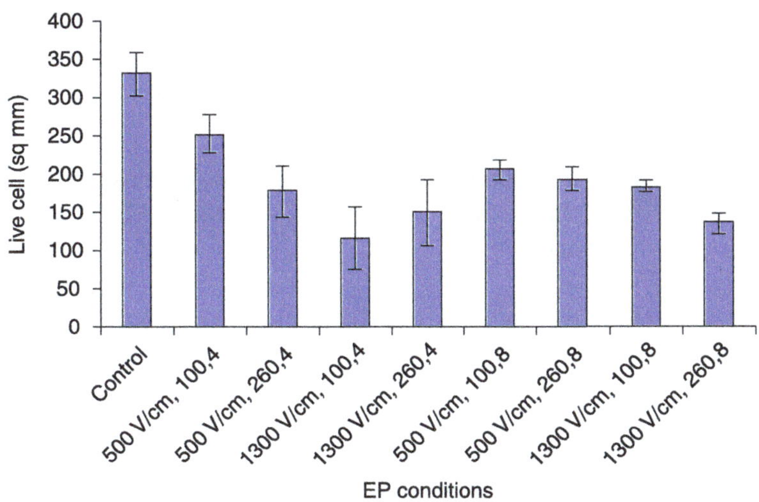

Figure 9.4 (b) The viability of cells at various EP conditions using light microscopy

Table 9.2 P values for various electroporation conditions

EP Condition	p value
500 V/cm, 260 μs, 8 pulses vs. 1300 V/cm, 100 μs, 8 pulses	0.5894
Control vs. 500 V/cm, 260 μs, 8 pulses	0.0022
Control vs. 1300 V/cm, 100 μs, 8 pulses	0.0012
Control vs. 500 V/cm, 100 μs, 4 pulses	0.0286
Control vs. 1300 V/cm, 260 μs, 4 pulses	0.0107
Control vs. 500 V/cm, 100 μs, 8 pulses	0.0033
Control vs. 1300 V/cm, 100 μs, 8 pulses	0.0005
All 10 0 μs vs. 260 μs	0.0009

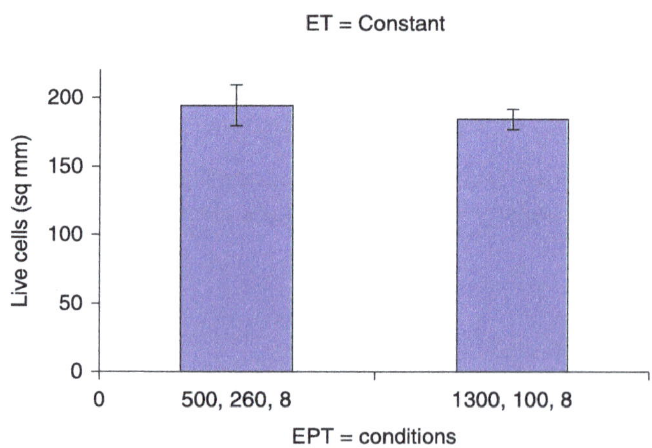

Figure 9.5 Comparison of 1300 V/cm, 100 μs, 8 pulses vs. 500 V/cm, 260 μs, 8 pulses, illustrating ET = K, a constant

Table 9.2 exemplifies the statistical significance of the results obtained. The 500 V/cm, 260 μs, 8 pulses are not statistically significant ($p = 0.5894$) from the 1300 V/cm, 100 μs pulses, as expected from ET = constant. Also, these conditions are statistically significant ($p = 0.0022$ and 0.0012, respectively) compared to the control illustrating their efficacy. Comparing all 100 μs pulses versus all 260 μs pulses illustrates that they are statistically significant, indicating that the length of pulses makes a difference in cellular porosity. In general, the overall results followed the expected behavior, that is, the higher the magnitude and longer the duration, the lower the viability (and hence higher permeability) and ET = constant is a useful hypothesis to adopt.

Figure 9.5 shows the comparison of 1300 V/cm, 100 μs, 8 pulses versus 500 V/cm, 260 μs, 8 pulses, illustrating the validity of ET = constant, and that the low intensity and long duration pulses could be used for effective proliferation control of cancer cells. The ratio of live cells under these two conditions is 0.95 ((500 V/cm, 260 μs, 8 pulses)/(1300 V/cm, 100 μs, 8 pulses)), indicating that the efficacy of 500 V, 260 μs and 8 pulses is close to that of 1300 V/cm, 100 μs and 8 pulses.

9.3.4 Flow cytometry

Figure 9.6 shows the flow cytometry results for the SKOV-3 cells. The pulse parameters applied were 500 V/cm, 260 μs, 6 pulses and 1300 V/cm, 100 μs, 6 pulses, respectively. The viabilities obtained are comparable for 500 V/cm (a) and 1300 V/cm (b) pulses, indicating again that ET = constant is valid.

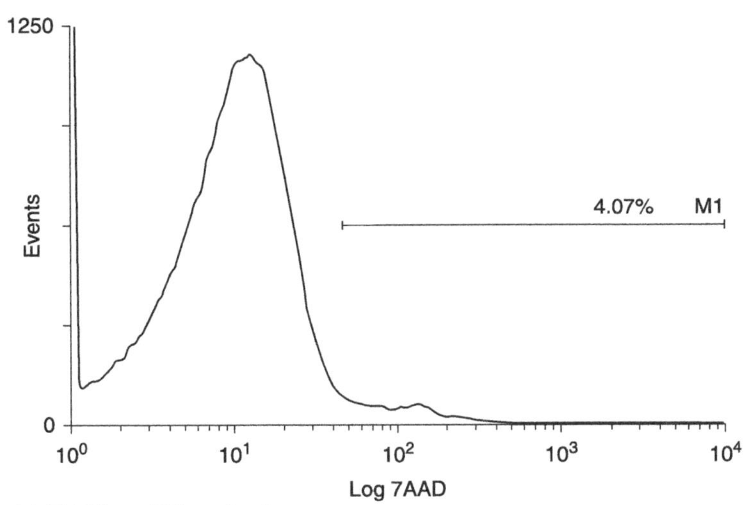

(a) 500 V/cm, 260 μs, 6 pulses

Figure 9.6 Flow cytometry results showing good correlation for 1300 V/cm, 100 μs, 6 pulses vs. 500 V/cm, 260 μs, 6 pulses

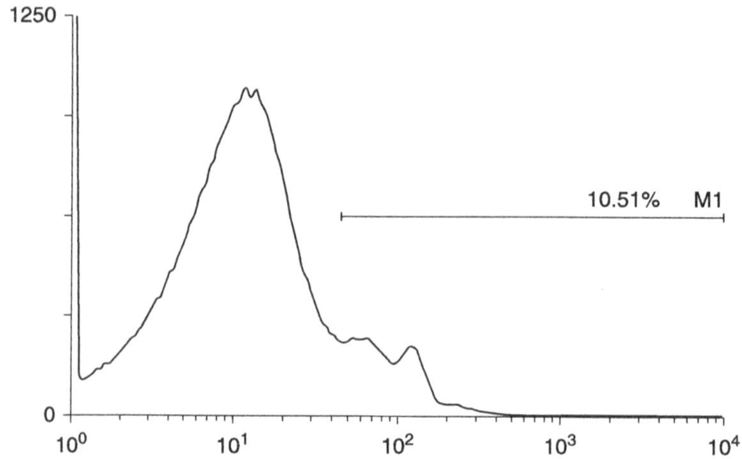

(b) 1300 V/cm, 100 μs, 6 pulses

Figure 9.6 Continued

9.3.5 *MTT assay*

Similar corroborating results were also obtained using MTT assay (data not shown).

9.4 Discussion and summary

In general, these results demonstrate that the proliferation of SKOV-3, the human ovarian cancer cells, could be controlled using both low and high voltage pulses of appropriate durations, following ET = K, a constant relation. This indicates that the higher the magnitude of the pulses, the lower the duration and vice versa. The viability is also controlled by the number of pulses. Figure 9.7 shows the influence of voltage and pulse durations on the number of live cells (cell density). It should be noted that the voltage is more dominant than pulse duration.

The SKOV-3 cell line was chosen in this research, because it has been routinely used in the xenograft orthotopic ovarian cancer animal model [22]. The main reasons for this are that SKOV-3 cells over-express HER-2 at least 100-fold higher than other cancer cell lines derived from the female genital tract, and can readily form tumors when intraperitoneally

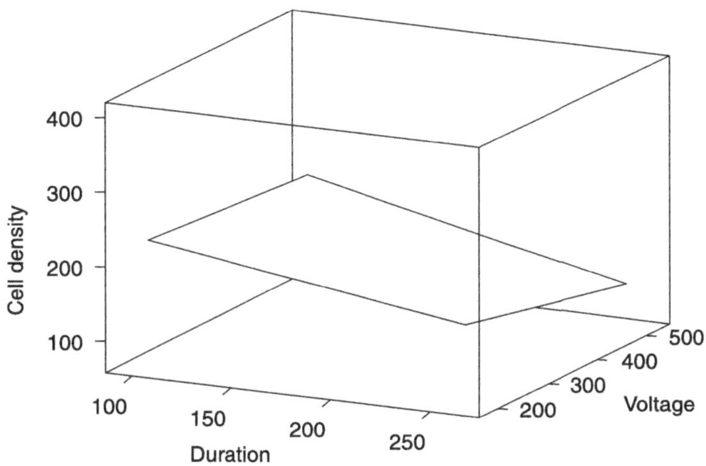

Figure 9.7 **Dependence of cell density (number of live cells) on duration of pulse and the applied voltage**

injected into the mice. Hence, preliminary work with this cell line is beneficial for future preclinical and clinical experiments.

After corroborating the results obtained by various assays, a model was constructed on the relationship between the various parameters used in this assay. This would lead to a better understanding of the controlling factors in efficient proliferation control. As illustrated in Figures 9.8–9.10, the data analysis of our experiments indicates that factor A (voltage magnitude) and BC (B: pulse duration, and C: number of pulses) interaction terms are significant. Factor B and C are included in the model in order to preserve the hierarchy. Analysis of the center point runs indicates that no curvature is present in our experiment. Thus a first-order model would be sufficient to represent the result. Residual plots show that the assumptions of normality and independence hold true.

The voltage applied (or the electric field strength, EV/cm) to the cells is a major parameter determining the number of living cells; the lower the voltage, the higher the number of living cells. The BC interaction graph shows that at the low level of pulse duration, the number of pulses applied would be important in determining the number of live cells (response). However, at the high level of pulse duration, the response is robust towards variation in the number of pulses.

These findings are of significance in the application of electrical pulses in treating cancer, by balancing the optimal value of electric field strength

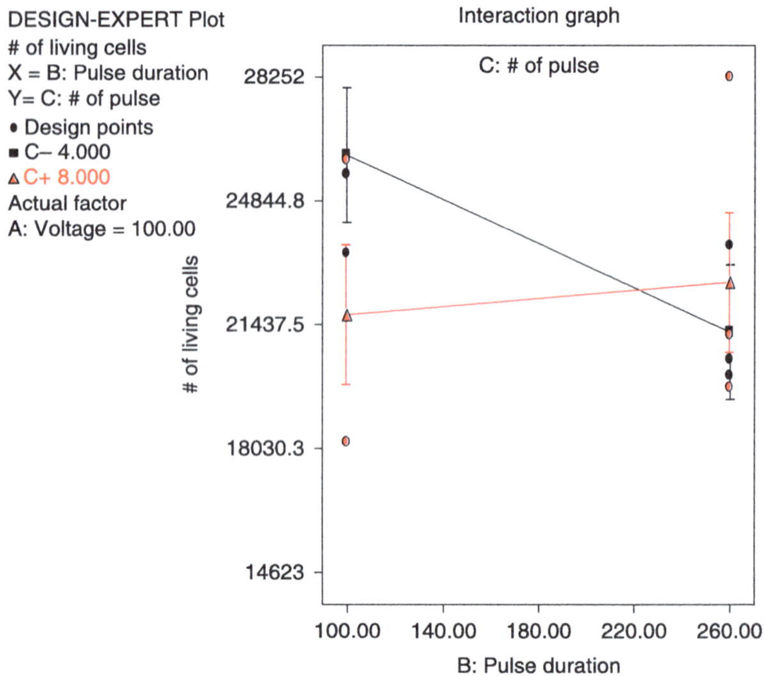

Figure 9.8 Interaction graph illustrating the relationship between voltage magnitude, pulse duration, and number of pulses at V=100

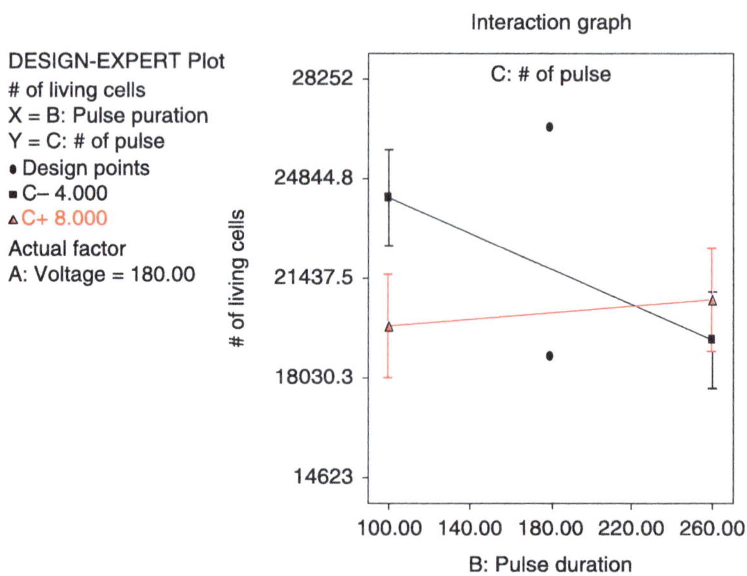

Figure 9.9 Interaction graph illustrating the relationship between voltage magnitude, pulse duration, and number of pulses at V=180

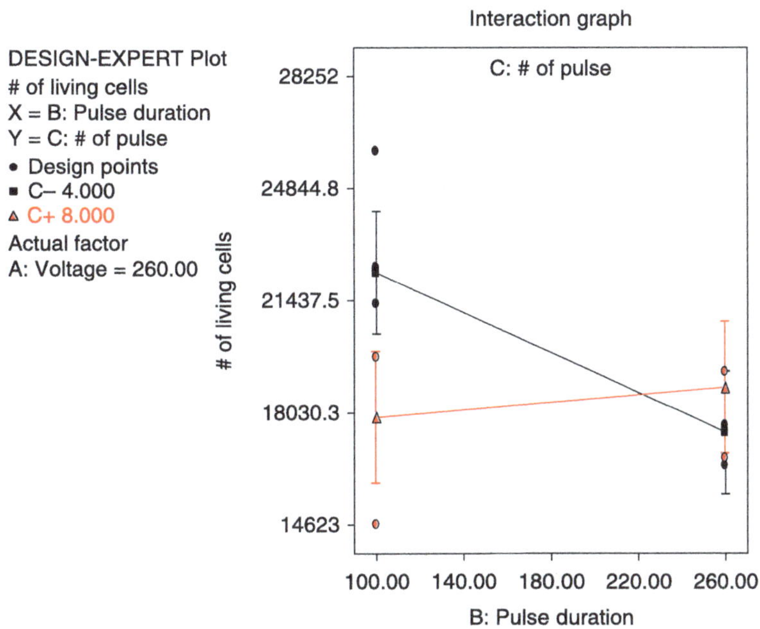

Figure 9.10 Interaction graph illustrating the relationship between voltage magnitude, pulse duration, and number of pulses at V = 260

and other pulse parameters so as to achieve the desired effect. EP, being a safe, local, site-specific physical technique, does not have any significant side effects as compared to conventional chemotherapy. Being a physical phenomena resulting from the interaction of the plasma membrane with the electric field, it can be applied to all histologies of solid tumors. In addition, all conventional cancer treatments, such as surgery, chemotherapy and radiation have major side effects that cause an overall weakness in the immune system. EP's safety, simplicity, versatility and low cost offer an attractive alternative to treat cancer. Its efficacy depends on various parameters, such as applied voltage (electric field strength), duration, number of pulses, shape of voltage (square or exponential), pulse interval, type and size of drug to be delivered, conductivity of the medium, temperature, among other things [23–27]. It is also cell/tissue specific, which means the pulse parameters have to be optimized for each cell or tissue [23].

We have delved into a detailed characterization of the effects of electrochemo-exposure on SKOV-3 human ovarian carcinoma cells and optimized the parameters to be used with this cell line. We explored the effect of voltage, duration and number of pulses using various combinations of these three parameters. Overall, our results indicated that the voltage applied (or electric field strength which is the linear ratio of voltage to the distance between the electrodes, V/cm) to the cells is a major factor in determining the number of live cells after electroporation; we can infer that the lower the voltage, the higher the number of living cells. While the other two factors, pulse duration (μs) and the number of pulses, do have significant effects individually, the interaction between the two factors is more significant and complex. At short pulse duration (100 μs), the number of pulses applied was important in determining the number of living cells. However, at longer duration (260 μs), the variation in the number of pulses appears to be more dominant than at 100 μs. In general, running the factors at the low voltages, which maximizes the cell recovery response (resealing of the cell membrane) is desired for basic reasons (safety, low cost, etc.), but must be strong enough to cause cell membrane permeability.

In summary, our results show the possibility of using low voltage, longer duration and the standard high voltage, short duration pulses for effective proliferation control of human ovarian cancer cells. Using very high magnitude pulses alone, we can completely eliminate the tumor (irreversible electroporation) and leave the healthy cells surrounding cells alone. Irreversible electroporation can be used as an alternative treatment for cryotherapy and/or thermal ablation, when they could not be applied. These results form the basis for taking this safe, simple and fast technique to the next level for clinical use.

9.5 Acknowledgments

The authors gratefully acknowledge the 2004 Arizona State University (ASU) East Faculty Grant-in-Aid (FGIA). Many thanks are due to Tim Curd of ASU for performing the experiments and Dr Paukin of ASU for assisting in conducting these experiments. Very many thanks are also due to Dr McHenry, Dean, ASU East for all his support and encouragement and to J S Faust of Wistar Institute (Flow Cytometry), PA for their help with flow cytometry.

9.6 References

1. Jemal A, Bray F, Center M M, Ferlay J, Ward E and Forman D (2011), 'Global cancer statistics', *CA Cancer J Clin*, **61**: 69–90.

2. Banks E (2000), 'The epidemiology of ovarian cancer', in: Bartlett J M S (ed.), *Ovarian Cancer – Methods and Protocols*, New Jersey: Humana Press.
3. Zweemer R P and Jacobs I J (2000), 'Familial ovarian cancer', in: Bartlett J M S (ed.), *Ovarian Cancer – Methods and Protocols*, New York: Humana Press.
4. Ramesh R, et al. (2000), 'Retroviral transfer of the Herpes Simplex Virus-Thymidine Kinase (HSV-tK) gene for the treatment of cancer', in: Walther W and Stein U (eds), *Gene Therapy of Cancer – Methods and Protocols*, New Jersey: Humana Press.
5. Hamilton T C (1992), 'Ovarian cancer, Part I: Biology', *Current Problems in Cancer*, Jan/Feb.
6. Bast R C Jr., Hennessy B and Mills G B (2009), 'The biology of ovarian cancer: new opportunities for translation', *Nat Rev Canc*, **9(6)**: 415–28.
7. Berger K and Bostwick J III (1998), *A Woman's Decision*, New York: St Martin's Griffin, p 90.
8. Love S M and Lindsey K (2005), *Dr Susan Love's Breast Book*. Cambridge MA, Perseus Publishing.
9. Heller R, Gilbert R and Jaroszeski M J (2000), 'Clinical trials for solid tumors using electrochemotherapy', in: Jaroszeski M, Heller R and Gilbert R (eds), *Electrochemotherapy, Electrogenetherapy, and Transdermal Delivery*, New Jersey: Humana Press, 137–56.
10. Rabussay D P, Nanda G S, and Goldfarb P M (2002), 'Enhancing the effectiveness of drug-based cancer therapy by electroporation (electropermeabilization)', *Technology in Cancer Research & Treatment*, **1**: 71–82.
11. Dev S B, et al. (2000), 'Medical applications of electroporation', *IEEE Trans Plasma Sci*, **28**: 206–23.
12. Campana L G, Valpione S, Mocellin S, Sundararajan R and Granziera E, et al. (2012), 'Superficial disease control in melanoma patients treated with electrochemotherapy in a palliative setting', *Brit J Surg*, **99**: 821–30.
13. Campana L G, Mocellin S, Basso M, Puccetti O, De Salvo G L, et al. (2009), 'Bleomycin-based electrochemotherapy: clinical outcome from a single Institution's experience with 52 patients', *Ann Surg Oncol*, **16**: 191–9.
14. Larkin J O, et al (2007), 'Electrochemotherapy: aspects of practical development and early clinical experience', *Ann Surg*, **245(3)**: 465–79.
15. Weaver J C and Schoenbach K H (2003), 'Biodielectrics', *Editorial IEEE Trans TDEI*, **10(5)**: 715–16.
16. Weaver J C (2003), 'Electroporation of biological membranes from multicellular to nano scales', *IEEE Trans TDEI*, 10(5), 754–68.
17. Yoon S H, Lee J M, Cho H I, Kim E K, Kim H S et al. (2009), 'Adoptive immunotherapy using human peripheral blood lymphocytes transferred with RNA encoding Her-2/neu-specific chimeric immune receptor in ovarian cancer xenograft model', *Canc Gene Ther*, **16(6)**: 489–97.
18. Sersa G, Kranjc S and Cemazar M (2000), 'Improvement of combined modality therapy with cisplatin and radiation using electroporation of tumors', *Int J Radiat Oncol Biol Phys*, **46(4)**: 1037–41.

19. Hui S W and Li L H (2000), '*In vitro* and *ex vivo* gene delivery to cells by electroporation', in: Jaroszeski M, Heller R and Richard Gilbert R (eds), *Electrochemotherapy, Electrogenetherapy, and Transdermal Delivery*, New Jersey: Humana Press, 157–71.
20. Greenlee R T et al. (2000), 'Cancer statistics 2000', *CA Cancer J Clin*, **50**: 7–33.
21. Al-Sakere B, André F, Bernat C, Connault E, Opolon P, et al. (2007), 'Tumor ablation with irreversible electroporation', *PLOS ONE*, **2**: on-line, 1–8.
22. Hung M, Yan D and Zhang S (2000), 'E1A-Mediated gene therapy', in: *Ovarian Cancer – Methods and Protocols*, Bartlett J M S (ed.). New Jersey: Humana Press.
23. O'Hare M J (2000), 'Electropermeabilization and electrosensitivity of different types of mammalian cells', in: Neumann E, Sowers A E and Jordan C A (eds), *Electroporation and Electrofusion in Cell Biology*, New Jersey: Plenum Press, 319–30.
24. Djuzenova C S, Zimmermann U, Frank H, Sukhorukov V L, Richter E and Fuhr G (1996), 'Effect of medium conductivity and composition on the uptake of propidium iodide into electropermeabilized myeloma cells', *Biochim Biophys Acta*, **1284(2)**: 143–52.
25. Sersa G (2000), 'Electrochemotherapy – animal model work review', in: Jaroszeski M, Heller R and Gilbert R (eds), *Electrochemotherapy, Electrogenetherapy, and Transdermal Delivery*, New Jersey: Humana Press, 119–36.
26. Rols M P and Teissie J (1998), 'Electropermeabilization of mammalian cells to macromolecules: control by pulse duration', *Biophys J*, **75**: 1415–23.
27. Mir L M and Orlowski S (2000), 'The basis of electrochemotherapy', in: Jaroszeski M, Heller R and Gilbert R (eds), *Electrochemotherapy, Electrogenetherapy, and Transdermal Delivery*, New Jersey: Humana Press, 99–117.

Irreversible electroporation: a drug-free cancer treatment

Raji Sundararajan, Therese Salameh, Ignacio G. Camarillo, R. Raja Prabu, Arutselvan Natarajan and Kavitha Sankaranarayanan

DOI: 10.1533/9781908818294.219

Abstract: Many potential drugs to treat cancer have found limited success, due to the lack of safe and efficient delivery systems that allow molecules to cross the hydrophilic and hydrophobic lipid bilayers of cell plasma membranes, which are normally impermeable. It would be ideal to have a cancer treatment that is drug-free, alleviating side effects, cost and drug resistance. Electroporation could be used for targeted and drug-less treatment, known as irreversible electroporation (IRE), where higher voltages and longer and more pulses, compared to conventional electroporation, are used to kill the tumor cells. This causes permanent permeabilization of the cell membranes and the consequent loss of cell homeostasis. This study evaluates the effects of various electroporation parameters on different cancer cell lines that were treated with various pulse parameters. The viability results indicate the differences in the viabilities and hence the aggressiveness of the cell lines experimented on and the IRE parameters needed to obtain the desired effect on proliferation control and complete tumor ablation.

Key words: irreversible electroporation, cancer, MCF-7 cells, MBA-MB-231 cells, 4T1 cells, A549 cells, chick embryo fibroblasts, viability, proliferation control, colony forming unit.

10.1 Introduction

Drug delivery has several issues, including route of administration, toxicity, efficacy, cost, resistance, side effects, to mention just a few [1]. Oral delivery drugs are the simplest and most straightforward, but they can cause several adverse effects. Medicines administered orally must circulate throughout the entire body before taking effect. This poses a potential threat to the patient, as the drug could take action in any part of the body causing adverse effects, as well as creating an opportunity for the drug to be broken down through enzymatic degradation. Chemo drugs for cancer have many side effects [2–5] and are costly, in addition to the normal drug delivery issues. An improved delivery system, such as electroporation, which will provide a more targeted pathway into the cell, will increase the cell's receptiveness to the drug and enhance the drug's capabilities while reducing systemic side effects.

While a few pills are useful in curing malaria, polio and communicable diseases, cancer is not so easily curable using chemo drugs. This is evidenced by the millions of cancer deaths each year and the poor response rates of many chemo drugs (Table 10.1) [3, 4]). The host cell defense mechanisms, which help in infections, are not doing so in cancers, as the cancer cells are the host cells. In addition, the cancer cells are our own cells, unlike microbes. Hence, drugs kill both normal and diseased cells. Besides side effects, there is also drug resistance, both primary and secondary. The various adverse effects of anticancer drugs include cystitis, stomatitis, pulmonary fibrosis, ototoxicity, oedema of hands, cardio toxicity, hepatotoxicity, neurotoxicity, peripheral neuritis, mental depression, pancreatitis, dermatitis, bone marrow depression, alopecia, vomiting, amenorrhoea, teratogenity, renal dysfunction, thrombocytopenia, muscle weakness, allergic reactions, diarrhea, etc. [2–5]. The various drug resistance mechanisms include decreased drug accumulation due to decreased drug influx, increased drug efflux, altered intracellular drug trafficking, increased inactivation of drug or toxic intermediate, increased repair of or tolerance to drug increased damage to DNA, protein and membranes, decreased drug activation, altered drug targets, etc. [2–4].

The socio-economic aspect of cancer drugs is another major issue, with chemo drugs costs being exorbitantly high. It is reported that in the US, 62% of bankruptcies and foreclosures are medical-related, including cancer [6]. In the low and middle income countries, it is unaffordable for the common people. There are also inoperable, and aggressive, recurrent tumors that need alternate/additional treatment modalities. Towards this end, it is desirable and ideal to have a cancer treatment using electrical

Table 10.1 **Breast cancer drug response rates [3]**

Drug	Response Rate
Doxorubicin	25–40%
Paclitaxel	17–54%
Tamoxifen	21–41%
Docetaxel	18–68%
Trastuzumab	10–20%

pulses with no drugs. Use of electrical pulses is also preferred, as the body is not only biochemical, but is also bioelectrical [7].

Electroporation is the technique of applying high intensity, short duration pulses to open up pores, allowing the passage of chemo drugs that are normally impermeable or less permeable through the cell plasma membranes [8–12]. It involves the rapid structural rearrangement of membranes, in response to the electric field. A noticeable effect is a rapid increase in electrical conductivity (permeability) attributed to the formation of pores in the bilayer lipid membrane. It leads to the transient or permanent rupturing of the membrane wall, which can be either reversible or irreversible.

Electroporation generally depends upon the intensity of the electric field, the duration of each pulse, the number of pulses, and the interval between them. Depending upon the magnitudes of these parameters, there are three effects on the cell membrane:

1. no changes on the cell membrane;
2. temporary opening of the cell membrane after which the cell can still survive (reversible electroporation); and
3. permanently open the cell membrane and the cell dies after irreversible electroporation (IRE).

Reversible electroporation is used in electrochemotherapy, wherein low dose chemo drugs and machomolecules are administered into the targeted area [8–12].

Electrodes are placed around the tumor area, generating reversible permeabilization that allows substances to pass through the cell membrane and into the cytoplasm. The upper limits to the range of electrical parameters that induce reversible electroporation are used for IRE. Thus, IRE is that which causes permanent permeabilization of the cell membranes and the consequent loss of cell homeostasis, when an electrical field is applied to cancer cells [13–17]. Advantages of IRE are that no drugs are used and it is a non-thermal mechanism of action, dependent on the blood flow, allows focal tissue ablation, and requires a short application time [13]. Also, IRE allows ablation of tumors in close proximity to other tissues. Other ablation mechanisms such as radiofrequency thermal ablation (RFA) release energy to the surrounding tissues and affect the vascularity and connective tissue structure [14]. RFA first uses ultrasound, computed tomography or magnetic resonance to locate the tumor and then inserts a small needle-electrode into the tumor [18]. Ionic vibration at the needle tip generates frictional heat that ablates the tumor.

The focus of this study is to determine safe and efficient pulse parameters for various cell lines that reduce or stop the proliferation, which could eventually be used as a basis for preclinical and clinical studies. Towards this, we studied the responses of different cancer cells and compare the aggressiveness of these cells and arrive at optimal pulse parameters for effective treatment.

10.2 Materials and methods

10.2.1 Breast cancer cell lines

Human breast cancer cells, MCF-7 and MDA-MB-231, and rat cancer cells 4T1, were cultured in RPMI-1640 medium with 10% fetal bovine serum and 1% penicillin/streptomycin stock and incubated in a 5% CO_2 atmosphere at 37 °C [19]. These cells were chosen as they represent major aspects of human breast cancer and their progression. For example, the MCF-7 cells were chosen because they are ERα+, non-invasive, human breast cancer cells, and a commonly used cell line for human breast cancer studies.

The MDA-MB-231 cells, obtained from a patient in 1973 at MD Anderson Center, were chosen as an ER negative cell line with invasive phenotype, *in vitro*. With epithelial-like morphology, it appears phenotypically as a spindle-shaped cell.

4T1 mammary gland cell lines were chosen as the tumor growth and the metastatic spread of 4T1 cells in BALB/c mice closely mimic human breast cancer. The 4T1 tumor is an animal model for stage IV human breast cancer. 4T1-induced tumors can be used as a post-operative model as well as a non-surgical model, because these tumors metastasize spontaneously in both models with similar kinetics. It is a metastatic breast cancer model which is an excellent mouse model for the study of metastatic progression of breast cancer in humans [20] respectively. Figures 10.1(a–c) show the morphology of these cells.

10.2.2 Lung adenocarcinoma (A549)

The A549 cell line was first developed in 1972 by Giard et al., through the removal and culturing of cancerous lung tissue in the explanted tumor of a 58-year-old Caucasian male [21]. They are human alveolar basal epithelial cells, are squamous in nature and responsible for the diffusion of substances, such as water and electrolytes, across the alveoli of lungs. They grow adherently, as a monolayer, *in vivo*.

The A549 cells (obtained from the Department of Biotechnology, University of Madras, India) were grown in DMEM with 5% FBS, 100 U/mL penicillin at 37 °C in a 5% CO_2 atmosphere [22]. For the experiments, confluent monolayer cultures (Figure 10.1(d)) were trypsinized using 0.25% Trypsin/EDTA and centrifuged at 1500 rpm for 7 minutes. The cell pellet was washed with sterile 1 × PBS and then the cells were suspended in fresh medium.

10.2.3 Chick embryo fiberoblast (CEF)

Chick embryos were obtained from 9- to 11-day-old eggs [21,23]. Briefly, the egg surface was thoroughly wiped with 95% EtOH and cracked at the top using forceps and placed in a 100 cm^2 sterile petri dish containing sterile PBS solution. The head, internal organs and appendages were removed. Tissues were isolated from the remaining parts under sterile conditions. Isolated tissues were minced and trypsinized using 3 mL of 37 °C trypsin/EDTA and incubated for 5 minutes and medium containing FBS was added to stop the action of trypsin. The tissues were removed and the cell solution was centrifuged at 1500 rpm for 7 minutes. The cell pellet was rinsed with 2–3 washes with PBS solution. The cells were then cultured in a 60 mm cell culture dish with DMEM containing 5% FBS and 1% penicillin/streptomycin-glutamine solution and incubated at 37 °C in a 5% CO_2 atmosphere. Healthy and confluent fibroblast cultures were used in the IRE experiments. Figures 10.1 (e–g) illustrate the various cell morphologies. [7–9]

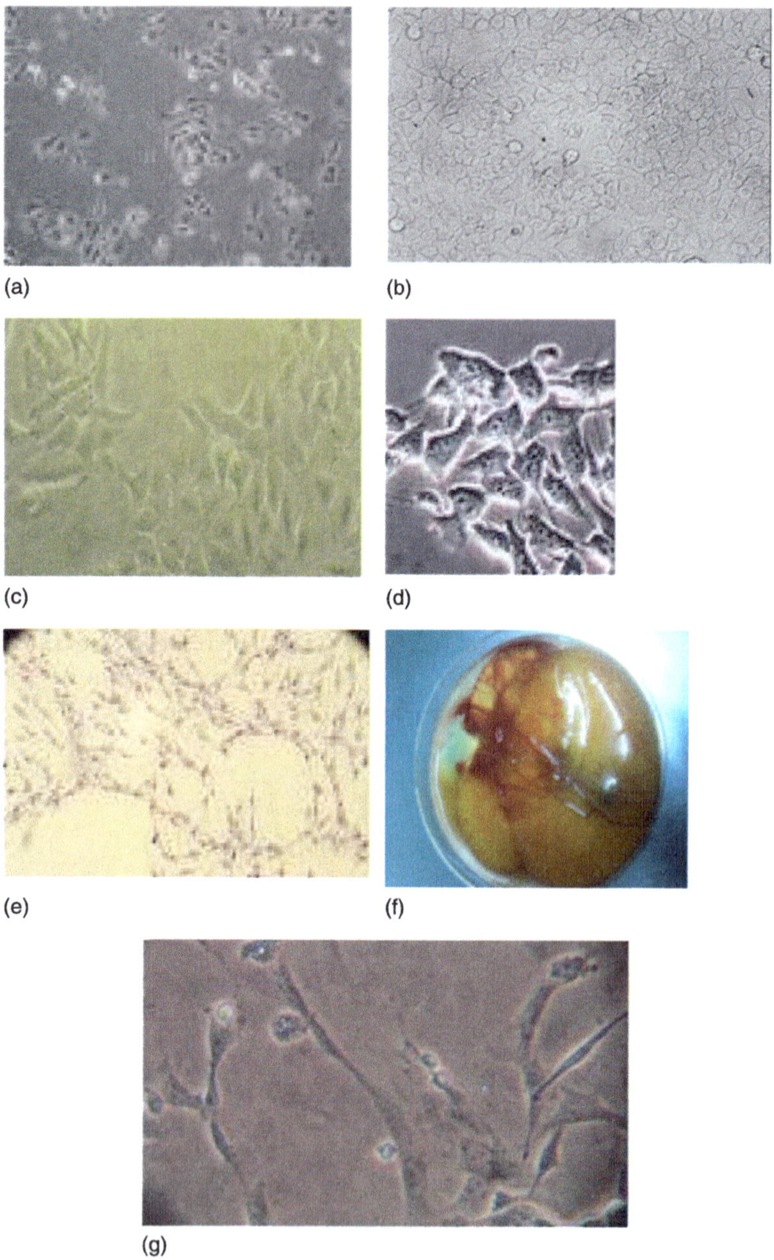

Figure 10.1 Cells studied for IRE [24–26]

10.2.4 Electroporation

Breast cancer cells

The cells were kept in RPMI-1650 medium with 10% fetal bovine serum and 1% penicillin/streptomycin stock and were dissociated from the flasks with 0.25% Trypsin/EDTA solution. A concentration of 1×10^6 cells/mL was used for the study; 700 μL and 400 μL of the cell solutions were added to 0.4 cm and 0.2 cm cuvettes, respectively.

IRE was carried out using six different parameters. Both high intensity field, short duration pulses and low intensity field, long duration pulses were used. The parameters applied are 100 V/cm, 24 ms; 500 V/cm, 24 ms; 500 V/cm, 1 ms; 1200 V/cm, 100 μs; 1500 V/cm, 300 μs; and 1500 V/cm, 2 ms. The pulses were applied in a single train of 8 pulses at a frequency of 1 Hz. Table 10.2 lists these pulses.

A549 cells

The variables studied in this research include electric field intensity, number of pulses and pulse length. The electric fields applied were 500 V/cm, 1500 V/cm and 2500 V/cm. The pulse length range and the number of pulses selected for the experiments were 300 μs, 250 μs and 200 μs, and the numbers of pulses applied are 60, 40 and 20, respectively. In the experiments with varying number of pulses, a single train has 10 pulses with frequency of 1 Hz and accordingly 2, 4 and 6 trains were used for 20, 40 and 60 pulses, respectively. An interval of 60 seconds was given between each train. Table 10.3 gives a summary of the pulse parameters used.

CEF blast cells

Two sets of studies were conducted using CEF. In the first case, both high intensity field, short duration pulses and low intensity field long duration pulses were used. The 6 parameters applied during electroporation are 100 V/cm, 24 ms; 500 V/cm, 24 ms; 500 V/cm, 1 ms; 1200 V/cm, 100 μs; 1500 V/cm, 300 μs; 1500 V/cm, 500 μs. The pulses were given in a single train of 8 pulses at a frequency of 1 Hz. The electroporation buffer used was DMEM supplemented with 10% FBS. The experiments were conducted in triplicate. In addition, cells were also subjected to a single pulse under the above conditions and their impact was studied. Table 10.4 gives the pulse conditions.

In the second case, pulse parameters same as for A549 lung cancer cells were studied.

A BTX ECM 830 (Harvard Apparatus, San Diego, CA), square wave electroporator was used for all the experiments (Figure 10.2); 4 mm cuvettes were used. The electroporation buffer used was DMEM supplemented with 10% FBS, for lung cancer cells and CEFs.

Table 10.2 Treatment parameters for breast cancer cells

Electric field intensity, V/cm	Pulse width
Control	–
100	24 ms
500	1 ms
500	24 ms
1200	100 μs
1500	300 μs
1500	2 ms

Table 10.3 Pulse parameters for A549 cells and CEFs

Electric field (V/cm)	Pulse length (μs)	No. of pulses
2500	200	20
1500	250	40
500	300	60

Table 10.4 Treatment parameters for CEFs

Electric field intensity, V/cm	Pulse width
Control	–
100	24 ms
500	1 ms
500	24 ms
1200	100 μs
1500	300 μs
1500	500 μs

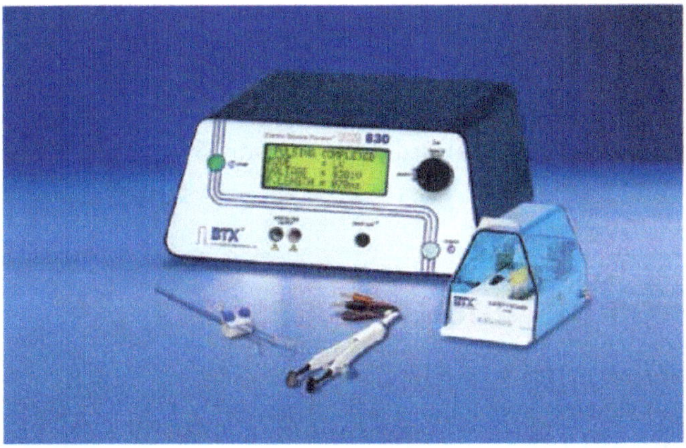

Figure 10.2 BTX ECM830 square wave electroporator used for pulsing

10.2.5 Viability calculations

After electroporation, cell viability was determined by Trypan Blue Exclusion assay, as explained in previous publications [19, 21, 23]. In the case of CEFs, the viability was found using phase contrast microscopy.

10.2.6 Colony forming unit assay for CEFs

The colony forming unit (CFU) is a measure of viable colonogenic cell numbers in CFU/mL [23]. A fixed number of isolated chick embryonic fibroblasts were plated in a 6-well plate along with 2–3 mL of DMEM medium. Then 100 μL of the sample + 700 μL of DMEM medium are taken in a 4-mm gap cuvette and electroporated using various parameters. In this case, only one pulse was applied. After electroporation, the cells were seeded onto tissue culture dishes in DMEM with antibiotics and maintained at 37 °C in a 5% CO_2 atmosphere. They were left undisturbed for 7 days until colonies of cells appeared. The dishes were washed with PBS and cells stained with 0.1% Crystal Violet in Methanol and incubated at room temperature for 10 minutes. The staining solution was discarded and the dishes observed under a microscope. A collection of cells of around 100 mm^2 (~50 cells) was counted as one colony.

Statistical analysis

The graphs were drawn using Microsoft Excel software, which show the percentage cell viability with error bars. The errors bars indicate standard deviation (breast cancer cells), std deviation/sqrt(3) and percentage error (A549 and CEFs).

10.3 Results and analyses

10.3.1 Viability studies

Human breast cancer cells

Figures 10.3 and 10.4 show the variation of viabilities for the ERα positive, non-invasive MCF-7 breast cancer cells and the invasive, ERα negative, MDA-MB-231 breast cancer cells with various IRE pulse parameters. The viabilities are similar for almost all parameters, except for the 1500 V/cm, 2 ms pulses and the 500 V/cm, 24 ms pulses in the case of MCF-7 cells. However, for the MDA-MB-231 cells, except for the 1500 V/cm, 2-ms pulses, the viabilities are almost similar. The viabilities of the MDA-MB-231 cells are greater than MCF-7 cells, especially under the 1500 V/cm, 2-ms pulse conditions.

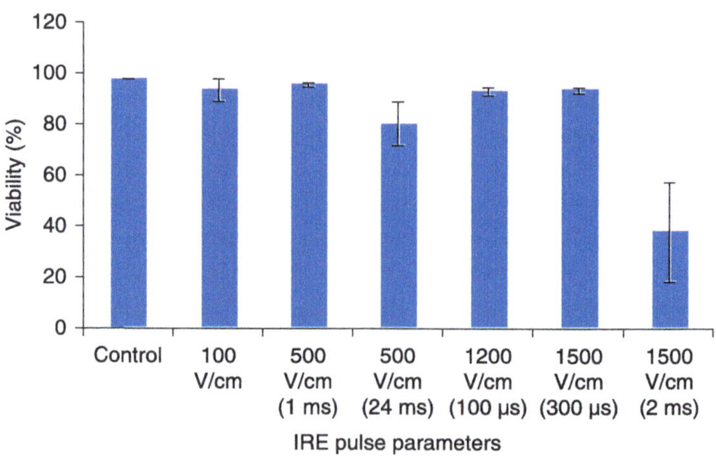

Figure 10.3 Viabilities of MCF-7 cells at various pulse parameters

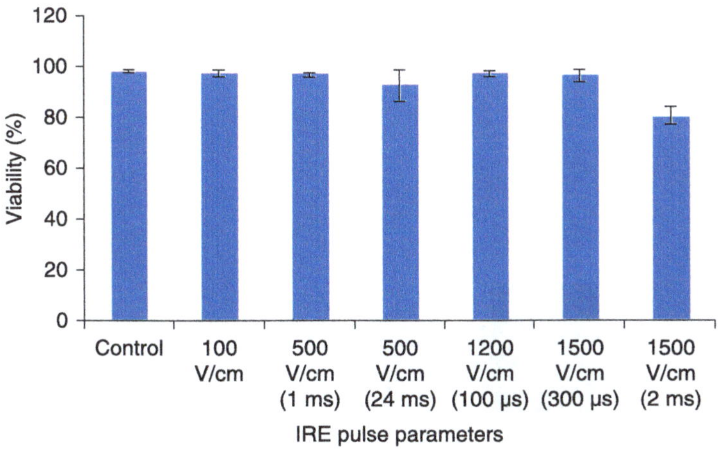

Figure 10.4 Viabilities of MDA-MB-231 cells at various pulse parameters

Under the pulse conditions 500 V/cm, 24 ms, 80% of MCF-7 cells survived, while under 1500 V/cm, 2 ms, the viability educed to 38%. For MDA-MB-231, the viability was about 80% at 1500 V/cm, 2 ms pulses. This indicates that more energy is needed to permeabilize the MDA-MB-231 cells than the MCF-7 cells. This could be due to the difference in the phenotype of these cells, as well as due to the difference in the plasma membrane biology and physiology of these two cells, which cause two entirely different cancer types that require different regimens of chemo treatment.

A comparison with previous work (Figure 10.5) shows good correlation with the trend of the results while the magnitudes are different [13,14]. They applied a single pulse (Figure 10.5(a)) and trains of 10 pulses (Figure 10.5(b)). The cell line used was human hepatocarcinoma (HepG2), which is obtained from liver hepatocellular carcinoma. Figure 10.5(a) shows the result obtained using those cells with a single pulse at 1 ms, 1.5 ms, 3 ms, 6 ms and 24 ms durations for the intensities of 2500 V/cm, 2000 V/cm, 1500 V/cm, 100 V/cm and 500 V/cm, respectively. The 1000 V/cm, 6 ms pulse has reduced viability more than 1500 V/cm, 3 ms pulses, possibly due to the reduced intensity, despite twice the duration.

It can be seen that with 30 (10 × 3) pulses at 1500 V/cm, 300 μs duration, there was complete cell death, indicating that it is possible to elicit complete cell death using an appropriate series of electrical pulses

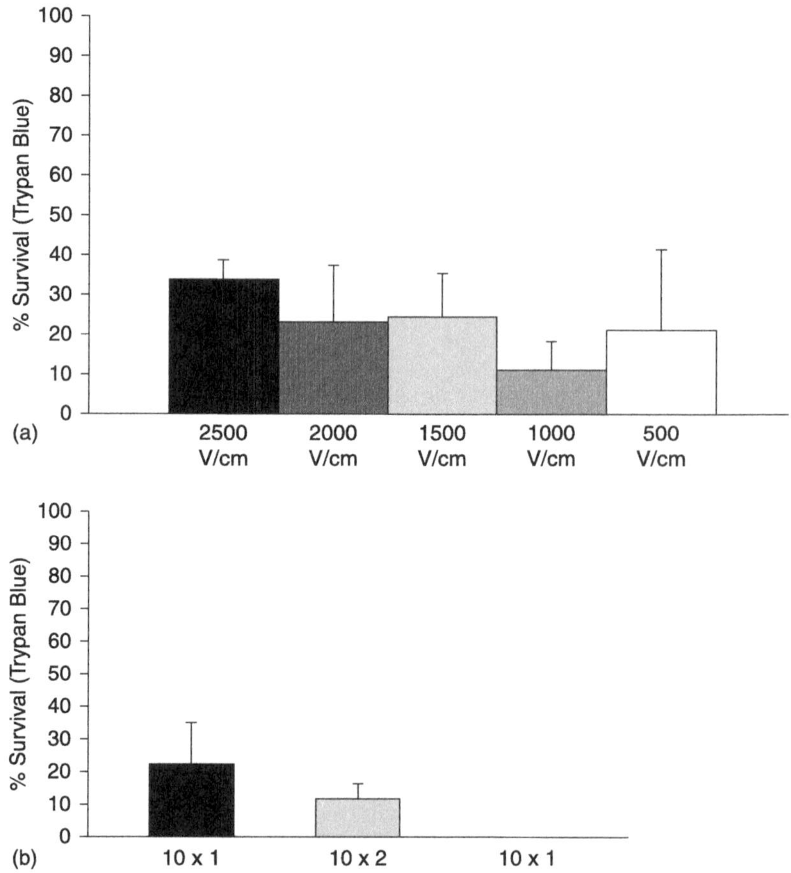

Figure 10.5 Viability of cells after IRE: (a) for various pulse parameters using single pulse, the durations of which are: 1 ms, 1.5 ms, 3 ms, 6 ms and 24 ms; (b) due to a number of pulses for 1500 V/cm, 300 μs pulses [13,14]

(Figure 10.5(b)). When we repeated 1500 V/cm, 300 μs, 30 pulse parameters using MCF-7 human breast cancer cells, there was notable viability (data not shown). We did not observe complete cell death. We could observe complete cell death using longer (ms) pulses at various intensities, including 500 V/cm, 1 ms pulses. This could be due to the differences in the cell sensitivities to the electrical pulses, based on their morphology, phenotype and other biological and physiological characteristics.

Rat 4T1 metastatic cells

Figure 10.6 shows the responses of 4T1 cells for a number of pulse conditions. The viabilities spread into a wider range than MCF-7 and MDA. The highest viability was 73.8% and the lowest was 0%. There were two parameters that killed all the cells (0% viability), at 500 V/cm, 24 ms and 1500 V/cm, 2 ms. This difference could also be due to the different cell biology and membrane structure of the 4T1 cells.

Figure 10.7 shows the viabilities at 3 × 10 and 8 × 10 pulses (train of 10 pulses each, with 150 ms between each pulse and 1 minute between each train of 10 pulses). Three intensities of the pulses were used: 1500 V/cm, 2500 V/cm and 100 V/cm, in that order, and all at 300 μs pulse length. With both 3 × 10 and 8 × 10 pulses, there is zero viability at 2500 V/cm intensity. It was 10.4% and 0% respectively for 1500 V/cm pulse intensity. At 100 V/cm, the viabilities were 32.5% and 3.9%, indicating, even at a lower intensity of 100 V/cm, that proliferation could be stalled using a large number of pulses, such as 80 pulses. Thus, for human clinical applications, we can use a large number of fast pulses at lower intensity to obtain a similar effect as that of higher intensity pulses.

Lung cancer cells

Figures 10.8 and 10.9 show the effect of the pulses on 549 human lung cancer cells. It can be seen that in these cells too, the viability follows a

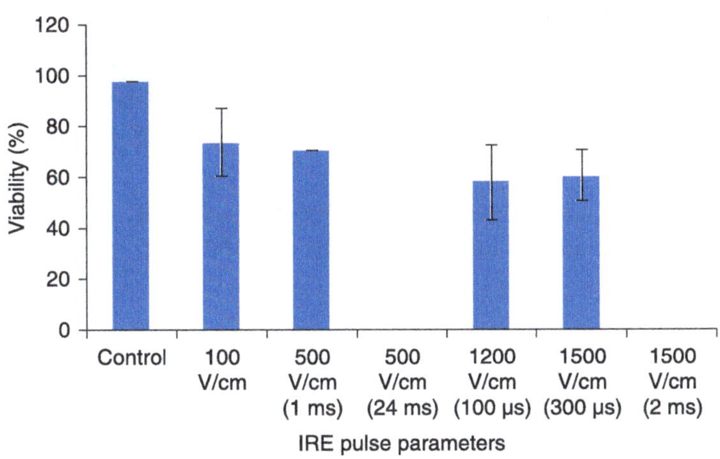

Figure 10.6 Viability of 4T1 cells after treatments with various pulse parameters

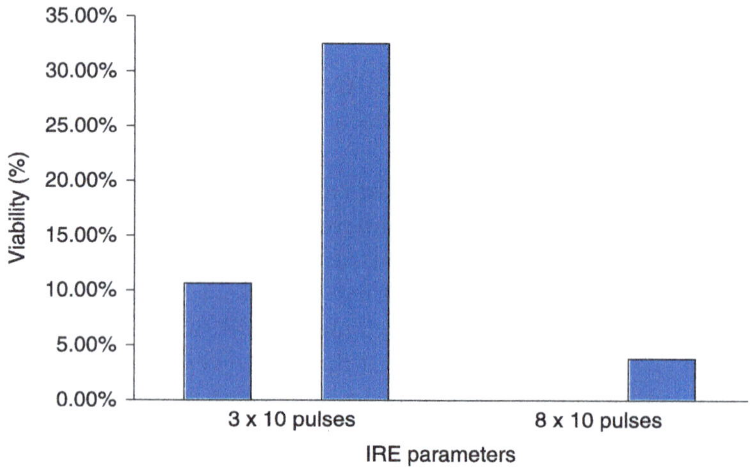

Figure 10.7 Viability of 4T1 cells after treatments with various pulse parameters

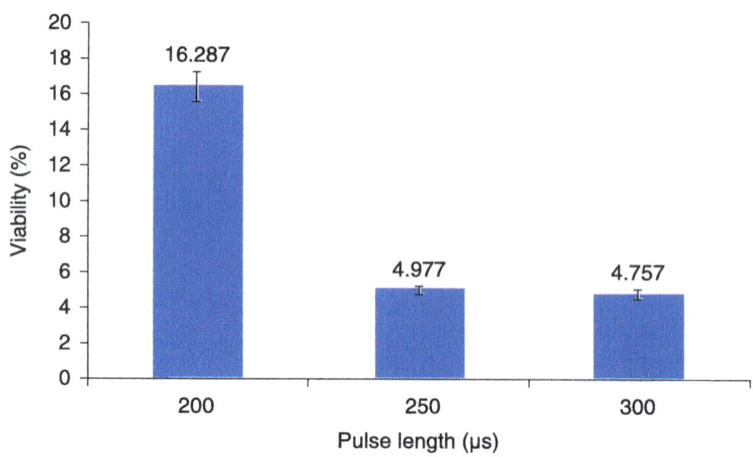

Figure 10.8 Influence of pulse length on the viability of A549 human lung cancer cells at 1500 V/cm

trend similar to that of MCF-7 and other cells discussed above. Figure 10.8 shows the variability in the proliferation control at 1500 V/cm using 200 μs, 250 μs and 300 μs pulses. The viability at 250 μs (4.977%) is marginally higher than 300 μs (4.757%) at the same electric field intensity of 1500 V/cm. The difference in viability percentage at 200 μs was much higher than 300 μs and 250 μs. For effective tumor ablation, the pulse

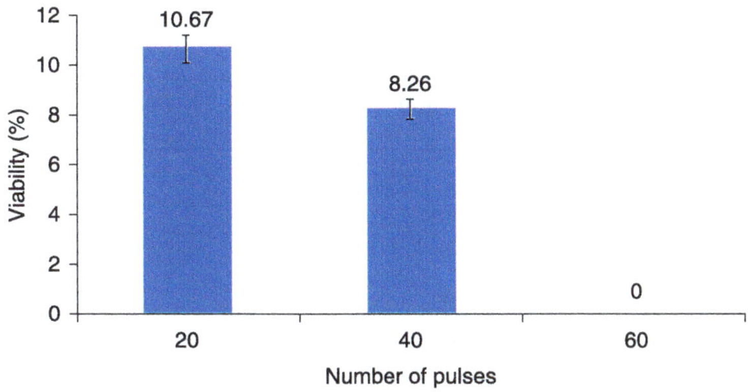

Figure 10.9 Influence of number of pulses on the viability of A549 human lung cancer cells at 1500 V/cm

length of 250 μs was chosen as the optimal parameter, since there was negligible difference between 250 μs and 300 μs. This indicates the non-linear nature of the cell behavior with the energy applied.

Figure 10.9 shows the effect of number of pulses for the same intensity of 1500 V/cm. The numbers of pulses applied were 20, 40 and 60. The viability for 60 pulses is completely zero and had only cell debris, but the viability at 40 pulses was 8.26% and at 20 pulses was 10.67%. These also show the nonlinearity or saturation effect of the cells with the applied energy.

Chick embryo fibroblasts

Table 10.5 shows the summary of various pulse conditions and their corresponding viabilities for the same intensities with 1 and 8 pulses.

Figures 10.10 and 10.11 illustrate the viabilities for the conditions studied. There are considerable differences in the viabilities influenced by the intensity and duration of the applied pulses. It is seen that 8 pulses at 500 V/cm, 24 ms, resulted in death of all cells (Figure 10.11, pulse condition A). For the same field with 1 ms pulse length, there is increase in viability, as it should be due to reduced energy applied. When the pulse length difference is not that significant, then there is no observed difference on viabilities. For example, at 1200 V/cm and 1500 V/cm, for a difference of 200 μs in pulse length, the viabilities are 7.36% and 7.12% respectively, showing that a significant change is obtained only when there is considerable difference in pulse length. It was also proved using

Table 10.5 CEF viabilities for a number of pulse conditions

Electric field intensity, V/cm	Pulse Length	% Viability	
		1 pulse	8 pulses
100	24 ms	12.50	10.19
500	24 ms	4.80	0.00
500	1 ms	36.60	25.56
1200	100 μs	12.80	7.36
1500	300 μs	9.60	7.12
1500	500 μs	8.90	9.07

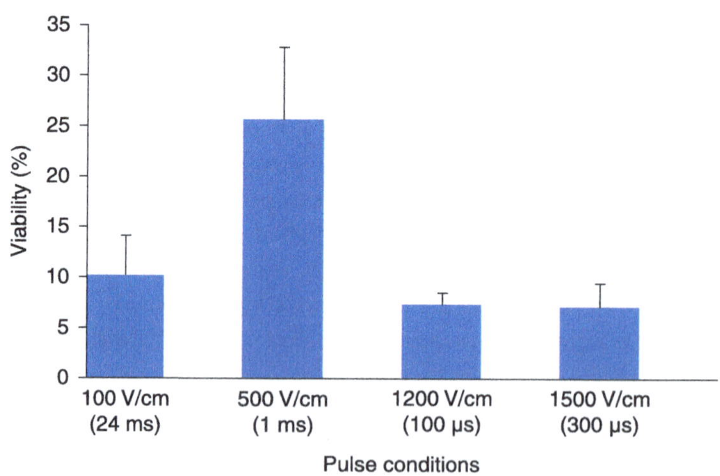

Figure 10.10 Chicken embryo fibroblasts viability for various IRE conditions: 100 V/cm, 24 ms; 500 V/cm, 1 ms, 1200 V/cm, 100 μs, and 1500 V/cm, 300 μs pulses. Error bars indicate std deviation/sqrt(3)

300 μs and 500 μs pulses at 1500 V/cm, which indicated that the difference was very little. Statistical analyses of these data indicated no significant difference ($p < 0.05$). This indicates that at 1500 V/cm, we can use either 300 or 500 μs pulses, and obtain similar cell death magnitudes. These results indicate that with fine-tuning the pulse durations for a given intensity, we can maximize the outcome and also show that there is a saturation or non-linear behavior with the energy applied.

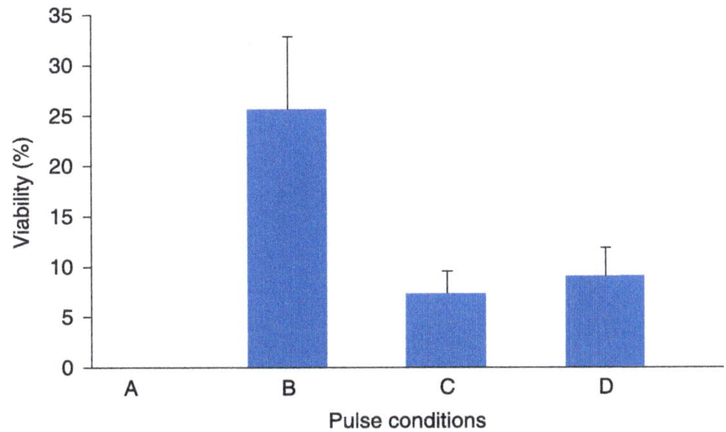

Figure 10.11 Chicken embryo fibroblasts viability for various IRE conditions: (A) 500 V/cm, 24 ms (zero viability); (B) 500 V/cm, 1 ms; (C) 1500 V/cm, 300 μs; and (D) 1500 V/cm, 300 μs. Error bars indicate std deviation/sqrt(3)

Figures 10.12 and 10.13 show the results for additional conditions for CEFs. Figure 10.12 shows the variation with pulse lengths. The pulse lengths were 300 μs, 250 μs and 200 μs with the constant electric field (1500 V/cm) and the number of pulses (80 pulses). The 80 pulses were given as 8 trains of 10 pulses, each with a train interval of 1 minute. It

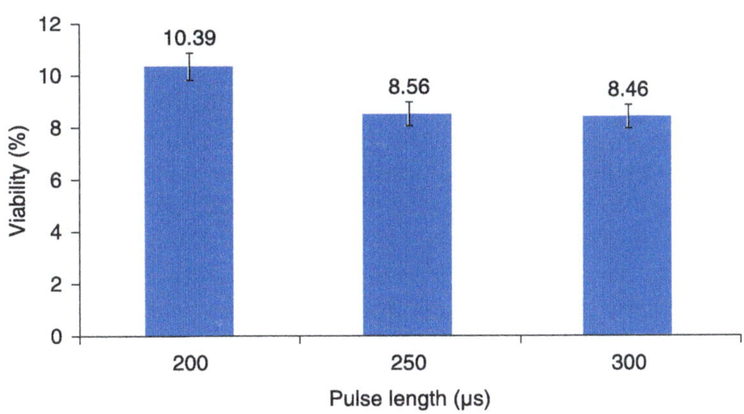

Figure 10.12 Influence of pulse length on the viability of CEFs at 1500 V/cm

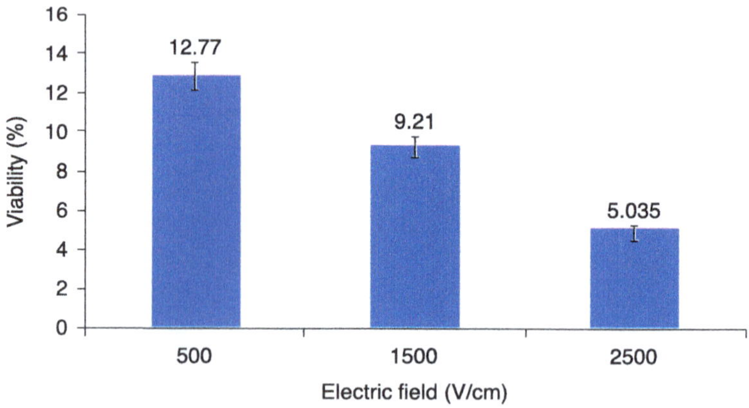

Figure 10.13 Influence of electric field intensity on the viability of CEFs

was found that there was very little difference in viability of cells for 300 μs and 250 μs, which showed viabilities of 8.46 and 8.56% respectively. This is similar to the results obtained using A549 lung cancer cells. Since the difference between the two pulse lengths is very small, 250 μs could be chosen as optimal, because it gave a viability level comparable to that of 300 μs. Also in the clinical aspect, the main objective would be to reduce the duration of the pulse as much as possible without compromising on the efficacy. Statistical analysis showed that the pulse length had 88.91% impact when the other parameters were kept constant.

Figure 10.13 shows the variation in proliferation for various pulse conditions listed in Table 10.6 and their corresponding viabilities. Among the three different electric fields applied, the viability of CEF was found to be the least with 2500 V/cm when compared with the other two field intensities, as to be expected being the highest energy supplied at this intensity, with while other conditions remaining the same. With constant number of pulses and varying electric field applied, the percentage cell viability was inversely proportional to the intensity of electric field applied.

With all the parameters varied in a combination, it was found that the forty 1500 V/cm, 250 μs pulses condition to have less viability percentage than 500 V/cm, 60 pulses, 300 μs. Although the number of pulses was higher in the second condition, the electric field intensity plays a major role in affecting viability over other parameters, as implied from Table 10.6. With higher electric field such as 2500 V/cm, complete cell death was obtained (Table 10.6 and Figure 10.14).

Table 10.6 Mean viability for combination of parameters used in IRE of CEF

S. No.	Voltage (V/cm)	Pulse length (μs)	No. of pulses	Viability (%)
1	2500	200	20	0.00
2	1500	250	40	12.11
3	500	300	60	12.59

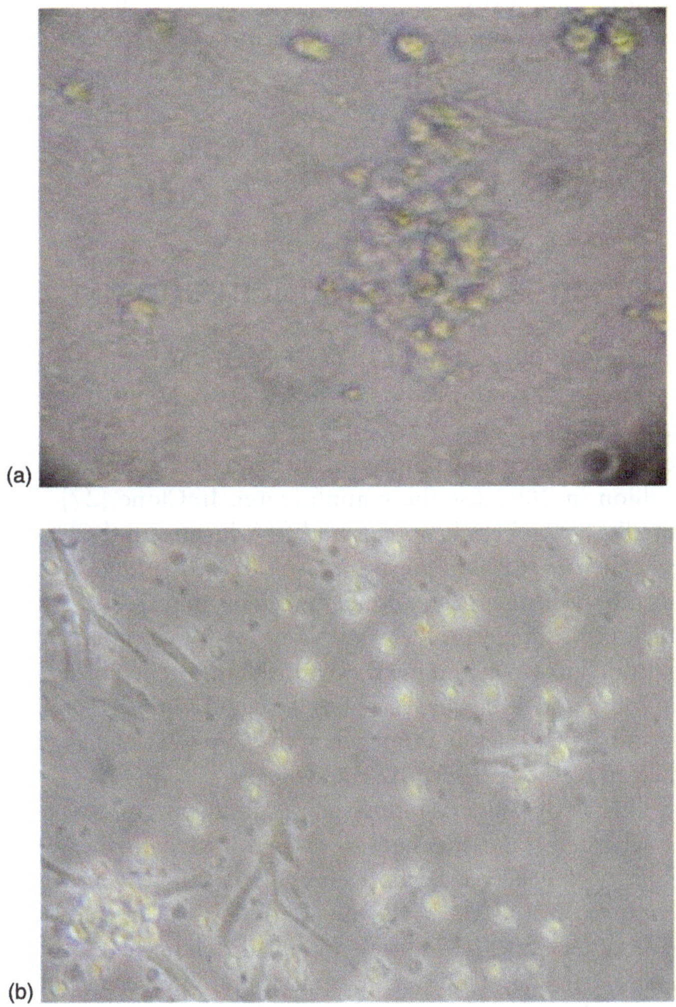

Figure 10.14 The chick embryo fibroblast cells before (a) and after electroporation at 2500 V/cm, 200 μs (b) CFU assay

The correlation co-efficient was found to be 99.8% for the electric field when compared to number of pulses or pulse length, which implies that the former is the major influencing factor in IRE. Figure 10.14 shows the morphology of the CEFs before and after electroporation at 2500 V/cm, 200 μs.

Figures 10.15(a–c) show the colonies formed by the various samples. Figure 10.15(a) shows the large number of colonies formed in the control sample (no pulses applied). In this case, the number of colonies before electroporation was found to be 86 CFU/ml, whereas after electroporation, the cell multiplication decreased, leading to reduced number of CFUs. The number of colonies at 1500 V/cm, 200 μs was 2 CFU/ml (Figure 10.15(b)). At 250 μs, only a few cells were observed (Figure 10.15(c)) and no CFUs were found. This could be due to the slightly increased pulse length, indicating the sensitivity of the column forming units in this case.

10.4 Discussion and conclusions

Oncology is one of the most profitable sectors in medicine. Cancer drugs cost from hundreds to thousands of dollars per month in the US. For example, Avstin (Genentech) costs up to $100000 a year and Erbitux, another cancer drug, costs around $10000 a month and pulled in revenues of $1.3 billion in 2007 for the manufacturer, ImClone [27]. Sixty-two percent of all personal bankruptcies and foreclosures in the US were due to medical expenses (69% of these are insured) [28]. 'Nobody talks about the financial impact of cancer, and it's huge', as indicated in a September 2008 article [29]. Single mothers with one or two children and breast cancer are too ill to work and nearly bankrupted by chemotherapy. Some lose their jobs, and even starve. They struggle with astronomical co-payments and deductibles at a time when they are too sick to work. These are in addition to the issues with drug delivery.

Once a disease of Western industrialized, affluent countries, cancer is now everywhere; but not the medication and the treatment [30]. Hence, there is a critical need to develop alternative, efficient and affordable cancer therapies. In this research, the efficacies of drug-free IRE in two human cancer cell lines exhibiting highly different phenotypes, a rat breast cancer cell line, a human lung adenocarcinoma cell line and in chick embryo fibroblasts, were examined. The results show the promise of completely ablating the cell proliferation using electrical pulses of both

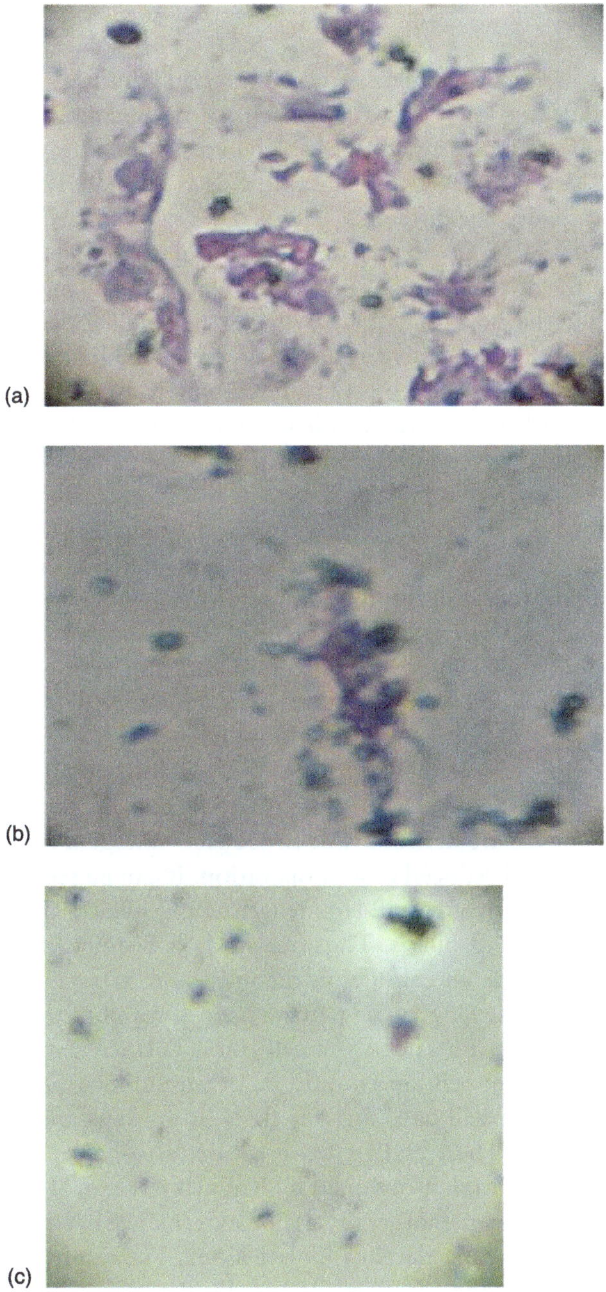

Figure 10.15 Colonies after the IRE: (a) colonies formed without any electroporation; (b) colonies formed after electroporating at 1500 V/cm, 200 μs; and (c) colonies formed after electroporating with one 1500 V/cm, 250 μs pulse

high and low intensities of micro- and milli-second durations. There were complete cell deaths in the case of 4T1 cells at 500 V/cm, 24 ms, 8 pulses, as well as 1500 V/cm, 2 ms, 8 pulses. The corresponding viabilities for MCF-7 and MDA-MB-231 for the same pulse parameters are 38% and 80.2% respectively. The difference between the two human breast cancer cell lines and the mouse breast cancer cell line could be due to the difference in the cell morphology, type of cell and the membrane.

Considering the two human cell lines, the MCF-7 is an estrogen receptor alpha (ERα) positive, weakly invasive, luminal epithelial-like human breast carcinoma cell line [31]. It is a widely used epithelial cancer cell line, derived from breast adenocarcinoma. MCF7 cells retain characteristics of differentiated mammary epithelium, including ability to process estradiol via cytoplasmic estrogen receptors. Although easy to propagate, the cells are generally slow-growing. However, MDA-MB-231 is an ERα negative, highly invasive, fibroblast-like cell line [31]. The considerable differences in response to protein kinases between these two human cell lines, MCF-7 and MDA-MB-231, could be due to their highly different phenotypes. Their cell architecture, morphology and motility, as well as their interaction with neighboring proteins and environment [31], could contribute to their difference in viabilities.

The experiments performed show that a longer pulse duration (millisecond) combined with a low intensity field gives a reduced viability. Low intensity fields are desirable due to reduced voltage levels and hence increased safety and reduced cost of operation. It can be seen that as low a magnitude of 100 V/cm at 24 ms length could elicit IRE effect with appropriate number of pulses. Thus our study is distinct from others in the use of low voltage magnitudes of the pulses.

In the case of pulse length, the percentage cell viability trend seems to be similar for both CEF and A549 cell lines. Further standardizations using different cancer cell lines would be of importance to structure an IRE protocol, which will pave the way for a promising cancer treatment in the near future. The trend of these results also correlates with the breast cancer cells and that obtained by Rubinsky's team [13, 14].

Irreversible electrochemotherapy is an extremely effective technique for ablation of cells using electrical pulses alone. The results of our study and previous studies indicate that electrical pulses can be used alone to provide a considerable reduction in viability of cells. This reduces the cost of chemo drugs, which vary from a few hundred dollars per month to several thousands per month, depending upon the type of drug. In the case of IRE, this drug cost is completely eliminated. Thus, there is high potential for this treatment for transfer to clinical applications.

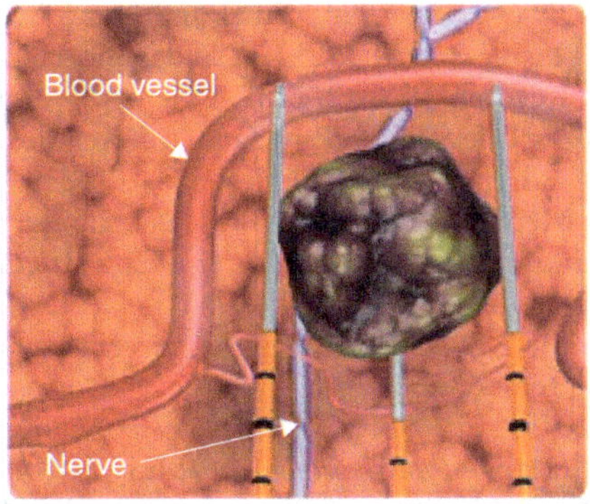

Figure 10.16 Medical application of IRE for treating an organ with cancer [32]

Figure 10.16 shows an application of microsecond pulse IRE for cancer in a human organ [32]. It can be applicable to cancer in the liver, kidney and other sensitive organs, without exposing the patient to traditional invasive surgical procedures, harsh chemotherapies or radiation treatments that might also damage nearby healthy areas.

10.5 Acknowledgments

The authors are very grateful to the students, Mindy Heish of Purdue University, R Snehalatha, K Santhosh, R Ramya and S Sajan of BS Abdur Rahman University, Chennai, India and K Priyadarshan, S Varsha and U Vimal Kumar of AU-KBC Research Centre, Anna University, Chennai, for their help with the experiments and S Venkatesh for editing.

10.6 References

1. Udayakumar P (2011a), 'Introduction and routes of drug administration', in: *Medical Pharmacology*. New Delhi: CBS Publishers & Distributors.
2. Udayakumar P (2011b), 'Cancer chemotherapy', in: *Medical Pharmacology*. New Delhi: CBS Publishers & Distributors.
3. Gonzalez-Augulo A M, Morales-Vasquez F and Hortobagyi G N (2007), 'Overview of resistance to systemic therapy in patients with breast cancer',

in *Breast Cancer Chemosensitivity*, Yu D and Hung M-C (eds), New York: Springer Science Business Media, LLC, Landes Bioscience, 1–22.

4. Lambert L and Keyomarsi K (2007), 'Cell cycle deregulation in breast cancer: Insurmountable chemoresistance or Achilles' heel?' in: *Breast Cancer Chemosensitivity*, Yu D and Hung M-C (eds), New York: Springer Science Business Media, LLC, Landes Bioscience, 52–69.
5. Love S M with Lindsey K (2000), *Dr Susan Love's Breast Book*, 3rd edition. New York: Capo Press.
6. McCarter, J (2013), 'Medical bills cost 62 percent of bankruptcies', *www.dailykos.com*. September 2013.
7. Haltiwanger S (2010), The electrical properties of cancer cells. Available at: *http://www.royalrife.com/haltiwanger1.pdf*, July 2010.
8. O'Sullivan G C (2011), Preface, *International Workshop on: Electrochemotherapy for Effective Cancer Treatment*, July 27–28, Chennai, India.
9. Rols, M P (2006), 'Electropermeabilization, a physical method for the delivery of therapeutic molecules into cells', *Biochimica et Biophysica Acta*, **1758**, 423.
10. Dev, S B, Dhar, D and Krassowska, W (2003), 'Electric field of a six-needle array electrode used in drug and DNA delivery *in vivo*: analytical versus numerical solution', *IEEE Transactions on Biomedical Engineering*, **50(11)**: 1296–300.
11. Larkin J O, Collins C G, Aaron S, Tangney M, Whelan M O et al. (2007), 'Electrochemotherapy: aspects of preclinical development and early clinical experience', *Ann Surg*, **245**: 469–79.
12. Campana L G, Mocellin S, Basso M, Puccetti O, Salvo G L et al. (2009), 'Bleomycin-based electrochemotherapy: clinical outcome from a single institution's experience with 52 patients', *Ann Surg Oncol*, **16**: 191–9.
13. Miller L, Leor J and Rubinsky B (2005), 'Cancer cells ablation with irreversible electroporation', *Technol Cancer Res Treat*, **4(6)**: 1–7.
14. Rubinsky B, Onik G and Mikus P (2007), 'Irreversible electroporation: a new ablation modality – clinical implications', *Technol Canc Res Treat*, **6(1)**: 1–12.
15. Rubinsky B (2007), 'Irreversible electroporation in medicine,' *Technol Cancer Res Treat*, **6(4)**: 255–9.
16. Tracy C R, Kabbani W and Cadeddu J A (2011), 'Irreversible electroporation: a novel method for renal tissue ablation', *BJU International*, **107(12)**: 1982–7.
17. Al-Sakere B, André, F, Bernat, C, Connault E, Opolon P, et al. (2007), 'Tumor ablation with irreversible electroporation', *PLOS ONE*, **2**: **1**–8.
18. NIH Clinical Center. 'Radiofrequency thermal ablation as tumor therapy,' *National Institute of Health*. N.p., 6 March. 2009. 10 May 2011. Available at: *http://clinicalcenter.nih.gov/drd/tumortherapy.html*
19. Heish M, Salameh T, Camarillo I G, and Sundararajan R (2011), 'Irreversible electroporation effects: a drug-free treatment for cancer', *Electrostatics Society of America (ESA) Annual Meeting on Electrostatics*.
20. 4T1 Metastatic Breast Cancer Model. (2012) Available at: *http://www2.healthsci.tufts.edu/saif/4T1.htm*

21. Sundararajan, R, Rajendran R, Shahid S, Santosh D K, Ramakrishnan K, et al. (2011), 'Effect of irreversible electroporation on cancer cells', *IEEE CEIDP*, 164–7.
22. *www.A549.com*. June 2011.
23. Sankaranarayana K, et al. (2011), 'Effect of irreversible electroporation on cell proliferation of fibroblasts', *ESA Annual Meeting on Electrostatics*.
24. MCF-7 cells (2011), Available at: *www.tgrbio.com/.../cell-models-mcf7-.php?id=45* and *cells.html*, May 2011.
25. MDA-MB-231 cells (2013), Available at: *icbp.lbl.gov/breastcancer/viewline*
26. 4T1 cells (2013). Available at: *radiology.rsna.org/content/248/3/936.full*
27. Arnst C and Capell K (2008), 'Lilly's risky rush into cancer drugs', *Business Week*, October 20, p 72.
28. Feder J (2009), 'Federal action is required', *US News & World Report*, February, p 6.
29. Phillips B J (2008), 'Up from under', *Readers Digest*, September, 17–18.
30. Kingsbury K (2007), 'The changing face of breast cancer', *Time magazine*, October 15.
31. Considerable differences between MCF-7 and MDA-MB-231 breast cancer cells in response to the protein kinase C (PKC) activator PMA (TPA). Available at: *http://www.bio.net/mm/cellbiol/2004-October/015589.html*, May 2011.
32. New in cancer treatment at Baptist Health: the nanoknife system. Available at: *http://www.baptist-health.com/technology/nanoknife.asp*, May 2011.

11

Targeted delivery of siRNA and other difficult to transfect molecules using electroporation: current status and future scope

Lisa M. Reece, Kavitha Sankaranarayanan, Priyadarshan Kathirvelu, Christy Cooper, Raji Sundararajan, and James F. Leary

DOI: 10.1533/9781908818294.245

Abstract: RNA interference (RNAi) is a natural phenomenon existing in plants, insects, and many mammals involved in gene silencing or the inhibition of gene expression. Small interfering RNAs (siRNAs) could be used as potential tools to implement this process that could lead to treatments in diseases such as cancer. However, the delivery of siRNA to eukaryotic cells is a challenge due to its lower cellular uptake and easy degradation by enzymes in blood, tissues, and cells. Electroporation-based siRNA delivery is an upcoming therapeutic tool to deliver various siRNAs, both *in vitro* and *in vivo*. Discussed in this chapter is the review of various successful deliveries of siRNA using electroporation.

Key words: RNA inference, siRNA, electroporation, cancer, gene silencing.

11.1 Introduction

RNA interference is an endogenous gene silencing mechanism recognized as a naturally occurring biological strategy for silencing alleles during development in plants, insects and many mammals [1, 2]. The RNAi mechanism plays different functions within a cell. RNAi confers immunity by degrading extraneous ribonucleic acids entering the cells and also promotes tissue specific gene expression patterns by modulating DNA conformation [3]. RNAi is initiated by a double stranded RNA that is homologous to the target mRNA. The double stranded RNA is cleaved by the enzyme, Dicer, into small duplexes of 21–25 nucleotides [1] with 2 nucleotide 3' overhangs. These small duplexes are small interfering RNAs (siRNAs) that are incorporated in to the RNA induced silencing complex (RISC), a large protein complex with argonaute as the major catalytic protein. The siRNA's are unwound into sense (passenger) and antisense (guide) strand by the activity of helicase in RISC. The antisense strand binds to the complementary target mRNA and mediates its degradation by cellular nucleases.

RNAi can be induced by dsRNAs of both endogenous and exogenous origins. The dsRNA of exogenous origin entering a cell, both through viral infection or experimental processing is cleaved and processed in RISC. The endogenous dsRNA are non-coding microRNAs (miRNAs) that regulate expression of specific genes. Till date there has been no report on endogenously expressed siRNAs in mammals [4]. The processing of target mRNA in RISC is based on the degree of complementarity exhibited by the siRNA or miRNA with the target. The siRNA shares complete complementarily with the target and results in mRNA degradation whereas miRNA binds partially and causes translational repression. This can be exploited for therapeutic uses in various diseases [4–6].

11.2 siRNA – a potential therapeutic tool for cancer treatment

The finding that synthetic dsRNA is capable of inducing RNAi mechanism is a promising strategy in treating many diseases including cancer. The siRNAs were initially used in gene knock down to study the function of specific genes. Unlike other antisense technology, siRNA promotes silencing of specific genes. The presence of an intact RNAi mediated gene silencing mechanism in a cancer cell [7] extends the application of siRNA in cancer treatment as well. The siRNA therapy for cancer is advantageous over chemotherapy in that it creates no side effects, no resistance and many siRNAs targeting multiple cancer candidate genes in a single treatment is a feasible alternative that results in cost reduction as well.

SiRNAs are readily synthesized with low production costs compared to protein or antibody therapies [5]. In addition, siRNAs have favorable pharmacokinetic properties and can be delivered to a wide range of organs [8].

11.3 siRNA- Gene targets in Cancer

Cancer is caused by a number of reasons including mutations, down regulation of tumor suppressor genes, socio economic factors, virus, etc. The genetic factors responsible for cancer are targeted in siRNA-mediated gene therapy. The genes promoting cell growth, proliferation, angiogenesis and metastasis in a cancer cell are potential targets for siRNA therapy as cancer is a multistage disease. Also the effects of existing drugs such as Imatinib and Rapamycin have been enhanced by silencing their resistance-associated gene BCRABL using RNAi technology [9].

The genes responsible for viability/cell survival in cancer cells can be identified using siRNAs through knock down approaches. Different biochemical, pharmacological, and histological assays have been used to determine the effects of siRNA inhibition of specific genes and to analyze the phenotypic changes in cells [5]. Currently, the use of siRNA to characterize gene function and potential therapeutic drug targets is a highly promising application for this technology [10]. Selective gene knockdown can be used to identify critical genes and/or pathways that can be targeted by siRNA, drugs and their combinations for treatment of cancers [5]. Conventional targets of traditional chemical drugs have been limited to certain classes of receptors, ion channels, and enzymes whereas a siRNA-based drug can target any mRNAs of interest, regardless of their cellular location of the translated proteins [11]. Thus with the use of siRNA almost any gene can be regulated.

The use of siRNAs is gaining momentum as a potential therapeutic tool for treatment of a variety of diseases including cancer, as they can induce RNA interference (RNAi) and inhibit the expression of target proteins, causing gene silencing or knock down [1–4]. However, cellular uptake of siRNA is extremely low due to its poly-ionic nature and it is easily degraded by enzymes in the blood, tissues, and cells, despite chemical modifications for improved stability. Thus, an effective delivery system is critical to induce RNAi at the site of action for both stabilization and effective transfection. Non-viral, physical techniques, such as electroporation (EP) could be used for improved targeted delivery compared to existing methods to achieve successful gene silencing in target cells *in vitro* and *in vivo will* enable the transition from bench to bed faster.

11.4 Delivery of siRNA

Molecular delivery is a great challenge in diseases like cancer treatment due to the impermeable nature of the bilayer lipid cell plasma membranes. The success of siRNA therapy depends on the effectiveness of siRNA delivery and design of the siRNA sequence [12, 13]. The various methods of siRNA delivery include viral, chemical, liposome-mediated, antibody conjugated, nanoparticle-mediated and non-viral, physical methods like electroporation (Figure 11.1). The various methods to upload are transfection using cationic liposomes, polymer nanoparticles and lipid conjunction, electroporation using voltage pulses and viral vectors [14–17]. An optimal delivery system should be biocompatible, non-immunogenic, inert (low toxicity), stable and tissue specific. More prolonged gene silencing has been achieved by expressing siRNAs from plasmid vectors that contain specific promoters [15–17].

Viral systems based on adenovirus, retrovirus, herpes simplex virus and non-viral delivery systems (liposomal formulations, mechanical

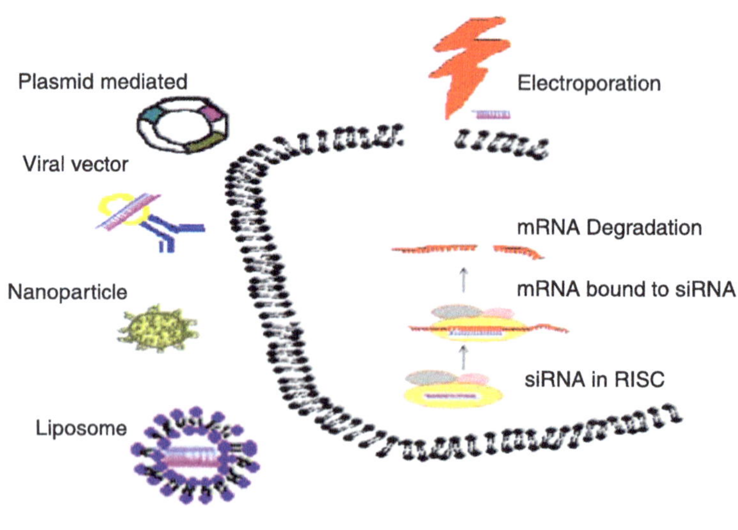

Figure 11.1 Various siRNA delivery systems and mechanism of siRNA mediated gene silencing in the cytoplasm.

administration) are being developed and used for delivery of siRNA and found to be more stable [18]. The genes, p53 mRNA in MCF-7 breast cancer and A549 lung carcinoma cells have been targeted with siRNA in a recombinant adenovirus system [19]. But the biosafety issues associated with the production of the viral vectors limits its use in RNAi mediated therapy. The inherent disadvantages of the viral delivery systems surpass their advantages of efficient delivery. The other techniques include lipid-based siRNA delivery, polymer-mediated siRNA delivery, nanoparticle-based siRNA delivery. Liposome mediated delivery is proven to show high transfection efficiency [20]. The liposomes are low toxic, less immunogenic and also prevent siRNA from nuclease degradation, thereby minimizing its renal clearance *in vivo*. However, nonspecific interferon response following administration of lipid-formulated siRNA molecules has been reported in mice [21, 22].

Major gene delivery systems exploiting siRNA strategies use either viral or non-viral vectors but some drawbacks are possible toxicity, immunogenicity, and possible inflammatory response [23]. For example, a major cause for concern over viral vectors is that there is a phenomenon known as insertional mutagenesis, in which the ectopic chromosomal integration of viral DNA disrupts the expression of a tumor-suppressor gene or activates an oncogene leading to malignant transformation of cells [24]. Another option for transfection of siRNA into cells is the use of Lipofectamine® 2000. Lipofectamine® 2000 Transfection Reagent is a cationic-lipid transfection reagent formulated by Life Technologies [25] for the transfection of plasmid DNA, small hairpin RNA (shRNA) and siRNA into eukaryotic cells. While this method has been successful since its launch in 1993, it does have a drawback in its application. One study that incorporated the transfection of SV40 large T-antigen into PKR/RNaseL-deficient cells, involved an overnight incubation with Lipofectamine® 2000 to assure cellular uptake [26]. Indeed, Life Technologies' protocol recommends harvesting cells 24–48 hours after transfection [27, 28]. While the successful transfection frequency using this procedure is high, the overall amount of time to complete the protocol may be an issue with some research, as the siRNA is mostly degraded in a very short time such that by 24 hours most of it would be gone. Thus getting it in fast is critically important.

For faster physical delivery electroporation is a promising tool and using electroporation to introduce siRNA into cells and tissues has been the most frequently applied technique using electrical pulses [29–40]. This has several advantages including delivery into difficult to transfect cells [14] and it does not have to induce nonspecific immune stimulations which might arise in some chemical delivery systems. In addition, due to their noninvasiveness and relative ease with which it can be used in clinical applications, it is gaining more interests.

11.5 Electroporation-based siRNA delivery

Physical, rather than chemical techniques are preferred for enhanced uptake of molecules including siRNAs. The physical method of siRNA delivery eliminates the adverse immune effects and biosafety issues associated with chemical delivery systems (viral or liposome).

Electroporation or electropermeabilization (EP) is a safe, non-viral, efficient gene delivery technique, wherein precisely controlled, high intensity, low duration electrical voltage pulses are applied to open up transient pores through semi-permeable membranes and tissues allowing targeted delivery of therapeutic molecules including genes, drugs, antibodies, and vaccines *in vitro* and *in vivo* including clinical trials [29–31]. EP has been proven for its efficacy *in vitro* and *in vivo* including clinical trials [5–8]. EP has been shown to offer a 100–1000 fold improved therapeutic benefit compared to drug alone in electro-chemo-therapy [30].

Kavitha et al. reported the use of electroporation alone and in combination with the cancer drug to reduce the viability of various cancer cell lines [32, 33]. The non-invasiveness and the relative convenience for clinical application further increase the use of physical method of delivery for siRNA [34]. Electroporation has also been shown to be the best method of choice for siRNA delivery *in vitro* and *in vivo*. In electroporation, the optimal electric field and pulses employed increases the permeability of cell membrane to siRNAs. For example, the siRNA targeting ERK1–2, responsible for cancer cell survival, delivered by electroporation in primary pancreatic cancer cells resulted in reduced cell growth [35]. In addition, EP has been used to transfect siRNA *in vitro* into human cell lines, such as human monocytes and dendritic cells (DC) that are refractive to current techniques [36]. The square wave electroporator, BTX ECM 830 (Genetronics, CA) was used for this purpose. RPMI 1640 media supplemented with 10% FCS and antibiotics was used. The pulse parameters used were: 500 V across 4 mm cuvette electrode gap, corresponding to an electric field intensity of 1250 V/cm, 2ms and one pulse. Around 5×10^6 DC cells were transfected with siRNAs with various sequences, such as 5′-GCCUCCUAUUUGGUUUATT-3′ (S-sense strand (passenger/target site) and 3′-TTCGGAGGAUAAAACCAAAUA-5′ (A-antisense (guide).

EP was found to be both a desirable and efficient method for siRNA delivery into monocyte-derived dendritic cells [37] while being able to resolve some of the cell-line specific limitations associated with liposome-based transfection. It has further been demonstrated that EP can transfect siRNA into hard-to-transfect cells [38].

Jordan et al. [39] optimized the electroporation pulse conditions for a number of primary and other difficult-to-transfect cells. The various cells studied include human umbilical vein endothelial cells (HUVEC), human primary fibroblasts (HPF), human neuroblastoma cells (SK-N-SH), mouse neuroblastoma cells (Neuro-2A), Jurkat, MCF-7, and others (Tables 11.1 and 11.2). Both square wave and exponential decay pulses were used to upload various siRNAs and plasmid DNAs. The results indicate that under appropriate voltage and pulse durations, it is possible to silence genes. They obtained impressive transfection efficiencies of 75% for Neuro-2A, 93% for HPF, and 94% for HUVEC cells. Figure 11.2

Table 11.1 SiRNA electroporation using square wave pulses [39]

Cell	Pulse Conditions
HUVEC	250V, 20ms
SK-N-SH	200V, 20ms
CHU K-1	250V, 20ms
CHO DG44	300V, 20ms
HeLa	250V, 20ms

Table 11.2 SiRNA electroporation using exponential decay pulses [39]

Cell	Pulse Conditions
HUVEC	350V, 500 μF, 1000Ω
Jurkat	250V, 500 μF, 1000Ω
Neuro-2A	250V, 500 μF, 1000Ω
HPF	250V, 500 μF, 1000Ω
MCF-7	250–300V, 500μF, 1000Ω
HeLa	250V, 350μF

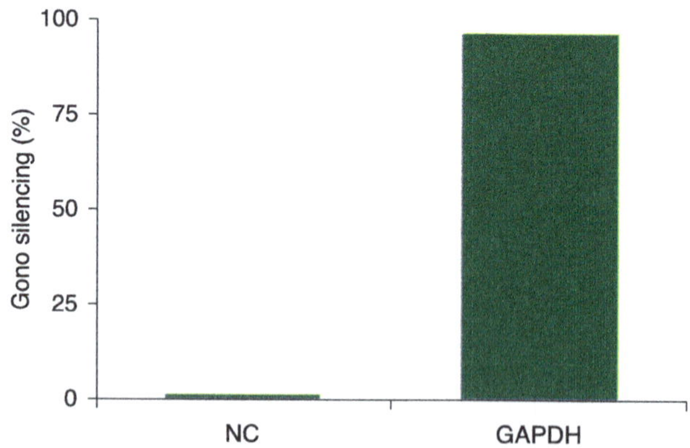

Figure 11.2 Electroporation of human primary fibroblast (HPF) cells with siLentMer™ siRNA using an exponential pulse of 250 V, 500 μF, 1000Ω. RT-qPCR results 24h post-electroporation of 100 nM GAPDH and negative control (NC) silent siRNA, expressed as percentage of gene silencing [39].

shows an illustration of the gene silencing in HPF cells with siRNA compared to GAPDH.

Electroporation was used to both efficiently and economically deliver DNMT1 siRNA into estrogen receptor (ER) negative breast cancer cell line, MDA-MB-468 to silence DNA methyl transferase mRNA [40]. The siRNA sense sequence was 5′-CGGUGCUCAUGCUUACAACTT-3′ and the antisense sequence was 5′-GUUGUAAGCAUGAGCACCGTT-3′. 19 pulse conditions were studied to optimize the most viable condition for siRNA delivery (Tables 11.3 and 11.4). 16 exponential and 3 square wave pulse forms were used and 220 V, 975 μF exponential pulse condition was found suitable with 97% viability. 2, 5, and 10 nM siRNA were delivered and their efficacy was studied using western blot analysis.

Table 11.3 SiRNA electroporation using square wave pulses into MDA-MB-468 cells [40]

Voltage, V	Pulse duration, ms
160	25
150	10
130	25

with kind permission [40]

Table 11.4 SiRNA electroporation using exponential decay pulses into MDA-MB-468 cells [40]

Sample	Voltage, V	Capacitance, μF
1	240	1000
2	155	1000
3	110	1000
4	360	975
5	340	975
6	300	975
7	280	975
8	220	975
9	210	975
10	200	975
11	180	975
12	140	975
13	160	750
14	200	500
15	160	500
16	140	500

with kind permission [40]

Figure 11.3 shows the viabilities of electroporated samples with the siRNA. It is identified that 10 nM siRNa with 74% viability as the optimal condition and the advantages of electroporation were realized including use of low dose siRNA, higher efficiency than liposomes and more viability.

HL-60 cells are noted for their difficulty to transfect. Electroporation was used to optimize the pulse parameters (the voltage, from 175 V to 250 V) as well the siRNA doses (25–200 nM) and the cell density (1×10^6 to 1×10^7) using Gene Pulser MXcell™ system [41].

Electroporation enabled efficient genetic transfection of *siRNA in vivo* into various organs and tumors without the aid of any infectious viral vector. Mazda et al. employed electroporation to deliver Epstein-Barr virus (EBV)-based plasmid vectors in order to increase viral transfection efficacy [42].

Vascular endothelial growth factor (VEGF) siRNA electroporation suppressed the growth of tumors exhibiting high VEGF expression to less than 10% of the control level [43]. In this study, using low intensity,

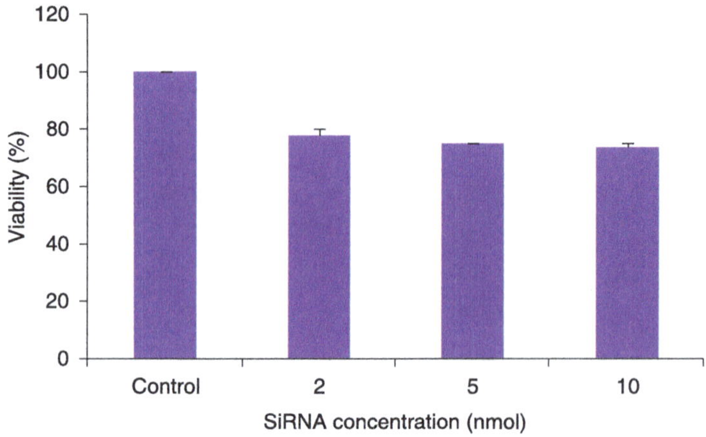

Figure 11.3 **DNMT1 siRNA delivery using exponential electroporation pulses at 220 V, 975 μF into MDA-MB-468 breast cancer cells.**
with kind permission [40]

long duration pulses of 91 V/cm, 50ms were applied, three each at positive and negative polarities at 100 ms interval between pulses into tumors grown in nude mice. For this purpose, vascular endothelial growth factor (VEGF) siRNA (5′-GGAGUACCCUGAUGAGAUCdTdT-3′ (sense) and 5′-AAAUUCCCGCGUUACGCGUdTdT-3′ (antisense)) and VEGF siRNA-SCR (5′-ACGCGUAACGCGGGAAUUUdTdT-3′ (sense) and 5′-AAAUUCCCGCGUUACGCGUdTdT-3′) were used. Using electroporation, they could achieve antitumor effect with only 1/4th the dose at twice the interval between the doses compared to atelocollagen-mediated method.

Golzio et al. uploaded the egfp22 siRNA (sense: 5′r(GCAAGCUGA CCC UGA AGU UCA U), ANTISENSE: 5′r(GAACUUCAG GGU CAG CUU GCC G) AND P76 SIrna (SENSE: 5′r(GCGGAGUGGCCU GCA GGU A)dTT, antisense: 5′r(UACCUG CAGGCC ACU CCG C)dTT into female C57Bl/6 mice [44]. Eight (4 each positive and negative) 800 V/cm, 5ms pulses were applied at 1 Hz frequency in their study. The results indicated that electric field could be used an efficient method for siRNA delivery for gene silencing into cells of solid tumors *in vivo* using non-systemic approach.

Other applications, such as naked siRNA transfection into muscle, retina, arthritic joints, and tumors have also been reported in [3].

SiRNA clinical trials indicate their applications as emerging biodrugs, due to their specific and potent RNAi triggering activity [45]. They are also applicable as therapeutic drugs due to their wide range of target proteins that can be used for gene silencing for treating diseases.

It must be noted, however, that naked therapeutic genes are rapidly degraded by nucleases, showing poor cellular uptake, non-specificity to the target cells and thus low transfection frequency [46]. Further, the high cost, intensive labor, and biosafety problems also limit the widespread application of siRNA as a therapeutic mechanism [39].

11.6 Summary and Future Scope

It is a great challenge to upload potent therapeutic molecules, such as siRNA, which is a promising therapeutic tool for specific gene silencing for various medical applications including cancer. However, enzymatic degradation, extremely low accumulation in the cells and difficult delivery, present challenges to *in vitro* and *in vivo* use of siRNAs. It is critical to use a delivery system that can achieve both stabilization and effective transfection in living cells. In addition, it should also be useful for improved targeted delivery into difficult-to-transfect cell lines.

Electroporation, whose pulse parameters and dosages have to be optimized for each cell/tissue is a good, non-viral, physical technique compared to other methods to achieve successful gene silencing in target cells *in vitro* and *in vivo* and enable the transition from bench to bed faster. The results of various researchers endorse the use of electroporation for effective and economical siRNA delivery. Thus, siRNA delivery using electroporation is a novel, next generation of physical delivery procedures having a wide scope for various applications. It is a reliable and upcoming promising strategy to deliver the synthetic siRNA into cells and tumors, enhancing the applicability of RNAi based therapy for cancer treatment.

11.7 References

[1] Higuchi, Y., Kawakami, S., and Hashida, M (Biodrugs 2010), 'Strategies for In Vivo delivery of siRNAs: Recent progress'. **24(3)**: 195–205.

[2] Storvold, GL., Andersen, TI., Perou, CM., and Frenzen, M (2006),'siRNA: a potential tool for future breast cancer therapy?'. *Critical Reviews™ in Oncogenesis*, **12(1–2)**: 127–150, 127–150.

[3] Lim, L. P., Lau, N. C., Garrett-Engele, P. et al.(2005), 'Microarray analysis shows that some microRNAs downregulate large numbers of target mRNAs.Nature'; **433**: 769–73.

[4] Abdelrahim, M., Safe, S., Baker, C. and Abudayyeh, A (2006), 'RNAi and cancer: Implications and applications'. *J RNAi Gene Silencing*; **2**: 136–145.
[5] Baulcombe, D. C (1996) 'RNA as a target and an initiator of post-transcriptional gene silencing in transgenic plants'. *Plant Molecular Biology*, **32**: 79–88.
[6] Cogoni, C., Romano, N., Macino, G (1994) 'Suppression of gene expression by homologous transgenes'. *Antonie Van Leeuwenhoek*, **65**: 205–209.
[7] Billy, E., Brondani, V., Zhang, H., Muller, U. and Filipowicz, W (2001), 'Specific interference with gene expression induced by long, double-stranded RNA in mouse embryonal teratocarcinoma cell lines'. *Proc Natl Acad Sci (USA)*; **98**: 14428–33.
[8] Braasch, D. A., Paroo, Z., Constantinescu, A. et al (2004), 'Biodistribution of phosphodiester and phosphorothioate siRNA'. *Bioorg Med Chem Lett*; **14**: 1139–43.
[9] Lu, P.Y., Xie, F.Y. and Woodle, M.C (2003), 'siRNA-mediated antitumorigenesis for drug target validation and therapeutics'. *Curr Opin Mol Ther*; **5**: 225–34.
[10] Karagiannis, T.C., El-Osta, A (2005), 'RNA interference and potential therapeutic applications of short interfering RNAs'. *Cancer Gene Ther*. **12**: 787–95.
[11] Chen, J., Wall, N.R., Kocher, K., Duclos, N., Fabbro, D., Neuberg, D. et al. (2004), 'Stable expression of small interfering RNA sensitizes TEL-PDGFbetaR to inhibition with imatinib or rapamycin'. *J Clin Invest*, **113**: 1784–91.
[12] McManus, M.T. and Sharp, P.A (2002), 'Gene silencing in mammals by small interfering RNAs'. *Nat Rev Genet*; **3**: 737–47.
[13] Bantounas, I., Phylactou, L.A. and Uney, J.B (2004). 'RNA interference and the use of small interfering RNA to study gene function in mammalian systems'. *J Mol Endocrinol*, **33**: 545–57.
[14] 'siRNA delivery methods into mammalian cells'. *Gene function study guide in stem cells. Pathways. Issue #9*, 2009. *SABiosciences*. www.SABiosciences.com
[15] Lee, N.S., Dohjima, T., Bauer, G. et al. (2002), 'Expression of small interfering RNAs targeted against HIV-1 rev transcripts in human cells'. *Nat Biotechnol*, **20**: 500–5.
[16] Paul, C.P., Good, P.D., Winer, I. and Engelke, D.R (2002), 'Effective expression of small interfering RNA in human cells'. *Nat Biotechnol*, **20**: 505–8.
[17] Sui, G., Soohoo, C., Affar, el. B. et al. (2002), 'A DNA vector-based RNAi technology to suppress gene expression in mammalian cells'. *Proc Natl Acad Sci (USA)*, **99**: 5515–20.
[18] Devi, G.R (2006), 'siRNA-based approaches in cancer therapy'. *Cancer Gene Therapy*, **13**: 819–29.
[19] Shen, C., Buck, A.K., Liu, X., Winkler, M., Reske, S.N (2003), 'Gene silencing by adenovirus-delivered siRNA'. *FEBS Lett* **539**: 111–14.
[20] Oh Y.K., Park T.G (2009), 'siRNA delivery systems for cancer treatment'. *Adv Drug Deliv Rev*, **6**: 1850–62.
[21] Ma, Z., Li, J., He, F., Wilson, A., Pitt, B., Li, S (2005), 'Cationic lipids

enhance siRNA-mediated interferon response in mice', *Biochem Biophys Res Commun*, **330**: 755–59.
[22] Hornung, V., Guenthner-Biller, M., Bourquin, C., Ablasser, A., Schlee, M., Uematsu, S., et al.(2005), 'Sequence-specific potent induction of IFN-alpha by short interfering RNA in plasmacytoid dendritic cells through TLR7'. *Nat Med*, **11**: 263–70.
[23] Rolland, A (2005), 'Gene medicines: The end of the beginning?'. Advanced Drug Delivery Reviews ,57: 669–673.
[24] Mao, S., Sun, W, Kissel, T (2010), 'Chitosan-based formulations for delivery of DNA and siRNA'. *Advanced Drug Delivery Reviews*. 62. DOI: http://dx.doi.org/10.1016/j.addr.2009.08.004.
[25] www.invitrogen.com, 2012.
[26] Gitlin, L., Karelsky, S, Andino, R (2002), 'Interfereing RNA confers intracellular antiviral immunity in human cells'. *Nature*. **418**: 430–434.
[27] Yu, J. Y., DeRuiter, S.L., and Turner, D.L (2002), 'RNA interference by expression of short-interfering RNAs and hairpin RNAs in mammalian cells'. *Procedures of the National Academy of Science*. **99(9)**: 6047–6052.
[28] Yu, J. Y (2005), 'Cotransfecting Plasmid DNAand RNAi into Mammalian Cells Using Lipofectamine™ 2000'. *L. Technologies*.
[29] Rabussay, D.P., Nanda, G.S., and Goldfarb, P.M (2002), 'Enhancing the effectiveness of drug-based cancer therapy by electroporation (electropermeabilization)', *Technology in Cancer Research & Treatment*, **1**:1.
[30] Larkin, J., et al. (2007), 'Electrochemotherapy: Aspects of preclinical development and early clinical experience'. *Annals of Surgery*, Vol. 245, No. 3, p. 469–479.
[31] Campana, L.G., et al.(2009), 'Bleomycin-based electrochemotherapy: clinical outcome from a single institution's experience with 52 patients'. *Annals of Surgical Oncology*, **16**: 191–199.
[32] R. Sundararajan, F. Xiao, T. Salameh, L. Reece, I.G. Camarillo, and L.G. Campana (2012). 'Effective proliferation control of human cancer cells using electrical pulses'. *IEEE Trans. Dielectrics and Electrical Insulation*, Vol. 19, No. 6.
[33] Sankaranarayanan K, Ramachandran RP, Kumar MS, Kumar V M, Vignesh S, Sadasivam P, et al. (2010), 'Viability of cancerous cell lines as a function of energy delivered'. *Electrical Insulation and Dielectric Phenomena CEIDP*, pg no:1–4.
[34] Novobrantseva, T.I., Akinc, A., Borodovsky, A., and de Fougerolles (2008), 'A Delivering silence: advancements in developing siRNA therapeutics'. *Curr Opin Drug Discov Devel*, **11**: 217–24.
[35] Salem, A.F., Funel, N., Pollina, L.E., Boggi, U., Del Chiaro M., Mosca, F (2008), 'Transfection of SIRNA for ERK1–2 in Primary Pancreatic Cancer Cells Cultures by Electroporation', *Reduce the Cell Growth In Vitro. J Pancreas*, **9**: 859–64.
[36] Mobergslien, A. and Sioud, M (2010), 'Optimized protocols for siRNA delivery into monocytes and dendritic cells', Chapter 5, in M. Sioud (ed.), *RNA Therapeutics, Methods in molecular Biology 629*, Springer Science + Business Media, LLC 2010.

[37] Prechtel, A. T., Turza, Nadine M, Theodoridis, Alexandros A., Kummer, Mirko, Steinkasserer, Alexander (2006), 'Small interfering RNA (siRNA) delivery into monocyte-derived dendritic cells by electroporation'. *Journal of Immunological Methods*. **311(1–2)**: 139–152'.

[38] Jiang, Q., Zhang, Zhaoqing, Bank, Alexander, Quellhorst, George, Huang, Jiaqiang, Medicus, Rudy (2009). SABiosciences Pathways. Technical Notes – siRNA Delivery Methods into Mammalian Cells (Gene Function Study Guide in Stem Cells). J. Hung, Sellappan, Shankar. Frederick, MD, USA, SABiosciences Corporation. 9.

[39] Jordan, ET., Collins, M., Terefe, J., Ugozzoli, L., and Rubio, T (2008), 'Optimizing electroporation conditions in primary and other difficult-to-transfect cells'. *J of Biomolecular Techniques*. **19**: 328–334.

[40] Arabsolghar, R., and Rasti, M (2012), 'Optimal electroporation condition for small interfering RNA trasnspection into MDA-MB-468 cell line'. *Iran J Med Sci*, Vol. 37, No. 3, pp. 187–193.

[41] Schaff, U., and Simon, S (2009), 'Electroporation parameters for transfection of HL-60 leukocytic cell line with siRNA using the Gene Pulser MXcell™ system'. *Tech note 5778*, Rev B.

[42] Mazda, O.,Kishida, T (2009), 'Molecular therapeutics of cancer by means of electroporation-based transfer of siRNAs and EBV-based expression vectors'. *Front Biosci*, **1**: 316–31.

[43] Takei, Y., Nemoto, T., Mu, P., et al. (2008), 'In vivo silencing of a molecualr target by short interfering RNA electroporation: tumor vascularization correlates to delivery efficiency'. *Mol Cancer Ther*, 7(**1**): 211–220.

[44] Golzio, M et al.(2007), 'In vivo gene silencing in solid tumors by tareted electrically mediated siRNA delivery'. *Gene Therapy*. **14**, 752–759.

[45] Oh, Y., and Park, TG (2009), 'siRNA delivery systems for cancer treatment'. *Advanced Drug Delivery Reviews*. **61**:850–862.

[46] Zhao, M., Yang, H., Jiang, X., Zhou, W., Zhu, B., Zeng, Ying, Y., Kaitai, R., (2008), 'Caiping. Lipofectamine RNAiMAX: An Efficient siRNA Transfection Reagent in Human Embryonic Stem Cells'. *Biotechnology*. **40**: 19–26.

12

Electric field distribution study of breast tumors

Raji Sundararajan, Poornima Agoramurthy, Gowri Sree Varadarajan and Luca G. Campana

DOI: 10.1533/9781908818294.259

Abstract: Electric fields at certain magnitudes and durations (frequencies) have the desired property of inhibiting the growth of cancer cells, which divide faster than normal cells. Thus, they disrupt the cancer cells more easily than normal cells, as evidenced by various experimental studies. Electric field distribution is critical for effective electroporation, the technique of enhanced drug delivery using electrical pulses. In this chapter, the electric field distribution is studied for a number of breast cancer models, using commercial, industry-standard software packages. Large electrode array geometries were also studied to treat larger areas, such as the chest. The cancer tissues have different electric field intensity compared to normal tissue. This could be due to the difference in the electrical properties, conductivity and permittivity of these cells compared to normal cells, and the type and placement of the electrodes. It is hoped that these results will help in improving electrical pulse-mediated drug delivery techniques for cancer treatment.

Key words: electric field distribution, needle array electrodes, plate electrodes, electroporation, electro chemotherapy, resistivity, conductivity, permittivity.

12.1 Introduction

Breast cancer is the cancer that originates in the tissues of the breast [1–5]. It is the most frequently diagnosed cancer in women, due to innumerable hormone variations exerted at various ages, from birth to puberty to pregnancy to menopausal [1–3]. There are mainly two types of breast cancer – ductal carcinoma (occurring in the ducts of the breast) and lobular carcinoma (starting in the milk-producing areas called the lobules) [5].

Standard treatments for treating cancer are surgery, chemotherapy, radiation therapy, immunotherapy and monoclonal antibody therapy [4]. These conventional therapies have major side effects and are expensive; 62% of all bankruptcies and foreclosures in the US are due to medical causes including cancers [6]. In order to reduce the death of a woman every 13 minutes in the US and 1 in every 50 seconds (400 000+ deaths) worldwide, there is a critical need for alternate, safe, effective and economical treatments. Towards this, electroporation is a promising concept that helps enhanced uptake of normally impermeable chemotherapeutic molecules across the lipid bilayer plasma membranes of the cells using high intensity, short duration electrical pulses [7–13]. Being a physical technique, it is applicable to all types of histologies of cancers. It induces transiently enhanced permeability of biological membranes by short pulses of electric fields.

Factors that effect electroporation are electric field strength, pulse length, shape of electrical pulse, number of pulses, interval between pulses, size of target cells, molecule to be uploaded, type and size of electrodes [7–14]. Electric field magnitude (and its distribution) is the most dominant factor, because it determines the response of cells and tissues to applied voltages. Thus, in order to efficiently electroporate cells and tissues, it is important to understand the electric field magnitude and distribution that occur within body tissue, because if the field is too high, there will be cell death and if it is too low, it will not open up pores to uptake the drugs.

This chapter deals with our study on the electric field magnitude and distribution study of various breast models and electrode geometries. For this purpose, finite element method-based, industrial standard, commercial software were utilized, as finite element analysis is an efficient tool that can be used to study electric, magnetic field distributions and stresses in materials when subjected to external stimuli. Two-dimensional models were studied for their electric field distribution and intensity for the effect of applied voltages in breast tissues for a number of electrode configurations, such as parallel plate electrodes, two- and multi-needle electrode arrays. These results will help improve the understanding of electroporation-mediated drug delivery techniques for cancers that are not receptive to conventional therapies.

12.2 Electric field concepts

The force exerted on a unit of positive charge is termed the electric field. For effective electroporation, it is essential to know the electric field intensity (magnitude) and its distribution in cells and/or tissues. Electric field intensity in V/m is more commonly used to express the strength of an electric field, and the direction of the field is in the direction of the force. The electric field in a region can be computed by using Maxwell's laws of electrostatics [15], as shown in Equations 12.1 and 12.2:

$$\nabla \cdot E = \frac{\rho}{\in_0} \quad [12.1]$$

$$\nabla \times E = 0 \quad [12.2]$$

where E is the electric field in V/m, ρ is the charge density and $\in_0$ is the permittivity of air. The electric potential (in volts), defined as the work done in bringing a unit of positive charge from infinity to that point, can be derived from Maxwell's equation and is related to the electric field, as shown in Equation 12.3;

$$-\nabla \phi = E \quad [12.3]$$

where ϕ is the electric potential in volts.

12.3 Electrical properties of cells

The electrical properties of cancer cells influence the magnitude and distribution of the electric field, which depends on the ion concentrations on either side of the cell membrane [16]. They are different, causing a voltage difference across the membrane. The cell membrane is composed of a bilayer of lipids, which acts as an insulator and restricts movement of ions in and out of the cell. Sodium and potassium ions are maintained at different concentrations on either side of the membrane due to the selective permeability of the membrane, thus causing a membrane potential difference. The membrane potential difference helps in maintaining cell membrane permeability to a large variety of nutrients and controls energy production and synthesis of macromolecules. It also creates a strong electric field around the cell membrane.

A healthy living cell has a membrane potential of –60 to –100 mV [17]. The electric field in a healthy cell can be as high as 100–200 kV/cm [18, 19]. When cancer arises, all regular mechanisms occurring in the cells are interrupted. The electrical properties of a cancer cell are much different than a healthy adult cell [16]. The various changes include reduction in cell energy production [16], change in charge distribution [17, 20], different lipid and sterol content [21], change in membrane potential and permeability (sodium and water move into the cell and potassium, calcium, and magnesium are transported out [22], with more water and sodium inside the cell and lower potassium than a normal cell) [23–25].

In general, the electrical properties of cancer cells differ from normal cells [16]. This is caused by the lower negative charges at the surface, and changes in the pH, mineral content, ionic species, structural membrane changes, membrane potential changes, extra cellular matrix changes and protein changes. In addition, there are changes in the genes and the sialic acid (a monosaccharide with a 9-carbon backbone) content. Cancer cells have lower potassium concentration and higher sodium concentrations than normal cells. As a result of these mineral changes, energy abnormalities and membrane charge distribution abnormalities, the capacitance changes and hence the relative permittivity. This leads to changes in the electric field intensities as observed in the simulation using different magnitudes for the relative permittivity and the conductivity (or the resistivity) for malignant tissue compared to the normal tissue. A clear understanding of these variables will enable us to obtain the optimal parameters of voltage to be applied for a given duration for effective electrochemotherapy applications in the clinic.

Due to the several changes that occur in a cancer cell, the electrical properties of that cell are altered. The electrical conductivity and permittivity of cancer cells were found to be greater than normal tissues [26]. The change in electrical properties of cancer cells makes them more susceptible to external electric fields. Thus, electroporation, the phenomenon based on electric fields, can be effectively used to treat cancer cells while normal cells remain less or unaffected. These electric fields have more of an affect on the rapidly dividing cells such as cancer cells than on normal cells [27].

The transmembrane voltage induced when an external field is applied is given by Equation 12.4, also known as steady state Schwan's equation [28]:

$$\Delta\psi_m = 1.5ERcos\theta \qquad [12.4]$$

where ψ_m is the voltage induced, E is the electric field, R is the radius of the cell and θ is the angle measured from the center of the cell with respect to the electric field. This is the voltage induced after the onset of the electric field. The transient behavior in the initial few microseconds can be described by the first-order Schwan's equation [29] as:

$$\Delta\psi_m = 1.5ERcos\theta(1-e^{\frac{-t}{\tau_m}}) \qquad [12.5]$$

where τ_m is the time constant of membrane charging:

$$\tau_m = \frac{R\varepsilon_m}{2d\sigma_i\sigma_e / (\sigma_i + 2\sigma_e + R\sigma_m)} \qquad [12.6]$$

where σ_i, σ_m and σ_s are the conductivities of the cytoplasm, cell membrane and extracellular medium, ε_m is the dielectric permittivity of the membrane, d is the membrane thickness and R is the cell radius.

12.4 Finite element modeling

Numerical methods are often used to study electric field distribution in many systems, including biological systems because of their complex and intricate nature. Finite element modeling is one such numerical method, which has been a powerful tool to explain the complex processes in biological systems [30]. The finite element method is an effective numerical technique used to find approximate solutions to partial differential equations. In this method, the entire region consists of a complex system of points called nodes, which connect together to form a grid. The grid, called the mesh, has defined properties and is created according to the anticipated stress in the materials. Higher stress regions will have a higher number of nodes and a compact mesh. The mesh is generated automatically by the software used.

Previous studies have indicated that numerical models of cell and tissue electroporation have been useful to describe the processes that occur in them and for evaluation and optimization of various parameters. Modeling and simulation has been used to improve electrode design and placement and in the planning of treatment [31, 32–35]. The effect of tissue characteristics and electrical parameters on electrical field distribution have also been studied and analyzed. New experiments can be designed and planned at lower costs, since experimental conditions can be evaluated by modeling [36].

Finite element analysis has proven to be beneficial in the computation of electric fields inside biological systems [37,38]. In this research, a finite element model of the human breast tissue along with electrodes was developed and electric field distribution was plotted for various models and varying different parameters. The entire model was treated as a domain divided into segments. A mesh was created and the electric field plotted for the entire region. The voltage and electric field plots were obtained after solving a set of partial differential equations. The partial differential equations governing this model are given below. The electric potential is computed by the partial differential equation:

$$\nabla \cdot \varepsilon_r \varepsilon_0 \nabla \phi(x,y) = -\rho(x,y) \qquad [12.7]$$

where $\phi\ (x,y)$ is the electric potential, ε_r is the relative permittivity, and can be different for each material, ε_0 is the permittivity of free space, 8.854×10^{-12}F/m and $\rho\ (x,y)$ is the charge density. After the solution for the potential is generated, the system automatically computes the E-field using the relation:

$$E = -\nabla \phi \qquad [12.8]$$

where E is the electric field and ϕ (x,y) is the electric potential.

Figure 12.1 shows a finite element model created for electric field studies on liver tissues [39]. Two needle electrodes are inserted for the application of voltage in a block of tissue. Figure 12.1(a) shows the model developed using FEMLAB software, and Figure 12.1(b) shows the mesh generated.

The same software was used by Larkin et al., to study electric field distribution in tumor tissues [7]. They compared electric fields in tissues for different electrode types, namely parallel plate and needle electrodes. Figure 12.2(a) shows the electric field distribution obtained in the case of parallel plate electrodes. Figure 12.2(b) shows the electric field in the case of needle electrodes, when a field of 1400 V/cm is applied. Plate electrodes generated a homogenous electric field, whereas needle electrodes showed a marked reduction in field away from the tips.

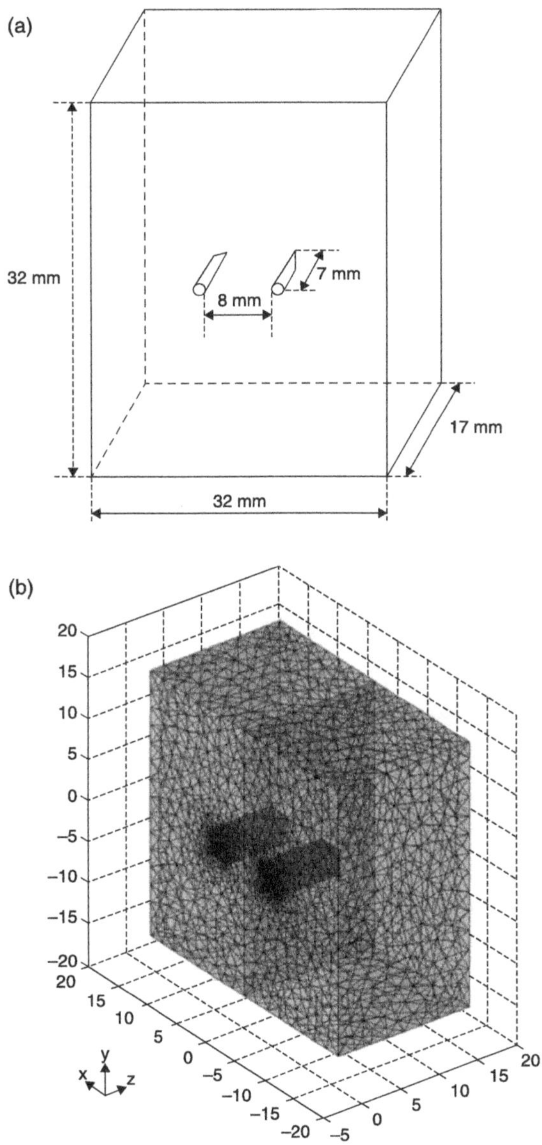

Figure 12.1 Finite element model (a) and mesh generated (b) for block of liver tissue with electrodes [39]

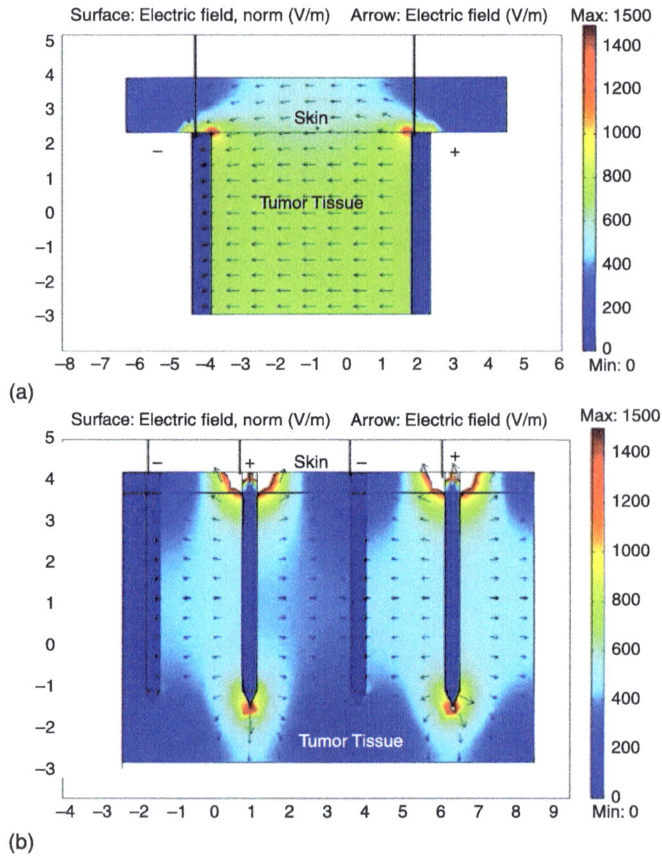

Figure 12.2 Electric field distribution of tumor tissue with parallel plate electrodes (a) and needle electrodes (b) [7]

12.5 Electric field intensity used in clinical electrochemotherapy

The electric field in the tissue depends on the voltage applied, inter-electrode gap and the electrode configuration. The field strengths that have been used in clinical trials vary between 800 and 1300 V/cm [7–14]. Eight square wave voltage pulses with a duration of 100 μs each at 1 Hz frequency are usually applied [7–13, 40–42]. When hexagonal needle electrode arrays are used, six pulses are applied [13].

12.6 Electrodes used

Different types of electrodes have been used in electrochemotherapy. The most common electrodes are parallel plate and needle electrodes. Parallel plate electrodes were the first electrodes introduced for tumor treatment, but proved to be disadvantageous due to their limited use. They can be used only for cutaneous tumors and cannot reach deeper tumors. Several configurations of needle electrodes were compared by Gilbert et al. [43]. They found that a circular array with 6 needle electrodes and pulses delivered with 2×2 needles was the most efficient. Gehl et al. developed an electrode configuration of a 2×4 needle electrode array with an inter-electrode distance of 4 mm [14]. They found that this configuration yielded a homogenous electric field and the applied voltage could be reduced [11]. Figure 12.3 shows the three most commonly used types of electrodes, developed by the IGEA Company [44].

(a) (b) (c)

Figure 12.3 Electrode types available for commercial use – parallel plate (a), 2×4 needle electrode array (b), circular array (c) with kind permission from [44]

12.7 Thermal effects of electroporation

Electroporation techniques primarily focus on optimizing pulse parameters, so that a maximum area of tissue is electroporated and minimum damage occurs to surrounding tissues [31]. While electroporating tissues and designing protocols, secondary effects arising from electrical currents need to be considered. One of the most important effects to be avoided is Joule heating [45], which is the process by which passing current through a conductor releases heat. For electroporation applications, a temperature rise below 50 °C in the tissue is considered harmless. The equation for determining temperature rise due to Joule heating is given by [46]:

$$\Delta T = \frac{tE^z}{\tau \rho c_p} \qquad [12.9]$$

where ΔT is the temperature rise in degree centigrade, t is the pulse length that causes the temperature rise, E is the electroporation voltage gradient, r is the resistivity of the medium, ρ is the density and c_p is the specific heat of the medium.

12.8 Simulation

Maxwell Student Version (SV), Version 3.1.04 [47] and Maxwell 13 [48], by Ansoft Corporation, were used to create lobular models and run two-dimensional simulations. The ductal carcinoma was studied using ANSOFT 13, Ansoft Corporation [49]. All simulations were run on standard PCs. The mesh generation was automatic, with every model consisting of thousands of nodes and elements. The mesh was generated in such a way that curved and higher stress regions have more nodes than other low stress areas.

12.9 Breast modeling

12.9.1 Anatomy

The breast is composed of three types of tissues: breast fat (or adipose tissue), glandular tissue and connective tissue (fibrous strands called Cooper's ligament) [50]. The breast lies between the second and sixth ribs of the human body and consists of 15–20 lobes that radiate from the nipple. Lobes are surrounded by fat and connective tissues. Each lobe further consists of several lobules, which are the milk producing regions [4]. Lobules are the basic structural units of the breast and have a lining of epithelial cells, from where most of the cancers develop. Milk is produced in the lobules and is carried to the nipple through ducts.

12.9.2 Common types of breast cancer

Mainly, there are two most common types of cancer: *In situ* (or non-invasive) and invasive. *In situ* cancers are those in which cancer cells remain within the basement membrane of the lobules and the draining lactiferous duct. Invasive cancers are those in which there is dissemination of cancer cells outside the basement membrane of the ducts and lobules into the surrounding adjacent normal tissue. The most frequently occurring breast cancers are ductal carcinoma *in situ* (DCIS), lobular carcinoma *in situ* (LCIS), invasive ductal carcinoma (IDC) and invasive lobular carcinoma (ILC). DCIS is a type of cancer in which cancerous cells are inside some of the ducts, but have not spread to other regions of the breast or body. LCIS is not a type of cancer *per se*, but in presence of this disease there are high chances of developing the disease. LCIS is characterized by changes in the cells within the breast lobes. Invasive IDC is the most common type of breast cancer (70–80% of breast cancer cases), and occurs in the epithelial cells that line the ducts of the breasts, which penetrate through the walls of the ducts and spread into surrounding breast tissue. ILC represents about 10% of breast cancer cases and occurs in the epithelial cells that line the lobules of the breast [4,5,51]. This type of cancer is very difficult to detect.

12.9.3 Model of a lobular carcinoma

Figure 12.4 shows the duct and lobular arrangement of the breast [52]. To simulate this, a novel 5-segment lobular model was developed to realistically represent the anatomy of the human breast lobule. Figure 12.5(a) shows this configuration. This model can be viewed as a representation of LCIS and ILC. The figure shows two lobules, lobule 1 and 2. For a cancer that originates at lobule 1, electrodes are placed on either side of the lobule and electric field distribution is studied after application of voltages at the electrodes. Electrode 1 is the positive terminal and electrode 2 is grounded. Dimensions of tissue and electrodes are given in Table 12.1. Layers of fat (20 mm) and skin (2 mm) are included in the model to make it closer to reality. The model with fat and skin layers included is shown in Figure 12.5(b).

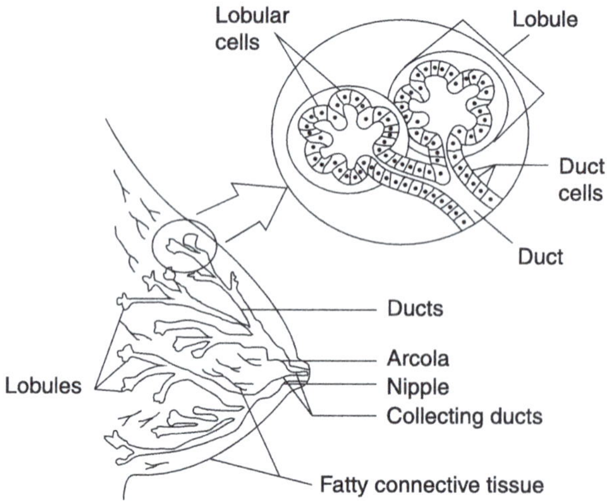

Figure 12.4 Human breast anatomy – duct and lobule arrangement [52]

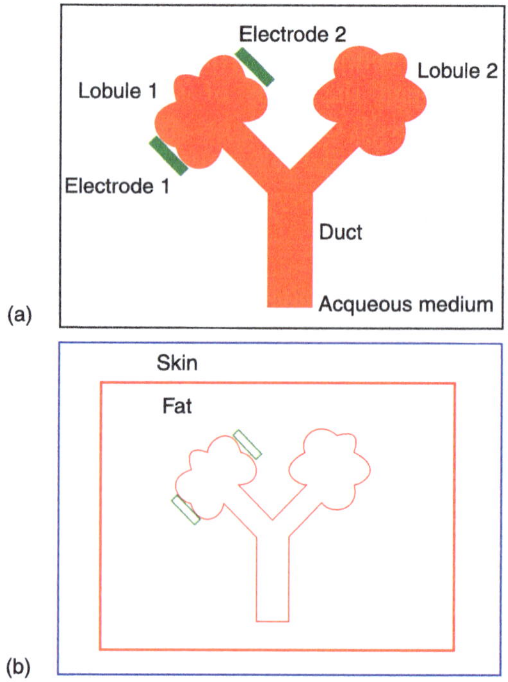

Figure 12.5 Electric-field model of 5-segment breast lobule tissue only (a) and with fat and skin (b)

Table 12.1 Dimensions of tissue and electrodes for lobule and duct model

Component	Dimension (mm)		
	Length	Breadth	Radius
Duct	12.0	4.0	–
Ductile	5.6	2.0	–
Lobule	–	–	4.24
Electrode	4.0	1.0	–

Table 12.2 Electrical properties used [53,54]

Tissue Type	Parameters	
	Rel. Permittivity	Conductivity (ohm.m)
Normal tissue	200.0	0.4
Cancer tissue	1000.0	2.0
Fat	15.0	0.2
Skin	33.5	0.7

12.9.4 *Model of a ductal carcinoma*

The cancer originating in the duct is called ductal carcinoma. Figure 12.6 illustrates two of the most common types of breast tumor, DCIS and IDC [55]. Figure 12.7 shows a model of the same. Here, the tumor is placed on the center of the duct. An 8-needle electrode array was inserted into the tumor. A high voltage of 480 V is applied to the left electrode, and the right electrode is grounded across an electrode gap of 4 mm, to give an electric field of 1200 V/cm [7–13] (Table 12.2). The dimensions of the model are shown in Table 12.3 and the electrical properties are shown in Table 12.4.

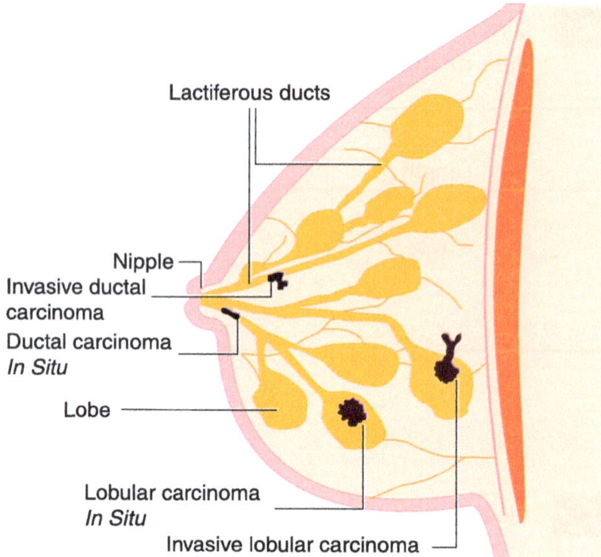

Figure 12.6 Breast with the representations of two of the most common types of breast tumor: *in situ* and invasive [55]

Figure 12.7 Model of ductal carcinoma

Table 12.3 Dimensions of the human breast used

	Length (mm)	Breadth (mm)	Radius (mm)	Thickness (mm)
Electrode	25.0	–	1.0	–
Fat	–	–	20.0	30.0
Skin	–	–	22.0	30.0
Lobule	–	–	3.2	30.0
Duct	10.0	5.7	–	30.0

Table 12.4 Electrical properties of the human breast used [53]

Tissue type	Parameters	
	Relative permittivity	Conductivity (S/m)
Normal tissue	52.0	2.2
Cancerous tissue	260.0	10.3
Fat	15.0	0.2
Skin	33.5	0.7
Stainless steel	1.0	$1.1e^{6}$

12.9.5 Model of breast carcinoma

Modeling of the entire human breast will help study electric fields in tumors that are present in the breast and have spread significantly. These can be any type of cancer – *in situ* or invasive, lobular or ductal (Figure 12.8 [56]). When cancerous cells have spread significantly and grown beyond their region of origin, tumor sizes increase and are more difficult to treat. Depending on the tumor size, appropriate electrode configurations are selected. Various tumor sizes and suitable electrode arrays are shown in this section. Dimensions of all model components are listed in Table 12.5.

A two-dimension model consisting of the human breast, tumor tissue and electrodes was developed to study electric field distribution (Figure 12.9). As shown in Figure 12.9, electrode 1 (positive) and electrode 2 (negative) are used to apply voltages. A voltage of 8400 V was

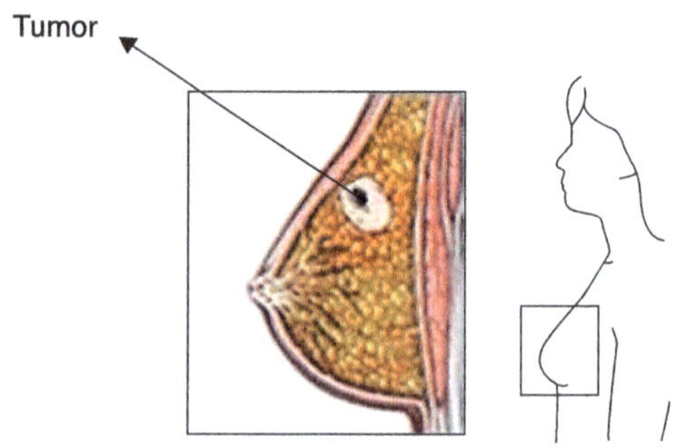

Figure 12.8 Human breast with tumor [56, 57]

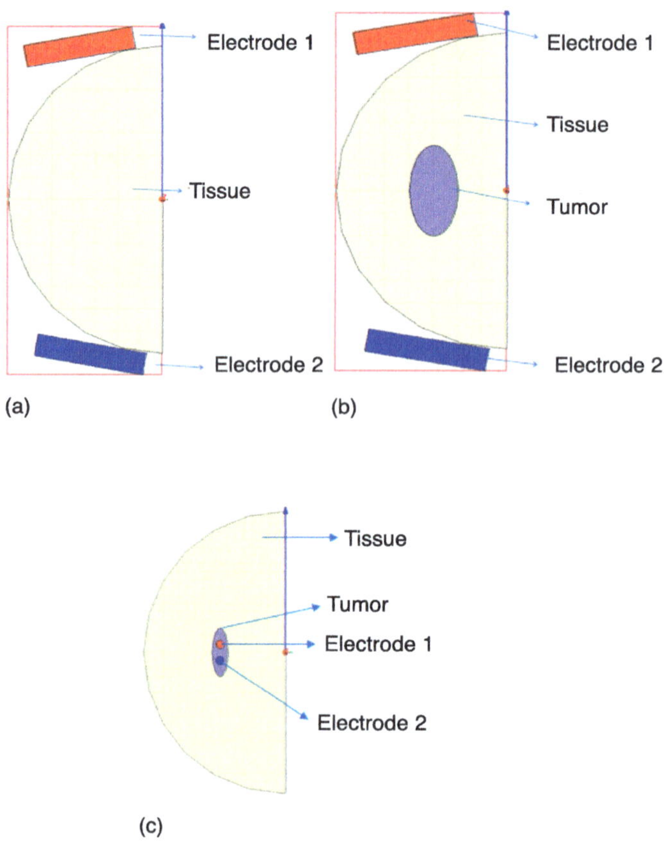

Figure 12.9 Finite element model of normal breast tissue with parallel plate electrodes (a); tumor within breast tissue with parallel plate electrodes (b); and tumor within breast tissue with needle electrodes (c)

applied to obtain a 1200 V/cm electric field intensity across the 7 cm diameter of the breast. The various models studied include the normal breast tissue (a) and breast tissue with a tumor (b) (Figure 12.10). Parallel plate electrodes (a, b) are used to apply voltages externally, whereas needle electrodes (c) are inserted directly into the tumor area (Figure 12.11). The dimensions of model components are listed in Table 12.6 and model parameters in Table 12.7.

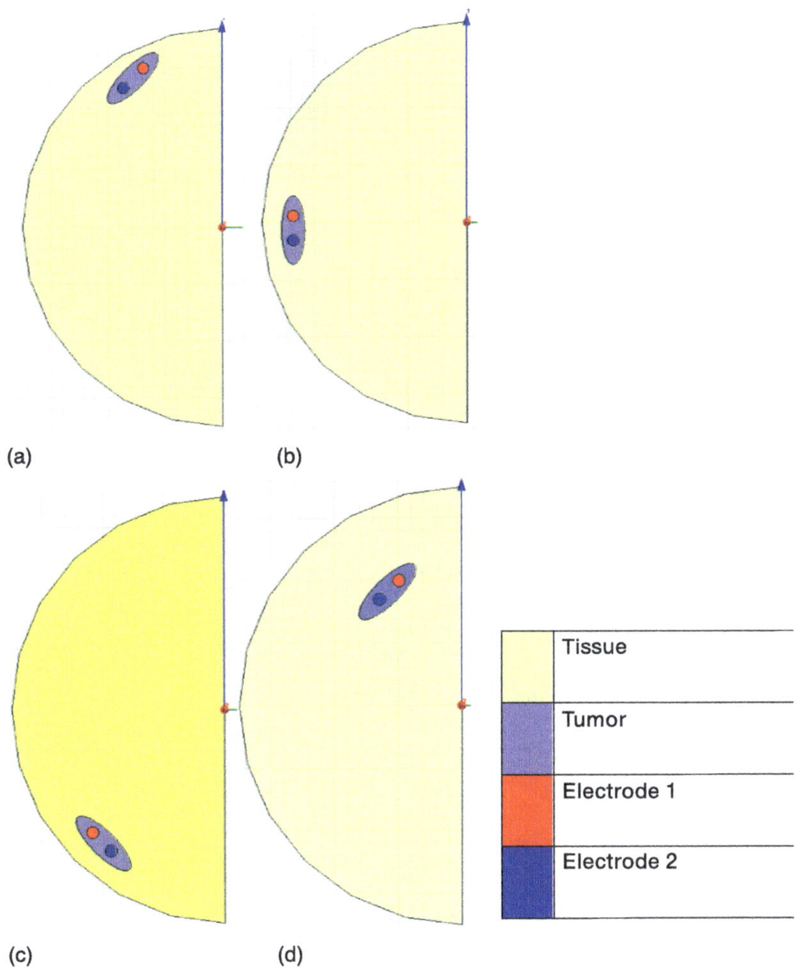

Figure 12.10 Models of tumor at various locations in the breast tissue: top (a); center (b); bottom (c); and top-center (d)

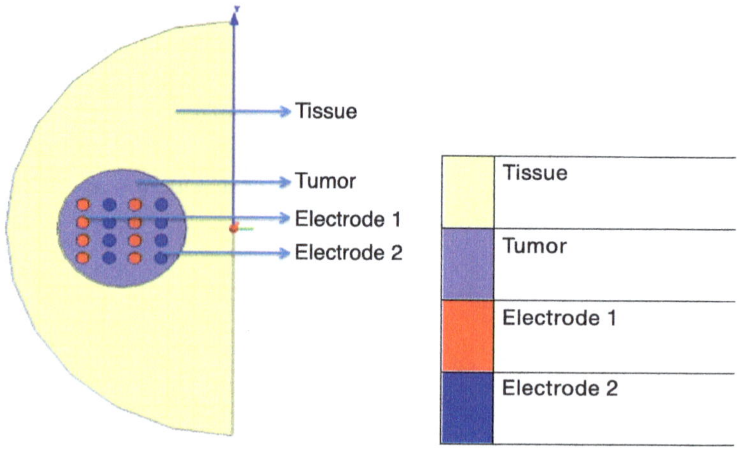

Figure 12.11 Big needle-electrode-array for larger areas

Table 12.5 Dimensions of model components

Model Component	Length × Breadth (cm × cm)	Diameter (cm)
Parallel plate electrode	2.5×0.5	–
Needle electrode	–	0.2
4×2 Needle Electrode array	1.4×0.6	–
4×2 Electrode array	1.4×1.4	–
Breast Tissue	–	7.0
Tumor	1.0×0.4 (ellipse)	–
Large tumor – 1	2.0×1.0 (ellipse)	–
Large tumor – 2	–	2.0

Table 12.6 Model parameters [53]

No.	Material	Properties
1	Tissue	$\varepsilon_r = 200$, $\sigma = 0.4$S/m
2	Tumor	$\varepsilon_r = 1000$, $\sigma = 2.0$S/m

12.10 Results and discussion

12.10.1 Effect of tumor on the electric field distribution of breast tissue using parallel plate electrodes

Figures 12.12(a and b) show the voltage distributions and Figures 12.13(a and b) show the electric field distribution of normal and malignant tissues under DC conditions. In lobule 1, which has the electrodes, the electric field is higher near the electrodes and reduces towards the center. The reduction inside the lobule could be due to the absorption of the electric

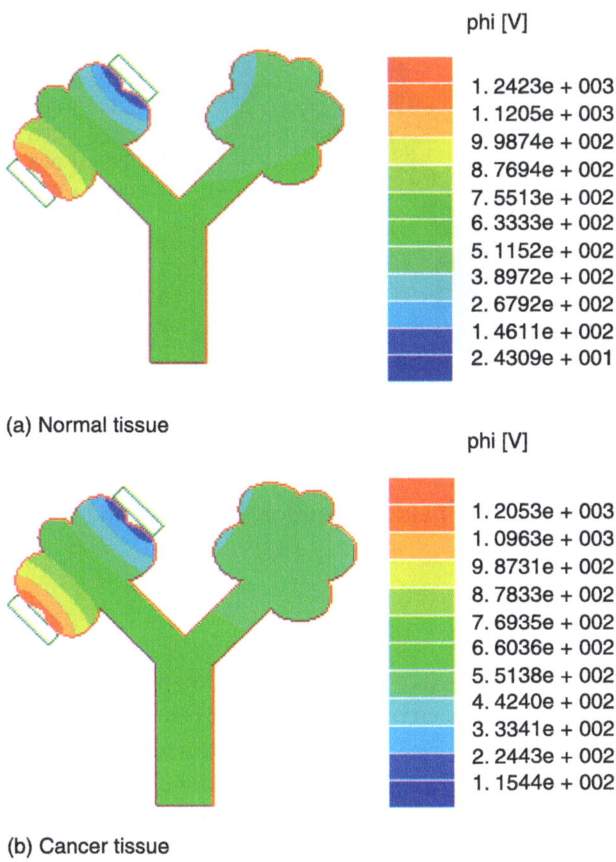

Figure 12.12 Voltage distribution normal and cancer tissues

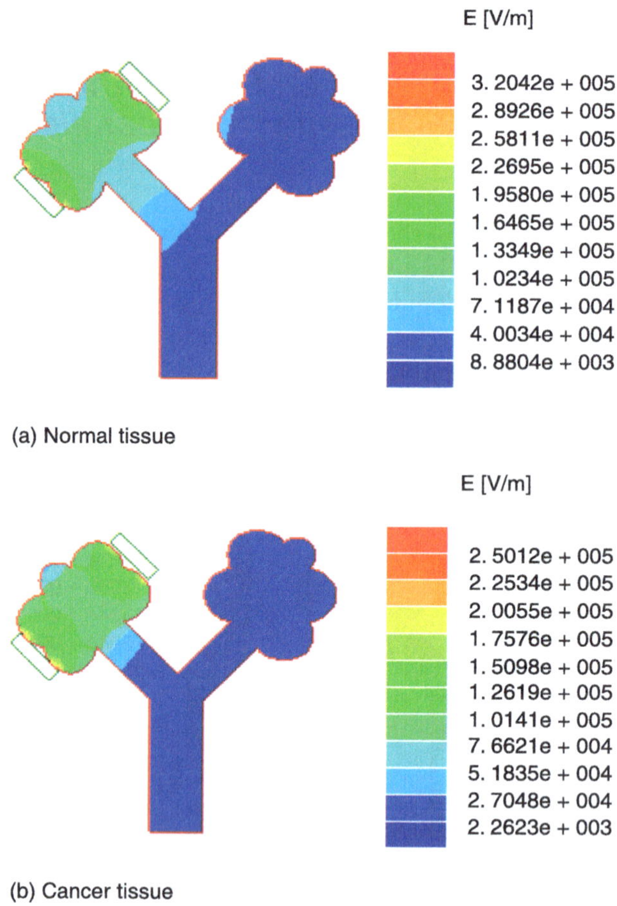

Figure 12.13 **Electric field distribution of normal and malignant tissues**

field inside the tissues, while it remains higher at the electrode ends. Similar results also have been reported by other researchers, who observed higher electric field only very close to the electrodes [58]. In adjacent lobule 2, which does not have the electrodes, the electric field magnitude is much lower. This illustrates that the electric field is mostly concentrated in lobule 1. A comparison of the malignant tissue electric field distribution with normal tissue indicates that the malignant tissue has slightly lower magnitude electric field intensity, while the trend is the same (Figure 12.14).

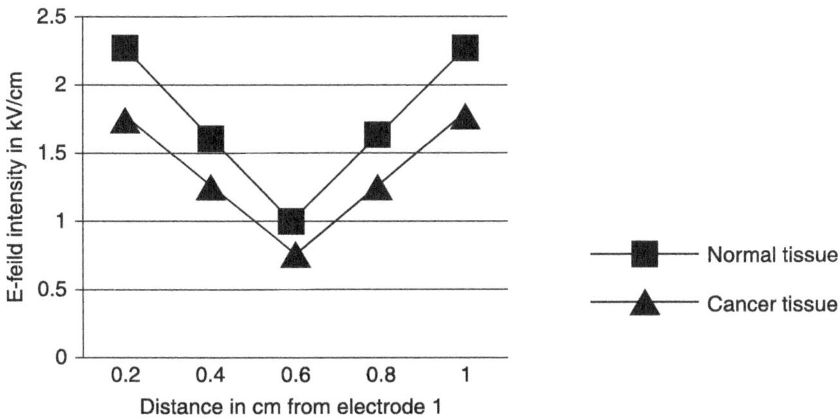

Figure 12.14 **Electric field distribution of normal and malignant tissues along the lobule**

This reduction in the malignant tissue could be attributed to the increase in the conductivity and permittivity of these cells, naturally lending themselves more susceptible to electrical pulse-mediated treatment techniques than normal tissues.

12.10.2 Effect of fat and skin layers

In order to simulate the tissue close to reality, fat (20 mm) and skin (2 mm) layers were introduced into the model. Figures 12.15(a and b) show the electric field distributions of normal and cancer tissue with the layers included. The electric field intensity is also lower in the case of cancer tissue than normal tissue in this simulation. The electric field distribution using tissue + fat only was also studied to investigate the electric field intensity without skin, simulating the inner part of the tissue.

Figures 12.16(a and b) show a comparison of the magnitudes of the electric field at various positions between the two electrodes for the normal and malignant tissues at the three different configurations studied, viz., tissue alone, tissue + fat and tissue + fat + skin. While the trend of variation is the same, there is a reduction in the electric field intensity of the malignant tissue compared to normal tissue in all three cases.

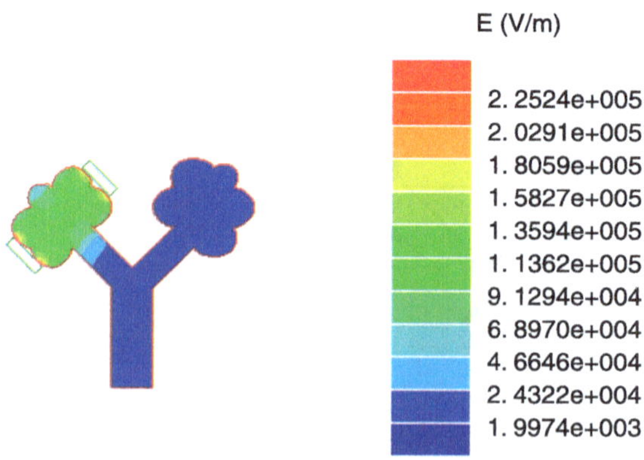

(a)

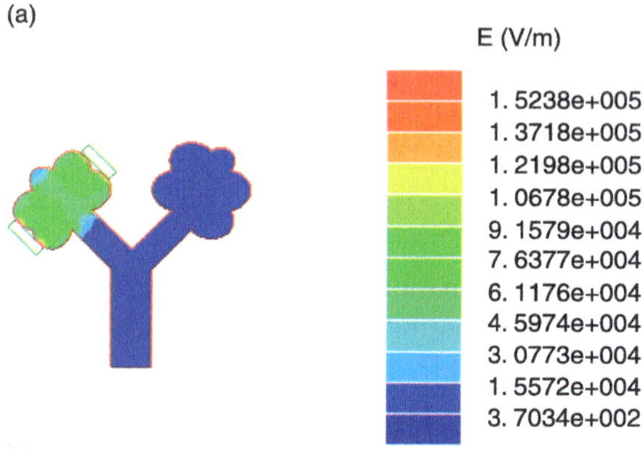

(b)

Figure 12.15 DC electric field distribution in normal (a) and malignant (b) tissues

12.10.3 Electric field distribution of ductal carcinoma

The electric field distribution of ductal carcinoma is shown in Figure 12.17. The electric field in the tumor area is 388 V/cm. The electric field between the needle electrode is 1170 V/cm. This correlates well with the applied field of 1200 V/cm.

Figures 12.18(a) and 12.18(b) show the electric field distributions of the ductal carcinoma for the malignant and normal tissues respectively. The 8-needle electrode is inserted into the duct, as illustrated in these figures. The electric field is homogenous in and around the electrode/

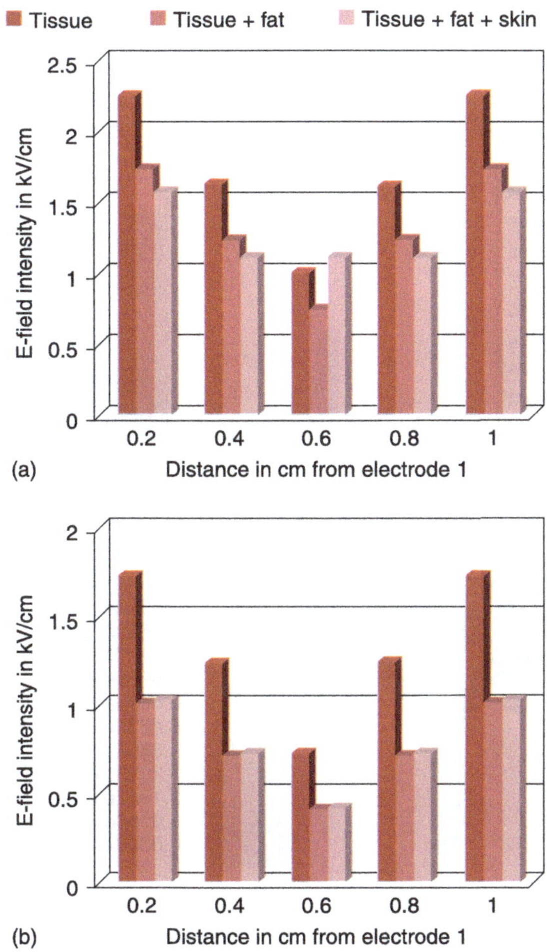

Figure 12.16 DC electric field distribution in normal (a) and malignant (b) tissues with fat and fat+skin layers

tumor area. The intensity of the electric field in the tumor tissue is about 329 V/cm, and about 287 V/cm in the normal tissue portion. In the adjacent areas, the electric field is higher near the electrode and is lower towards the tissue.

It is observed that the electric field magnitude of ductal carcinoma is 11.5% higher than that of the normal tissue. This contradicts that of lobular carcinoma, where the malignant tissue has a slightly lower magnitude than the normal tissue.

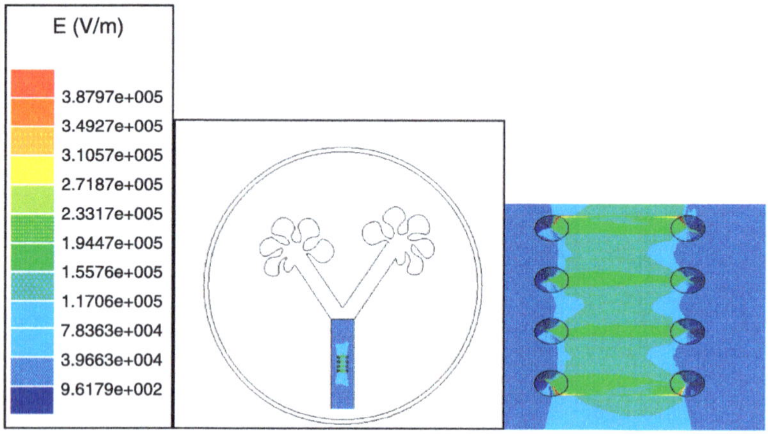

Figure 12.17 Electric field distribution of ductal carcinoma

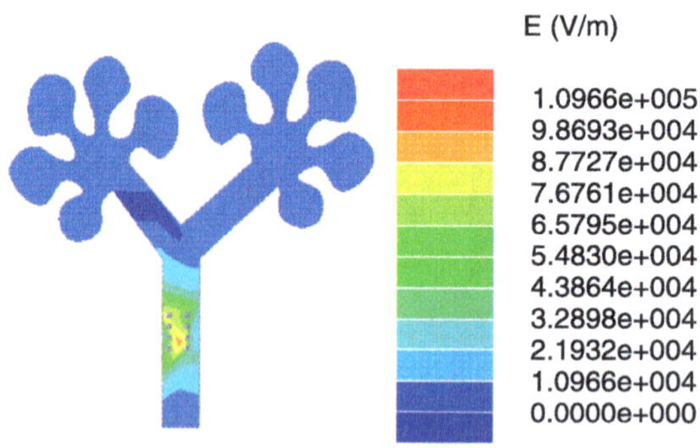

(a) Ductal carcinoma

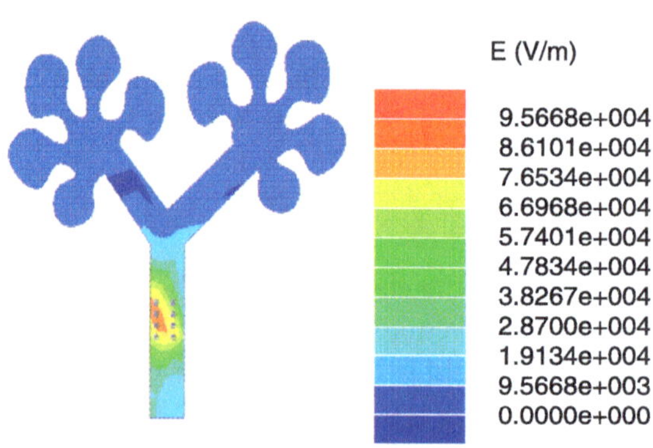

(b) Normal tissue

Figure 12.18 Electric field distribution of ductal carcinoma and normal tissue

12.10.4 Electric field distribution of breast model

A comparison of normal (a) and tumor (b) breast tissue is shown in Figure 12.19. Parallel plate electrodes were used in this case. With parallel plate electrodes, it is observed that the normal tissue has higher electric field intensity than tumor tissue.

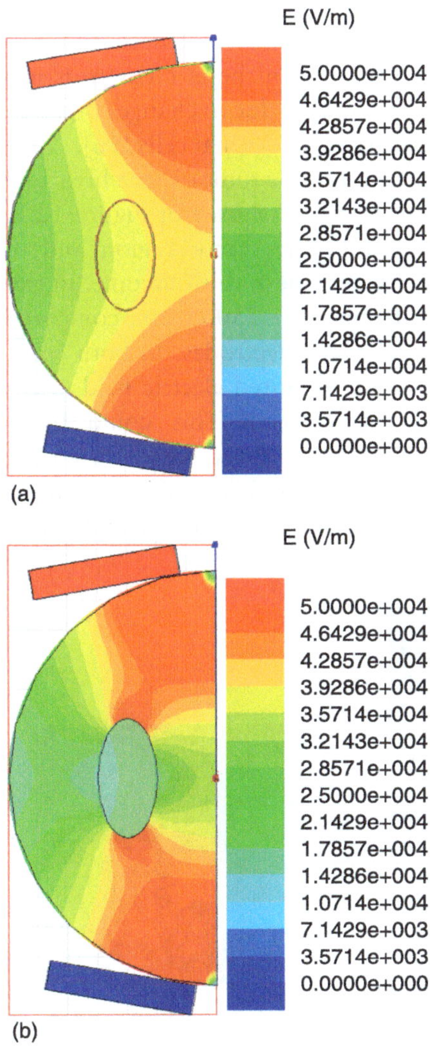

Figure 12.19 Electric field distribution of normal breast tissue (a) and tumor within breast tissue (b)

A voltage of 8400 V (across a 7-cm gap to produce the desired electric field of 1200 V/cm) causes an electric field of magnitude between 350 and 500 V/cm in the normal breast tissue (Figure 12.19(a)). A rectangular block of tissue of 7-cm gap with 8400 V applied showed that the field was 1150 V/cm, close to the desired value of 1200 V/cm. Thus, the reduction in electric field in the case of breast tissue could be due to the shape of the tissue and the difficulty in placing plate electrodes exactly parallel to the tissue (due to the limitation of the software and exploiting its features). The electric field in the tumor is about 146 V/cm (Figure 12.19(b)), which is 60% lower than that of the electric field in the normal tissue. The difference in the electric fields in the normal tissue surrounding the tumor could be due to changes in the electrical properties of malignant cells compared to normal cells [16].

The voltage distribution of the tumor in the breast model, using needle electrodes with a 4 mm gap, is shown in Figure 12.20. The maximum voltage is concentrated in the tumor region and did not affect the surrounding normal tissue, a desirable outcome. In this case, the electric field intensity varies from 2100–2500 V/cm, compared to the 146 V/cm obtained using parallel plate electrodes. In both cases, suitable voltage was applied to obtain the required electric field intensity of 1200 V/cm (assuming uniform distribution). This could be attributed to the enhancement of the field at the sharp edge of the needle. However, it is interesting to note that all the electric field is concentrated in between the electrodes, which happens to be the tumor and there is very little field

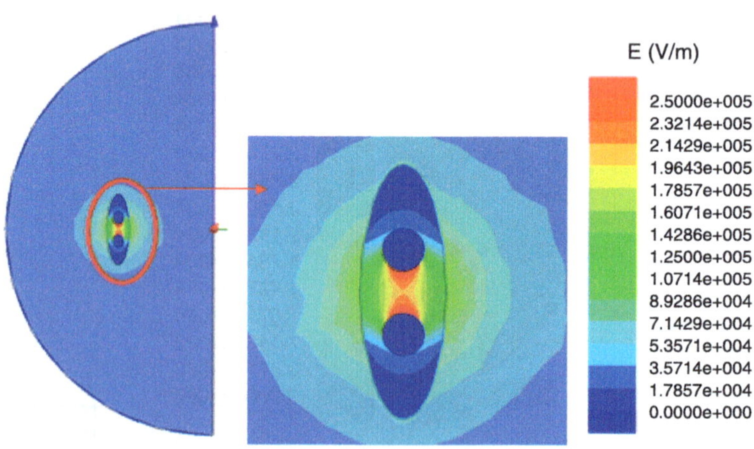

Figure 12.20 Electric field distribution of tumor in breast tissue with needle electrodes

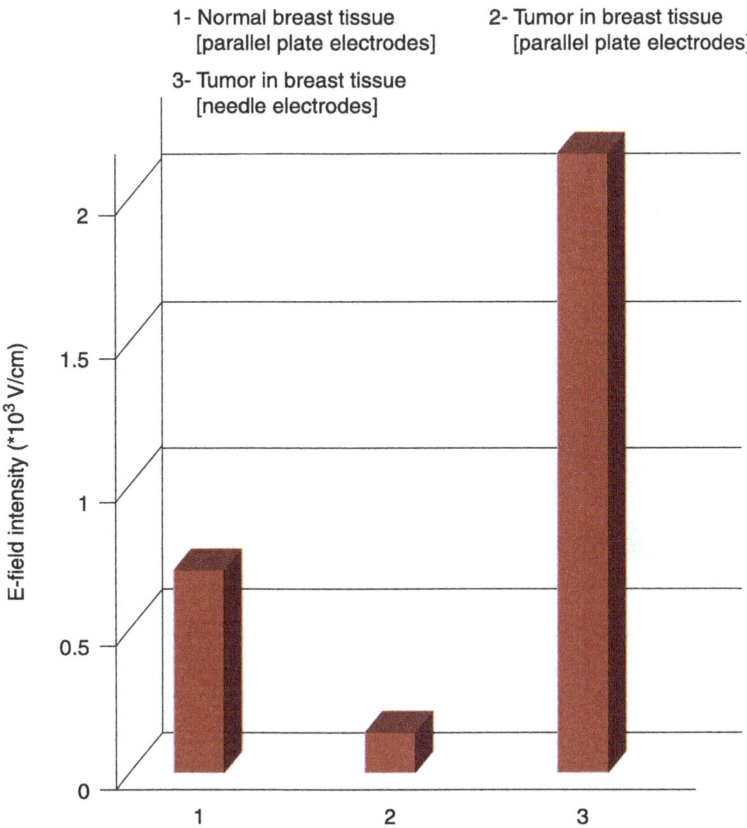

Figure 12.21 **Comparison of electric fields using various electrodes**

affecting the surrounding normal tissue, again a desirable outcome, as we do not want the surrounding normal tissue to be affected. Figure 12.21 shows a comparison of these magnitudes.

12.10.5 Large multi-needle electrodes for larger tumors

Typically in the clinic, a needle array electrode of 4 mm electrode gap is used. To treat larger areas, such as in the case of chest wall breast carcinoma (Figure 12.22), the surgeon has to remove and insert the electrodes repeatedly multiple times to cover the entire area. This is not only more time-consuming but also requires that the patient be sedated for longer durations and hence higher doses, etc., and the treatment time also

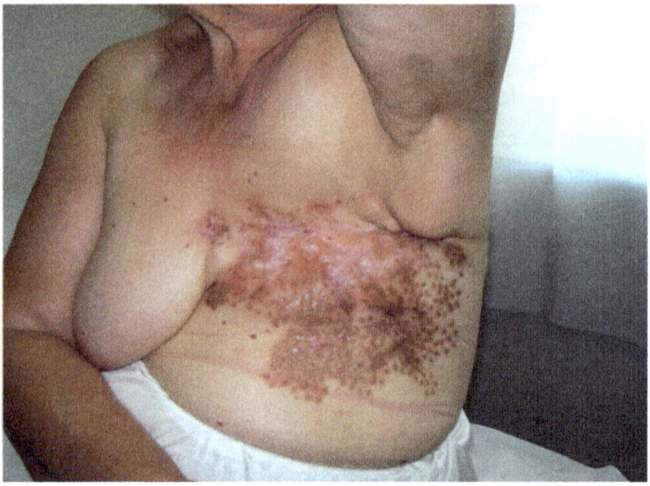

Figure 12.22 Electroporated larger chest area

increases for the oncological surgeon. Hence, there is a critical need for larger needle array electrodes. A novel, larger, multi-needle electrode array, which would cover larger areas, would be more efficient and convenient.

Figure 12.23 shows a larger tumor area with 16-needle electrodes. The electric field is again duplicated across the electrode array and the same

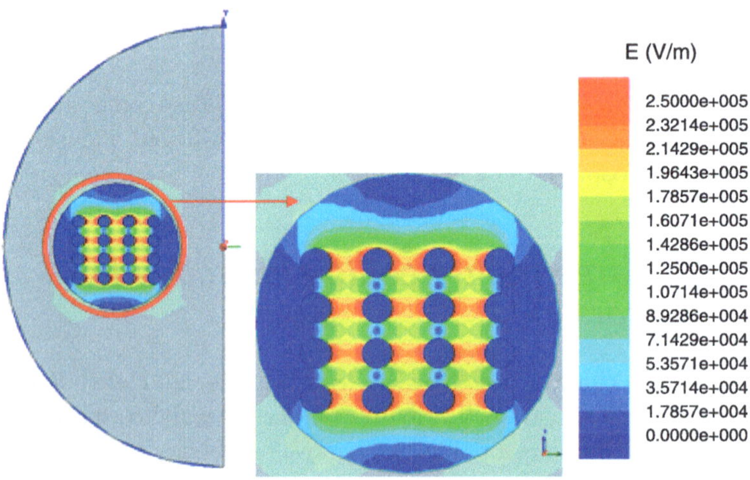

Figure 12.23 Electric field distribution of tumor with 16-needle electrodes

magnitude is maintained. The electric field distribution reveals that the electric field magnitude is the same as that of a single pair of needle electrodes with a 4-mm gap (2100–2500 V/cm) and is duplicated across each of the 4 pairs of electrodes. This shows that designing an electrode array with multiple electrodes would help fast and efficient treatment of invasive cancers using electrical pulses.

12.10.6 Thermal effects of electroporation

Using Equation 12.9, the temperature rise was calculated for various pulse parameters that are used in standard electroporation protocols. Table 12.7 lists the various pulse duration and voltages used and the corresponding temperature rise. There is not much heat dissipated for the pulse parameters used at present. The maximum rise is around 3 °C. Similar studies by Weaver indicated that temperature rise is around 1 °C per pulse, which correlates with our results [59]. Care should be taken during design and planning, so that there is no significant damage to tissues because of Joule heating. Previous studies have indicated that a temperature rise above 50 °C can cause thermal damage, and electroporation parameters have always been chosen to produce minimal heat dissipation [45, 46, 60].

Table 12.7 **Temperature rise due to electroporation**

Pulse duration	Voltage	Temp rise (°C)
300 ns	10 kV	0.102
10 ns	300 kV	3.061
100 μs	1200 V	0.490
25 ms	125 V	1.329
10 ms	200 V	1.360

12.11 Conclusions

- Electric field distributions of novel breast lobular, ductal and other models have been conducted using industry standard ANSYS and MAXWELL software.
- The 5-segment lobular models showed 8% reduction in malignant tissue compared to normal, using plate electrodes.

- In the case of ducts, there was an 11.5% increase for malignant tissue than normal.
- These variations depend on the shape, electrode geometry and the electrical parameters, including permittivity and conductivity.
- There is no noticeable temperature rise (max. 3 °C), hence electroporation is a non-thermal phenomenon and does not cause tissue heating and burning.
- Novel, large, multi-needle electrode arrays were also studied to treat larger areas such as chest.

In electrochemotherapy, electric field distribution plays an important role to inactivate the tumor. Simulation of human breast tissue will help in improving electroporation techniques for many types of breast cancers. Results obtained from comparison of parallel plate and needle electrodes and thermal calculations will help in electrode design. Multi-needle electrode arrays will help improve current methods significantly. These results will aid in performing effective electrochemotherapy, using the pulse-mediated drug delivery technique for cancers that are not receptive to current standards of cure.

12.12 References

1. Haagensen C D (1986), *Diseases of the Breast*, 3rd edition. Philadelphia, PA: W B Saunders Company.
2. Andolina V F and Lille S L (2011), *Mammographic imaging: A Practical Guide*, 3rd edition. Baltimore, MD: Wolters Kluwer – Lippincott Williams & Wilkins.
3. Gonzalez-Augulo A M, Morales-Vasquez F and Hortobagyi G N (2007), 'Overview of resistance to systemic therapy in patients with breast cancer', in: *Breast Cancer Chemosensitivity*, Yu D and Hung M-C (eds), New York: Springer Sci Business Media, 1–22.
4. Love S M and Linsey K (2005), *Dr Susan Love's Breast Book*. Cambridge MA: Perseus Publishing.
5. King R J B and Robins M W (2006), *Cancer Biology*, 3rd edition. New York: Pearson, Prentice Hall.
6. *Readers Digest*, December 2012.
7. Larkin J O, et al. (2007), 'Electrochemotherapy aspects of preclinical development and early clinical experience', *Ann Surg*, **245(3)**: 459–79.
8. Campana L G, et al. (2009), 'Bleomycin-based Electrochemotherapy: Clinical outcome from a single institution's experience with 52 patients', *Ann Surg Oncol*, **16**: 191–9.

9. Campana L G, Valpione S, Mocellin S, Sundararajan R, Granziera E, et al. (2012), 'Superficial disease control in melanoma patients treated with electrochemotherapy in a palliative setting', *Brit J Surg*, **99**: 821–30.
10. Gehl J and Geertsen P F (2000), 'Efficient palliation of hemorrhaging malignant melanoma skin metastases by electrochemotherapy', *Melanoma Res*, **10**: 1–5.
11. Gothelf A, Mir L M and Gehl J (2003), 'Electrochemotherapy: results of cancer treatment using enhanced delivery of Bleomycin by electroporation', *Cancer Treat Rev*, **29**: 371–87.
12. Mir L M, et al. (1991), 'Electrochemotherapy, a novel antitumor treatment: first clinical trial', *C R Acad Sci Paris*, **313**: 613–18.
13. Heller R, Jaroszeski M J, Glass L F, Messina J L, Rapaport D P, et al. (1996), 'Phase I/II trial for the treatment of cutaneous and subcutaneous tumors using electrochemotherapy', *Cancer*, **77**: 964–71.
14. Gehl J, Skovsgaard T and Mir L M (1998), 'Enhancement of cytotoxicity by electro-permeabilization: an improved method for screening drugs', *Anticancer Drugs*, **9**: 319–25.
15. Haddad A and Warne D (eds) (2007), *Advances in High Voltage Engineering*. Herts UK: The Institution of Engineering & Technology (ITE).
16. Haltiwanger S (2010), The electrical properties of cancer cells. Available at: *http://www.royalrife.com/haltiwanger1.pdf*, July.
17. Cure J C (1991), 'Cancer an electrical phenomenon', *Resonant*.
18. Reilly J P (1998), *Applied Bioelectricity: from Electrical Stimulation to Electropathology*. New York: Springer.
19. Brown G (1999), *The Energy of Life: The Science of What Makes Our Minds and Bodies Work*. New York: The Free Press.
20. Cure J C (1995), 'On the electrical characteristics of cancer', in: *The Second International Congress of Electrochemical Treatment of Cancer*, Florida: Jupiter.
21. Revici E (1961), *Research in Pathophysiology as Basis for Guided Chemotherapy with Special Application to Cancer*. Princeton NJ: D Van Nostrand Company.
22. Seeger P G and Wolz S (1990), Successful biological control of cancer: by combat against the causes. Available at: *Gesamtherstellung,NeuwiederVerlagsgesellschaftmbH*
23. Cone C D (1970), 'Variation of the transmembrane potential level as a basic mechanism of mitosis control', *Oncology*, **24**: 438–70.
24. Cone C D (1975), 'The role of surface electrical transmembrane potential in normal and malignant mitogenesis', *Ann NY Acad Sci*, **238**: 420–35.
25. Cope F W (1978), 'Medical application of the Ling association-induction hypothesis: the high potassium, low sodium diet of the Gerson cancer therapy', *Physiol Chem Phys*, **10(5)**: 465–8.
26. Foster K R and Schepps J L (1981), 'Dielectric properties of tumor and normal tissues at radio through microwave frequencies', *J Microwave Power*, **16**: 107–19.
27. Palti Y. Tumor-curing electric fields. Available at: *www.rexresearch.com/palti/palti.htm*

28. Schwan H P (1957), 'Electrical properties of tissue and cell suspensions', *Adv Biol Med Phys*, **5**: 147–209.
29. Pauly H and Schwan H P (1959), 'Über die impedanzeiner suspension vonkugelformigen teilchenmiteiner cchale', *ZNaturforsch B*, **14**: 125–31.
30. Rubinsky B (2010), 'Irreversible electroporation', in *BIOMED*, on-line, 183–202.
31. Miklavčič D, Semrov D, Mekid H and Mir L M (2000), 'A validated model of *in vivo* electric field distribution in tissues for electrochemotherapy and for DNA electrotransfer for gene therapy', *Biochimicaet Biophysica Acta*, **1523**: 73–83.
32. Belehradek J, Orlowski S, Poddevin B, Paoletti C and Mir L M (1991), 'Electrochemotherapy of spontaneous mammary tumors in mice', *Eur J Cancer*, **27**: 73–6.
33. Domengeet C, Orlowski S, Luboniski B, De Baere T, Schwaab G, et al. (1996), 'Antitumor electrochemotherapy. New advances in the clinical protocol', *Cancer*, **77**: 956–63.
34. Rudolf Z, Stabuc B, Cemazar M, Miklavčič D, Vodovnik L and Sersa G (1995), 'Electrochemotherapy with Bleomycin: The first clinical experience in malignant melanoma patients', *Radiol Oncol*, **29**: 229–35.
35. Miklavčič D, Corovic S, Pucihar G and Pavselj N (2006), 'Importance of tumour coverage by sufficiently high local electric field for effective electrochemotherapy', *Eur J Cancer Suppl*, **4**: 45–51.
36. Semrov D and Miklavčič D (2000), 'Numerical modeling for *in vivo* electroporation', *Mol Med*, **37**: 63–81.
37. Semrov D, Karba R and Valencic V (1997), 'DC electrical stimulation for chronic wound healing enhancement. Part II: Parameter determination by numerical modeling', *Bioelectrochem Bioenerg*, **43**: 271–7.
38. MiklavčičD, Semrov V, Valencic G, Sersa G and Vodovnik L (1997), 'Tumor treatment by direct electric current: Computation of electric current and power density distribution', *Electr Magnetobiol*, **16**: 119–28.
39. Selet D C, Cukjati D, Batiuskaite D, Slivnik T, Mir L M and Miklavčič D (2005), 'Sequential finite element model of tissue electropermeabilization', *IEEE Trans Biomed Engin*, **52(5)**: 816–27.
40. Semrov D and MiklavčičD (1998), 'Calculation of the electrical parameters in electrochemotherapy of solid tumors in mice', *Comp Biol Med*, **28**: 439–48.
41. Brandisky K and Daskalov I (1999), 'Electrical field and current distributions in electrochemotherapy', *Bioelectrochem Bioenerg*. **48**: 201–8.
42. Dev S B, Dhar D and Krassowska W (2003), 'Electric field of a six-needle array electrode used in drug and DNA delivery *in vivo*: analytical versus numerical solution', *IEEE Trans Biomed Engin*, **50(11)**: 1296–300.
43. Gilber R A, Jaroszeski M J and Heller R (1997), 'Novel electrode designs for electrochemotherapy', *Biochimicaet Biophysica Acta*, **1334**: 9–14.
44. Cliniporator Techincal sheet (2010), IGEA, March.
45. Davalos R V, Rubinsky B and Mir L M (2003), 'Theoretical analysis of the thermal effects during *in vivo* tissue electroporation', *Bioelectrochemistry*, **61**: 99–107.
46. Miller L, Leor J and Rubinsky B (2005), 'Cancer cells ablation with irreversible electroporation', *Technol Cancer Res Treat*, **4(6)**: 699–705.

47. Maxwell Student version (SV) (2009), Version 3.1.04, Ansoft Corporation.
48. Maxwell 13 (2010), Ansoft Corporation, Pittsburg.
49. Ansoft 13, Ansoft Corporation.
50. Jossinet J (1998), 'The impedivity of freshly excised human breast tissue physiological measurement', *Physiol Meas*, **19(1)**: 61–75.
51. Cancer Help UK (2007), Breast cancer section overview. Available at: *www.cancerhelp.org.uk/help/defaultpage=3270*
52. *www.Saint Francis Breast health services*, July 2011.
53. Chaudary S S, Mishra R K, Swarup A, and Thomas J M (1984), 'Dielectric properties of normal and malignant human breast tissues at radiowave and microwave frequencies', *Ind J Biochem Biophys*, **21**: 76–9.
54. Wilke N, et al. (2005), 'Influence of electrode design on electric field distribution during electroporation', *The 3rd European Medical and Biological Engineering Conference*, 2005.
55. Gorey T, Kennedy J, Flanagan F and Murray M (2006), *The Breast in Health and Illness* – an informational guide for patients and their cancers, AstraZeneca Pharmaceuticals Ltd.
56. *http://trialx.com/curebyte/2011/07/10/what-does-breast-tumor-look-like/*, 2011.
57. *A.D.A.M.* Medical encyclopedia. Fibroadenoma of the breast (n.d.).. Retrieved on August 2011.
58. Wu L, McGough R J, Arabe O A and Samulski T V (2006), 'An RF phased array applicator designed for Hyperthermia Breast Cancer treatments', *Phys Med Biol*, **51(1)**: 1–20.
59. Weaver J C (1994), 'Electroporation: a general phenomenon for manipulating cells and tissues', *J Cell Biochem*, **51(4)**: 426–35.
60. Edd J F, Horowitz L, Davalos R V, Mir L M and Rubinsky B (2006), '*In vivo* results of a new focal tissue ablation technique: irreversible electroporation', *IEEE Trans Biomed Engin*, **53(7)**: 1409–15.

Index

CPSIA information can be obtained at www.ICGtesting.com
Printed in the USA
BVOW10*0830020814

360613BV00004B/2/P

9 781907 568152